THE INVENTION OF PHYSICAL SCIENCE

BOSTON STUDIES IN THE PHILOSOPHY OF SCIENCE

Editor

ROBERT S. COHEN, *Boston University*

Editorial Advisory Board

THOMAS GLICK, *Boston University*
ADOLF GRÜNBAUM, *University of Pittsburgh*
SAHOTRA SARKAR, *Boston University*
SYLVAN S. SCHWEBER, *Brandeis University*
JOHN J. STACHEL, *Boston University*
MARX W. WARTOFSKY, *Baruch College of the City University of New York*

VOLUME 139

ERWIN N. HIEBERT

THE INVENTION OF PHYSICAL SCIENCE

Intersections of Mathematics,
Theology and Natural Philosophy
Since the Seventeenth Century

Essays in Honor of Erwin N. Hiebert

Edited by

MARY JO NYE
University of Oklahoma, Norman, U.S.A.

JOAN L. RICHARDS
Brown University, Providence, R.I., U.S.A.

and

ROGER H. STUEWER
University of Minnesota, Minneapolis, U.S.A.

KLUWER ACADEMIC PUBLISHERS
DORDRECHT / BOSTON / LONDON

Library of Congress Cataloging-in-Publication Data

```
The Invention of physical science : intersections of mathematics,
  theology, and natural philosophy since the seventeenth century :
  essays in honor of Erwin N. Hiebert / edited by Mary Jo Nye, Joan L.
  Richards, Roger H. Stuewer.
       p.   cm. -- (Boston studies in the philosophy of science ; v.
  139)
    Includes index.
    ISBN 0-7923-1753-X (HB : acid free paper)
    1. Physical sciences--History.  2. Mathematics--History.
  3. Theology--History.  4. Physics--History.  5. Hiebert, Erwin N.,
  1919-     .  I. Hiebert, Erwin N., 1919-    .  II. Nye, Mary Jo.
  III. Richards, Joan L., 1948-     .  IV. Stuewer, Roger H.  V. Series.
  Q174.B67  vol. 139
  [Q158.5]
  001'.01 s--dc20
  [500.2'09]                                                    92-12520
```

ISBN 0-7923-1753-X (HB)

Published by Kluwer Academic Publishers,
P.O. Box 17, 3300 AA Dordrecht, The Netherlands.

Kluwer Academic Publishers incorporates
the publishing programmes of
D. Reidel, Martinus Nijhoff, Dr W. Junk and MTP Press.

Sold and distributed in the U.S.A. and Canada
by Kluwer Academic Publishers,
101 Philip Drive, Norwell, MA 02061, U.S.A.

In all other countries, sold and distributed
by Kluwer Academic Publishers Group,
P.O. Box 322, 3300 AH Dordrecht, The Netherlands.

Printed on acid-free paper

All Rights Reserved
© 1992 Kluwer Academic Publishers
No part of the material protected by this copyright notice may be reproduced or
utilized in any form or by any means, electronic or mechanical,
including photocopying, recording or by any information storage and
retrieval system, without written permission from the copyright owner.

Printed in the Netherlands

TABLE OF CONTENTS

EDITORIAL PREFACE	ix
ROGER H. STUEWER / A Personal Appreciation. Erwin Nick Hiebert. The Wisconsin Years	xi
JOAN L. RICHARDS / A Personal Appreciation. Erwin Nick Hiebert. The Harvard Years	xix
MARY JO NYE / Introduction	xxv

PART I. NATURAL THEOLOGY, NATURAL PHILOSOPHY AND THE CERTAINTY OF MATHEMATICS

SARA SCHECHNER GENUTH / Devils' Hells and Astronomers' Heavens: Religion, Method, and Popular Culture in Speculations about Life on Comets	3
LORRAINE J. DASTON / The Doctrine of Chances without Chance: Determinism, Mathematical Probability, and Quantification in the Seventeenth Century	27
JOAN L. RICHARDS / God, Truth, and Mathematics in Nineteenth-Century England	51

PART II. PROBLEMS OF CONTINGENCY, COHERENCE, AND TRUTH

FREDERICK GREGORY / Theologians, Science, and Theories of Truth in Nineteenth-Century Germany	81
SKULI SIGURDSSON / Equivalence, Pragmatic Platonism, and Discovery of the Calculus	97

PART III. THE AIMS AND FOUNDATIONS
OF PHYSICAL SCIENCE: THE CASES OF
ELECTRICAL PHYSICS, PSYCHOPHYSICS,
AND PHYSICAL CHEMISTRY

JED Z. BUCHWALD / The Training of German Research Physicist Heinrich Hertz — 119

RICHARD L. KREMER / From Psychophysics to Phenomenalism: Mach and Hering on Color Vision — 147

DIANA KORMOS BARKAN / A Usable Past: Creating Disciplinary Space for Physical Chemistry — 175

PART IV. EXPLANATION AND DISCOVERY:
THE CLAIMS OF CHEMISTRY, PHYSICS,
AND FORTRAN

MARY JO NYE / Physics and Chemistry: Commensurate or Incommensurate Sciences? — 205

PETER GALISON / FORTRAN, Physics, and Human Nature — 225

APPENDIX I. Erwin N. Hiebert's Doctoral Students and Directed Dissertations — 261

APPENDIX II. Erwin N. Hiebert. Selected List of Publications — 263

NOTES ON CONTRIBUTORS — 267

NAME INDEX — 271

EDITORIAL PREFACE

In the early decades of our century, the French novelist Jules Romains wrote a series entitled *Men of Good Will*. How distressed he was by the European situations at the turn of 1900 and thereafter, by the interplay of persons and their places in those situations, by the sad contrast between human potential and human reality. Then his years of writing came to a climax and a nadir in 1938 with Volumes 15 and 16, *Prélude à Verdun* and *Verdun*, published in English in the ominous year, 1939, simply as *Verdun* (in one book, N.Y., Knopf). And yet, his tales leading to the horror, the pathetic patriotism, and the hypocrisy of that First World War of our time were tales of men and women who deserved no such fates, of many who were worthy in their different ways. Not just a minority, not merely a 'saving remnant', but living people deserved better despite their weaknesses, and deserved respect despite their inadequacies. In his almost Hegelian revelation of humanity through 'concrete universals' of character in practice, Jules Romains was the optimist of essential human decency. Look deeply *and connect*, he seemed to believe, and you will find men of good will.

Watching, hearing, reading Erwin Hiebert in America and in Europe and Asia for years, listening to so many he has touched and taught and criticized and inspired, I think of his having the same essential optimism about human decency, to be found everywhere. Professor Hiebert is a man of good will himself. *Hiebert connects*. He shows good will through his lovely combination: demanding work of high quality while offering personal bonds of collaborative effort. His own scholarship has both qualities; from him I learn that there is no contest between the rigor of the internalist's historical study of chemistry and physics and the equally rigorous humane interpretation of the externalist's psychology and social contexts of the living chemists or physicists.

Hiebert's research students are well represented in this book, and I bet he will be proud of them and of their essays in the book. But we may also note the many others: undergraduates by the score, colleagues in faculties and academic societies, and those in his world-wide career, in France and Germany, indeed in both Germanies, in China and all over Central and

Eastern and the rest of Europe, in the Soviet days when he reached out again and again, always helping, and always expecting critical standards and self-respecting scholarship. How did he succeed? Perhaps the Mennonite spirit? Perhaps the ability to see plain 'good will'?

The dread waste and shambles of Verdun had nearly faded for Erwin Hiebert's and my generation in our years. We witnessed deliberate genocide and the tactical slaughter bombing at Hiroshima and Nagasaki and Dresden. Hiebert, through this mid-century and after, is historian of the very nuclear science that is intertwined with both the wasting and the successes of our technical civilization. Is he an optimist about the individual? about scientists of good will in an immoral historical society? We may hope for more to come from his care and his insight.

June 1992   ROBERT S. COHEN

ROGER H. STUEWER

A PERSONAL APPRECIATION: ERWIN NICK HIEBERT. THE WISCONSIN YEARS

Erwin Hiebert has played a unique role in my life as teacher, colleague, and friend. My world line has now been joined to his for almost three decades, and over the years, as we shared so many experiences and exchanged so many ideas, my affection and respect for him has only grown deeper and more profound. This I know is not unique, and if I try to recapture here some of the ways in which he has influenced me, I know that many others share my feelings. I hope that they will find some echoes of their own warm experiences with Erwin in this account of my own.

I first met Erwin during the academic year 1963–64 at the University of Wisconsin. I was then a second-year graduate student completing my master's degree in physics and working as a research assistant in nuclear physics in the laboratory of H. H. Barschall, another teacher and scholar who has influenced me greatly. Committed as I was to physics, I nevertheless decided to branch out a bit that fall and take a course on the history of early modern science taught by Robert C. Stauffer. That eventually changed my career. As the course progressed, Professor Stauffer saw that I had a true interest in the history of science, and he suggested that I talk to Erwin about it. I set up an appointment with him toward the end of the semester.

At that time I was unaware of Erwin's high stature in the profession, but I did know that graduate students in the history of science liked him a lot. My first impression of him, however, was not too encouraging. As he invited me to sit down in his office in the old departmental space at 820 Irving Place, he struck me as a harried, nervous person (he even smoked a pipe or occasional cigarette in those days) who seemed to have a million things on his mind and was not terribly interested in adding another graduate student to his responsibilities. That too struck me as somewhat odd, because those were the days when I thought that the life of a graduate student was the most stressful one possible, and that of a professor the epitome of serenity. On his part, Erwin probably saw me as a typically naive young physicist. For some reason, however, probably simply to test my meddle, he suggested that I join his seminar on Mach the following semester, and I took the plunge.

That seminar had already begun in the fall semester. On the bibliography that Erwin prepared for it was the announcement: "The first seminar meeting will be held 8 P.M. October 10, 636 N. Frances Street." A serious seminar held in the evening in the professor's home? And, as I soon learned, with the students being served refreshments by wife Elfrieda and being greeted by children Cathy, Margaret, and Tom, and dog Peter. Never before had I experienced anything similar. Moreover, the bibliography for the seminar seemed patently ridiculous to a graduate student in physics who knew that a course requires one and only one textbook. Yet here was a list of five translated works by Mach plus ones by Meyerson, Duhem, and Stallo for every student to buy and read. Furthermore, there followed the instruction: "Each participant in the seminar should choose as soon as possible one of the volumes on Mach's work and thought listed below. Begin analysis of the work for later report." Five of the six volumes were in German, the other one in French. The bibliography concluded by listing "Four other works of similar scope but not in the UW Libraries" – all of which were in German.

That was not the end. By the time I joined the seminar in the second semester, Erwin had expanded the works to be discussed to include Alexander's *Sensationalism and Scientific Explanation*, Lenin's *Materialism and Empirio-Criticism*, as well as excerpts from Hertz, Peirce, Planck, Einstein, Frank, Cassirer, Poincaré, Campbell, and Enriques. The students then included Carolyn Merchant Iltis, Arthur Donovan, Robert McRae, Thomas Hawkins, and Edward Daub. Each of us had to type our reports on ditto masters, run off multiple copies in the departmental office for distribution to everyone else, and lead the discussion. It dawned on me that semester that not only physics graduate students had to learn how to swim to survive.

Erwin's seminar was an entirely new intellectual experience for me, and during the next academic year 1964–65 I became more and more convinced that my true research interests lay in the history of physics rather than in physics itself. Ultimately, in the fall of 1965, after completing my course work and passing my preliminary Ph.D. examinations in physics, I was awarded a fellowship and transferred into the history of science.

Only someone who has actually confronted such a basic career decision can fully appreciate the conflicting emotions it arouses, and only later did I learn that Erwin had made a similar decision at a similar time in his life. He received his bachelor's degree in chemistry and mathematics from Bethel College in Newton, Kansas, in 1941 and his master's degree in chemistry and physics from the University of Kansas in 1943 – the same year that he and Elfrieda were married. He then was employed as a research chemist for the Standard Oil Company and on the Manhattan Project in Chicago from 1943–46, served as Assistant to the Chief of the Scientific Branch of the War Department General Staff in Washington, D.C., from 1946–47, and worked as a research chemist in the Institute for the Study of Metals of the University of Chicago from 1947–50. Meanwhile, Elfrieda had completed bachelor's and master's degrees in 1945 and 1946 at the University of Chicago, and Erwin had completed a second master's degree in physical chemistry there in 1949.

But during this period Erwin's interests too were changing more and more into the history of science. The University of Chicago was a stopping-off place for an international constellation of scholars in all disciplines after the war, and while working on his master's degree in physical chemistry he also took a broad range of courses in history, philosophy, and sociology. The one that made the most lasting impression on him was a course given by Alexandre Koyré on "Scientific Thought in the Age of Newton." He began to read everything Koyré had written, and encouraged especially by the physical chemist Farrington Daniels, he decided to pursue his interests in the history of science further at the University of Wisconsin. There no one exerted a stronger influence on his career and approach to teaching and research in the history of science than Marshall Clagett, from whom he took no less than five seminars on medieval science (held evenings in the Clagett home). In those seminars Erwin had to hold his own with medieval specialists such as Edward Grant and John Murdoch, and all had to meet Clagett's rigorous scholarly standards, mastering Latin, analyzing texts in detail, placing them within their historical context, and exploring their ramifications in discussions and term papers. At the same time, Erwin did not neglect further scientific study. He completed his Ph.D. degree at Wisconsin in 1954 with a double major in physical chemistry and history of science.

By that time Erwin's academic career was already launched. He taught as assistant professor of chemistry at San Francisco State College from 1952–54. He then went as a Fulbright Lecturer to the Max Planck Institut für Physik in Göttingen for the academic year 1954–55, where his hosts asked him, since he had studied with Clagett, to lecture on medieval mechanics. Simultaneously, Elfrieda was a Fulbright Scholar in musicology at the University of Göttingen. During their year in Göttingen, Erwin received and accepted an offer as instructor in the history of science at Harvard University. He spent the academic years 1955–57 at Harvard and then joined the faculty of the University of Wisconsin, becoming assistant professor of the history of science in 1957, associate professor in 1960, and professor in 1963. During these years he also worked as a scientist on the Wisconsin Geophysical Expedition in the Canadian Arctic (summer of 1959), and he also held visiting appointments at Kabul (summer of 1961), the Institute for Advanced Study in Princeton (1961–62), Tübingen (summer semester 1965), and Harvard (first semester 1965–66).

Erwin thus was very much the peripatetic professor, and he has never changed in this respect. His broad and deep knowledge of the history and philosophy of science is known to everyone in the field, and he has been constantly in demand as a lecturer at universities and conferences throughout the world. Even as I write, he is spending, in his retirement, the winter semester 1991–92 in Göttingen as visiting professor. On this trip, as on many of his earlier ones, Elfrieda has accompanied him, pursuing her own scholarly interests. She is an accomplished pianist who received her Ph.D. degree in musicology from the University of Wisconsin in 1970 with a dissertation on Beethoven's Piano Trios. All three of their children now too have doctoral degrees, and like their parents all are accomplished musicians.

Erwin's absence from Wisconsin during part of the 1965–66 academic year did not trouble me, since I had my hands full with other courses in the history of science. I had already taken Erwin's course on the structure of matter 1895–1935 and his seminar on the early quantum theory; now I enrolled in other departmental offerings and Erwin's year-long course on the history of nineteenth-century physical thought, which Robert McRae, then an advanced graduate student, taught during the first semester while Erwin was at Harvard. In October 1966 I passed my preliminary Ph.D. examinations in the history of science and began to work on my dissertation on the history of the Compton effect.

Even then, however, I could not resist taking a seminar that Erwin offered that fall on Helmholtz's musical acoustics. I was never alone in this respect. Far more graduate students took seminars from Erwin and eventually completed their doctoral degrees under him at Wisconsin and Harvard than anyone else in the field (see Appendix I). My first impression of Erwin was accurate: he did indeed have so many demands on his time that he could afford to take on only the most promising of the graduate students who wished to work under him. At the same time, he always insisted that his students possess a high degree of competency in some field of science as a necessary foundation for historical research. Some, of course, chose to analyze philosophical, sociological, or religious dimensions of science. Nonetheless, they too understood that Erwin always required a thorough and detailed knowledge of some field of science as an essential prerequisite for illuminating these broader concerns.

I selected my dissertation topic through consultation with Erwin, and once chosen he responded to it in typical fashion: he allowed me the freedom to pursue it as I saw fit. During the course of the year, as my research and writing progressed, I met every now and then with Erwin, bringing him up to date. A favorite time was over lunch, and a favorite place was the Brathaus on State Street, consuming first beer and a brat or steak sandwich and then oodles of coffee. We parted intellectually invigorated and slightly tired, I to Memorial Library to my student carrel, he to a class, another meeting, or also to Memorial Library to his faculty study to work or to catch a few mid-day winks – he is an early riser, and one essential piece of furniture in his library study has always been a couch or cot. The system worked well, and by the spring of 1967 I was sufficiently far along on my dissertation to go onto the job market.

Since I had followed Erwin's model by completing a double Ph.D. major in history of science and physics, I received a spectrum of offers, but the one I accepted I owe directly to Erwin. Earlier, at a conference at Notre Dame, Erwin had met Herbert Feigl, Director of the Minnesota Center for Philosophy of Science, and Erwin now telephoned Herbert (actually tracking him down in Hawaii) and told him about me. In early April 1967 Herbert invited me for an interview, introduced me to Morton Hamermesh, Head of Minnesota's School of Physics, and by the end of the month the two had arranged a joint appointment for me as assistant professor in their departments. My family and I left Madison for Minneapolis in July, during the summer and fall I finished writing my dissertation, and in January 1968 I returned to Madison to take my final oral examination, the readers of my dissertation being Erwin, H. H. Barschall, and David Lindberg. Under Erwin's umbrella, I had acquired an assistant professorship and a Ph.D. degree, in that order.

The warm relationship that Erwin and I had established by that time soon grew into a close personal one through a fortunate set of circumstances. At the end of the summer of 1968, between August 25 and September 15, three conferences took place in Europe one after the other: the International Congress of the History of Science in Paris, the International Congress of Philosophy in Vienna, and a conference commemorating the 100th birthday of Arnold Sommerfeld in Munich. Erwin and I attended all three conferences, and for three weeks we walked the streets of Paris, Vienna, and Munich together, discussing practically everything under the sun. Erwin seemed to know everyone, and in the halls and at the receptions he went out of his way to introduce me, an unknown beginner, to many of the leaders of the various professions. It thus came as no surprise to me when Erwin was elected as Vice President of the Division of History of Science of the International Union of the History and Philosophy of Science in 1974 and eventually became President of the Division in 1982. He probably knows personally and has had intellectual contacts with at least as many foreign scholars and students as ones in the United States.

But most gratifying to me in 1968 was the close personal relationship that grew out of our many talks in Paris, Vienna, and Munich. We came to realize much more fully than before that our backgrounds were quite similar in significant respects. Erwin was born in Saskatchewan, Canada, on May 27, 1919, and attended Faraday Grade School and Isaac Newton High School in Winnipeg. Both of us had worked on farms in our youths, had gained a healthy respect for the know-how of farmers, and enjoyed their earthy sense of humor. One of our favorite aphorisms became, "We can always shock grain!" His forebears spoke *Plautdietsch*, mine *Plattdietsch*. He, like I, had grown up in a religious family, and I could understand how his Mennonite tradition could remain as a strong influence in his personal and family life. Both of us had read widely beginning at an early age, and when we later witnessed religious and ethnic bigotry were equally repelled by it. Both of us, still later, had made a similar transition from scientist to historian of science, and we had a similar educational philosophy. Both of us, in a word, had acquired many of our convictions in a kindred way, and we shared each other's trust.

After those weeks in Europe together, Erwin and I saw each other frequently on other trips and kept closely in touch by mail and telephone. In March 1969, while he was again on leave at the Institute for Advanced Study in Princeton, he introduced me to Helen Dukas, Einstein's former secretary, whose helpfulness and kindness I shall never forget. In September 1969 Erwin first visited Minneapolis after my arrival there to take part in a conference exploring the relationship between the history and philosophy of science. Among those also attending was Imre Lakatos, who became and remained a good mutual friend until his untimely death in 1974.

That conference remains vividly in my memory for other reasons as well. When I was asked to edit its proceedings as Volume 5 of the *Minnesota Studies*, I wrote to all of the contributors, including Erwin, setting forth a timetable that began by establishing October 31, that is, a date one month after the conference, as the date of receipt of the "definitive version" of all of the manuscripts. I am now positively astonished that I was then so naive about scholarly behavior. However, I am equally astonished that on November 11 I again could write to Erwin, opening with the sentence, "Guess what – you and Bob Cohen are my only delinquents." I eventually received Erwin's manuscript in February of the New Year – but he had a good reason for the delay. He had just received several telephone calls from John Murdoch at Harvard trying to persuade him to leave Wisconsin and join the Harvard department. That, as Erwin told me in a letter of January 29, precipitated a flurry of activity at Wisconsin, with all sorts of people trying to persuade him to remain there. He was "just bewildered" as to what to do, but he and Elfrieda were flying to Boston the next day to investigate things. He closed with a postscript, quoting Stephano in Shakespeare's *Tempest*: "Prithee, do not turn me about; my stomach is not constant." In early June he sent me a clipping from the Madison *Capitol Times* with the headline, "Science Historian Hiebert Leaving U.W. for Harvard," and above it was the caption, "Wife Gets Musicology Ph.D."

Erwin and Elfrieda's move to Cambridge with their family opened up a new phase in their lives, and increased our geographical separation, but did not diminish our close ties. We have continued to write and call each other, to see each other at meetings and conferences, and to work with each other closely, as for example for more than three years after 1971 while he was Vice President and President, and I Secretary, of the History of Science Society.

During these years as well, in common with all of Erwin's graduate students, I asked Erwin to write letters of reference for me on various occasions. This sort of activity should not be forgotten. The mind boggles, in fact, over the many hundreds of letters of reference that

Erwin has written for his graduate students and colleagues during the course of his career. Such letters, as everyone who has written them knows, require a great deal of thought and can easily take several hours apiece to formulate, write, and rewrite. Erwin has always responded graciously and willingly to these requests. He has always known that the many people he has helped in this way have deeply appreciated his support, but the ultimate beneficiary of Erwin's efforts has been the discipline of the history and philosophy of science. His success in placing his students and others who have turned to him for help in academic and other positions has resulted in the founding of new programs in the field and the strengthening of old ones throughout the country, amplifying his own influence enormously.

The Wisconsin years left a deep imprint on the lives of Erwin, Elfrieda, their children, and Erwin's graduate students. In the fall of 1978, several of us knew that Erwin was about to celebrate his sixtieth birthday, and Edward Daub and I conceived of a way to capture Erwin and Elfrieda once again in Madison, at the annual meeting of the History of Science Society. We arranged a morning session there composed of papers given by eighteen of Erwin's graduate students, four of which were subsequently published by Joseph Dauben in *Historia Mathematica* in 1980. We also arranged to take over a restaurant in nearby New Glarus one evening to celebrate the event properly. No less than twenty-five of Erwin's graduate students from Wisconsin and Harvard turned out that evening, many accompanied by their wives and husbands, to enjoy a wonderful dinner, toast Erwin and Elfrieda, and present Erwin with a packet of letters expressing our appreciation to him for the many ways in which he helped us over the years. It was a happy, warm, memorable occasion, and since then Erwin and Elfrieda have received – and as always have responded wonderfully to – many more expressions of gratitude and affection from their students, colleagues, and friends throughout the world.

University of Minnesota

JOAN L. RICHARDS

A PERSONAL APPRECIATION: ERWIN NICK HIEBERT. THE HARVARD YEARS

I knew of Erwin Hiebert for several years before I knew him. He came to the Harvard History of Science Department when I was a senior there, but I neither took any courses with him nor even met him that year. I did hear of him though, and what I heard was somewhat daunting; a kind of awed intensity surrounded everyone's references to the new professor from Wisconsin. Throughout the process of writing my senior thesis on Poincaré there lurked in the back of my mind, the shadowy figure of Erwin Hiebert who would, in the end, evaluate my efforts.

When I finally got my comments, I was relieved and somewhat surprised. Whereas I had thought he would have little time for my efforts, this newcomer read my work with remarkable care; he commented on everything from the translations, to the importance and persuasiveness of the thesis, to the chapter titles (or lack thereof). His remarks were insightful, helpful, somewhat humorous and most of all interesting. They were not enough to prevent me, a rather disillusioned radical of the sixties, from leaving the history of science, academe, and the United States for two years upon my graduation. They were enough, however, to make me excited to find that Erwin had agreed to be my advisor upon my return.

What I had glimpsed in the remarks of the unmet reader who later became my advisor was Erwin's total engagement with his work and that of his students. From afar this translates into an intensity which can be daunting to the uninitiated; closer up it is a quality of caring which marks all of Erwin's activities. For most, this quality was first manifested in the classroom where Erwin's involvement in his material was wonderful to see. Year after year his courses changed, even when presented under the same title. Year after year lectures seemed to emerge as the product of some great battle with the material. Nothing was glib, no point was pat or fixed. All of the issues were confronted and examined almost physically, as Erwin paced back and forth, occasionally knocking his glasses from his nose with the force of his gestures.

As a seminar leader Erwin was equally captivating; for years undergraduate and graduate students worked with him in his seminars, including the beloved "The Scientist as Philosopher of Science." Though the title remained the same the material covered did not, and every time it was offered Erwin and his students grappled with different issues in the original texts of scientific thinkers like Mach, Duhem, Helmholtz, Durkheim, James, Einstein, Bohr and others. As an anonymous student writer reported in the Harvard *Confidential Guide* of 1981, Erwin was "not a man intimidated by broad subject matter," and in his classes he moved briskly and self-confidently across the fields of physics, chemistry, psychology, religion, philosophy and whatever else might illuminate the material. It is no wonder that the same confi-guide author stated that "most students find Hiebert's perspective on science and philosophy one of the most refreshing they encounter at Harvard."

Erwin's perspective might be described as broad, inter-disciplinary, or, perhaps, historical, but a better term than all of these is humane because it includes Erwin's personal as well as academic style in its purview. With Erwin the two are always inseparable. Often his seminars were held in his house in Belmont where the conversation was no less intense than in the institutional surroundings of the department. In his warm and comfortable living room they were enlivened by the ambience of his family, often in concrete ways. I well remember a seminar on Helmholtz one fall. Through the chilly dark evenings we concentrated particularly on *Die Lehre von den Tonempfindungen*, wherein Helmholtz developed a physiologically-based theory of music. As we followed the twists and turns of Helmholtz's thought, Elfrieda illustrated acoustical points on the piano and occasionally Erwin pitched in with his clarinet. The best evenings of all were the ones when Elfrieda could be persuaded to close the evening for us with a piece of piano music.

The unique combination of Erwin and Elfrieda as physicist and musician displayed in this seminar, flowered for undergraduates in the context of the Harvard House system. The saga began one fall when Mather House acquired a harpsichord kit, which a group of students were to assemble for class credit under the tutelage of its maker, Frank Hubbard. Unfortunately, Hubbard died suddenly just as the term was starting. At the very last minute Erwin and Elfrieda were willing to step into the breach, adding the class to their already busy schedules. Week after week they considered with the students both the physical and the musical properties of the instrument they were constructing together. By the end of the term Mather House had its harpsichord.

In addition, there had been planted the seeds of the chamber music program which has thrived under Elfrieda's careful tutelage ever since. Erwin, for his part remained ever-true to his original commitment to Dunster House, which is next door to Mather. Over the years, the influence of each of the Hieberts has spilled liberally onto both of their adjoining Houses, bringing warmth, thought, and music into the every day lives of their students.

The same was true in the history of science department. When, in the spring of 1989, the department planned a day-long gathering of scholars to honor Erwin, some of the papers here collected were originally given. Also an integral part of the program was the music played by Elfrieda and a number of Erwin's students; this was but one in a long series of *Musikabends* the department enjoyed with the Hieberts.

Music was just one of the ways that Erwin managed to bring out the best in the extraordinarily diverse group of people who worked with him over the years; he was equally effective in seminars. There, even as he shared and explored his particular concerns with the class, he insisted that we do the same. My work on non-Euclidean geometry, a far cry from physiological acoustics, began as a paper for the Helmholtz seminar. Several of the papers in this volume reflect interests generated in a similar fashion from seminars with Erwin.

Erwin's embrace of diversity did not extend merely to the ideas of his students, but to their being as well. One of the striking aspects of the appended list of those who wrote dissertations with him is the large number of women among them, with dissertations dating all of the way back to 1967. In a time of transition for women in academe Erwin offered unambivalent support to those of us lucky enough to work with him. Our experience as graduate students could not have been more different than that of many of our contemporaries who were trying to be taken seriously. Never did Erwin question either our intellectual or our personal competence to pursue our chosen careers.

This is not to say Erwin did not always and at all times vigorously question our views as well as those of all of his students. Though watching him lecture might at first have been entertaining, and listening to him struggle with issues could be entrancing, Erwin's intellectual engagement was never a spectator sport. I well remember my surprise when, early in my first year as his student he asked me my position within a controversy we were considering. I had carefully studied the different positions and meticulously analysed the perspectives which fed into each person's approach, but was wholly unprepared to venture a stand of my own. Erwin had little patience with that kind of intellectual ducking; I realized on that day that he cared about my opinion which meant that first I had to have one. After that day I always did.

I then was in the position to learn a second big lesson from Erwin. He insisted that I take the material seriously and commit myself, which meant that there was now the possibility that he and I would disagree. This could be frightening at first, but soon became heady and stimulating. Erwin is a master at serious disagreement without personal cost; he will argue passionately against your point of view, even while keeping it clear throughout that it is not at your expense that you are disagreeing.

The power of his approach was graphically demonstrated for me at the 1989 meeting of the history of science society held in Gainesville. There I presided over a session where Diana Barkan and Rich Kremer delivered earlier versions of the papers which appear in this volume. In the discussion period, Erwin rose up and vigorously challenged a number of the points which had been made. Instantly the panelists were engaged in sharp debate – defending, clarifying, arguing – it was a powerful flashback to the accumulated years of seminars with Erwin. Afterwards, several people in the audience who did not know Erwin personally, came to commiserate with me on chairing such a difficult session. I could only reply that it was not difficult at all. On the contrary, it was exhilarating in a way all of Erwin's students easily recognized.

Erwin did not merely model for his students how to interact with him but, also, how to work with each other. I well remember in my first year, his constant glowing references to other graduate students – their strengths and achievements. At first I was jealous, but soon it became clear to me that this was an inappropriate and unnecessary response. Erwin's support was broad enough to encompass us all and there was no need to struggle against others for his attention or approval. This recognition, shared among his students, freed us to learn from and enrich each other. As we became more advanced, the ideas and questions planted in Erwin's seminars were frequently addressed and developed independently of his oversight. My dissertation, like many others, was brought to completion in the context of weekly lunches among Erwin's finishing students. By the time we had finished we recognized our validity to be not only as individual scholars but as part of an intellectual community.

Erwin, himself, modeled for us many of the kinds of responsibilities attendant on participation in the larger community of the history of science. His driving interest has always been on the kinds of issues he pursued with his classes. Nonetheless following this interest has meant Erwin has been caught up in many projects, groups, and organizations. So, for example, he was one of the editors of the *Dictionary of Scientific Biography* with special responsibility for the chemistry articles. In addition to membership in the American Chemical Society, and Sigma Xi, Erwin is a member of the Association of Members of the Institute for Advanced Study in Princeton and an elected fellow of the American Academy of Arts and Sciences.

Erwin is not just a passive participant in such groups, and he has gamely taken on the administrative responsibilities these involvements have thrust upon him. He was chairman of both the Wisconsin and Harvard History of Science Departments (1960–75 and 1977–84, respectively), President of the Midwest History of Science Society (1967–68), and Chairman of Section L of the American Association for the Advancement of Science (1982). As President of the History of Science Society (Vice-President 1971–72, president 1973–74) Erwin chaired its Council meetings, implemented its decisions, helped arrange annual meetings, corresponded with officers of other scholarly societies, and carried out a host of other duties as required. Particularly noteworthy was his pushing for a society *Newsletter*, to be published several times each year to keep its members informed of noteworthy events, job openings etc. Ever since an essential society publication, the focus of the *Newsletter* on constructive communication is a mark of the community-building spirit which marks all of Erwin's interactions.

These have always reached far beyond the communities or even the countries in which he has worked. Erwin's students are almost as remarkable for their international diversity as they are for their female representation. His ties to the international history of science community have always been strong and warm. He is a member of the British Society for the History of Science, the Canadian Society for the History and Philosophy of Science, and was elected as a Fellow of the Academie International d'Histoire des Sciences, as an Overseas Fellow of Churchill College, Cambridge, as Honorabilis Sodalis of the Czechoslovak Society for the History of Science and Technology, and as Auswärtiges Mitglied of the Sächsische Akademie der Wissenschaft zu Leipzig. He served for eight years as Vice-President (second and then first), another five as President and an additional three as ex-officio member of the Council of the "Division of the History of Science of the International Union of the History and Philosophy of Science" of the International Council of Scientific Unions (ICSU).

The performance of these duties, as well as the many visiting appointments he has accepted all over the world, has put Erwin in touch with a whole constellation of international scholars, a large number of whom were from the Soviet Union and former East-bloc countries. Perhaps because of his ancestral roots in the same part of the world, Erwin was able to develop warm relations with these people, based on the same alert tolerance his students so valued. Sensitive to the difficulties of the political, intellectual, religious, ethnic, and scholarly conditions under which they were forced to work, Erwin was always ready to respond generously with whatever help he could, usually gifts of books. The ties he established over the years form an important part of the backdrop to the attempt with Hans Wussing from Leipzig to launch the ""Science Networks" series of Historical Studies as a collaborative East-West venture. All that has taken on a new face recently but the series is still going strong with the Birkhäuser Verlag in Basel.

This effort is but one small piece of the larger picture of Erwin and his work at present. Now that he is officially retired, Erwin finally can reclaim some of the time and energy he so long focused on his students, concentrate on his own projects, and strengthen his ties with colleagues around the world. As always, his interests are many and diverse. He has long been working on a study of Lisa Meitner and her circle in the period between the two world wars. Another, even more longstanding interest is in the interplay of physics and chemistry, especially in the area of physical chemistry. In addition, Erwin is still actively pursuing, for publication this time, his interest in scientists as philosophers as well as the relations between science and religion.

Brown University

MARY JO NYE*

INTRODUCTION

The modern physical sciences are characterized by overlapping domains of specialized scientific disciplines which include classical mechanics and electromagnetism, thermodynamics and physical chemistry, particle physics and quantum chemistry, cognitive science and neural psychology. These contemporary scientific domains have evolved as the result of social and institutional dynamics in particular places at particular times. The physical sciences also have evolved as the result of conceptual negotiations among generations of mathematicians, experimentalists, philosophers and theologians who have been concerned to establish priorities in understanding physical phenomena and the natural world.

Some matters of epistemological discussion and debate have been perennial issues since the seventeenth century: whether the basis of knowledge lies in sensation or ideas, in facts or theories, in a correspondence between concepts and the world or in successive theories leading to a coherent picture of the world. In the seventeenth century, traditions of natural philosophy, natural theology, and Euclidean geometry were so closely intertwined that all often were pursued and professed by the same learned cleric or layman. By the late nineteenth century, the demarcation between science and religion was much clearer than in the seventeenth century, but many scientists remained strongly concerned with questions of metaphysics and epistemology, and some theologians explored the implications of scientific epistemology for the rational claims of religious doctrine.

The essays in this volume address developments in the physical sciences over the past three centuries with particular attention to epistemological issues underlying the conceptual development of the diverse sciences constituting "physical science." Of special interest are discussions at the intersection of mathematics, theology and natural philosophy about the nature and end of physical explanation. Should, and would, its results demonstrate chance or determinism, probability or certainty, structure or function, purpose or relation, commensurability or incommensurability, unity or diversity, the ordinary or the extraordinary in the laws and relationships governing the physical world?

Several overlapping themes are to be found in the sections of this volume. One theme in all four parts explores the role of mathematics in understanding the physical world: is mathematics a tool or a goal, a contingent representation or a universal truth in scientific law and physical theory? Has our perception of the function of mathematics in physical theory or the nature of mathematics as a fundamental science changed since the seventeenth century?[1]

A second theme involves the changing relationship between natural theology and natural philosophy, focussing on implications for theories of scientific and mathematical truth. In the seventeenth century, theologians came to recognize that the explanation of physical phenomena was to constitute the domain of the new scientists. However, theology was still to play a crucial role in the formation of this new science,[2] for example, convincing the practitioners of physical science that the laws they discovered must be deterministic in structure.

By the late nineteenth century, as physical theories, like evolutionary theories, increasingly emphasized chance, uncertainty, and even the contingency of the very scientific theories which demonstrated contingency, scientists, philosophers and theologians began reflecting on the history of science and reformulating definitions of contingent and necessary truths.[3]

A third theme in the volume concerns this ongoing definition and redefinition of the conceptual foundations and boundaries of the physical sciences as they proliferated into specialized disciplines. In nineteenth-century Germany and Austria, perhaps more than anywhere else, there was considerable determination still to ground the physical sciences conceptually in metaphysics and mathematics, if no longer in theology, despite institutional trends toward professionalized, specialized laboratory and engineering science.[4]

One result was the reaffirmation of the high place of theoretical science as abstract and idealized knowledge. Responding to David Hilbert's proof by "contradiction" of the existence of a finite basis for certain invariants, the Erlangen mathematician Paul Gordan was reported to have exclaimed, "Das is nicht Mathematik, das ist Theologie."[5] Response to the general theory of relativity and to non-classical wave mechanics sometimes took the form "Das is nicht Physik, das ist Mathematik."[6]

In reexamining the empirical, conceptual, and mathematical foundations of traditional mechanical and chemical science, physical scientists of the late nineteenth and the twentieth centuries worked out the foundations of newly defined fields like electrical science, biological physics and psychophysics, thermodynamics and physical chemistry, quantum chemistry and nuclear physics. But the new specialization by no means resulted in the complete loss to the new physical sciences of the epistemological and indeed the metaphysical controversies typical of natural philosophy.

Natural Theology, Natural Philosophy, and the Certainty of Mathematics

In the first essay, Sara Genuth outlines a transformation in theories of comets in the seventeenth century. Theologically-minded philosophers combined theories of the plurality of worlds with Biblical teachings of salvation and damnation in order to portray comets as wandering hells of wicked worlds, rather like cosmological penal colonies. In contrast, natural philosophers succeeded in capturing the literature on comets in the eighteenth century with physically-inspired speculations about temperature effects, electrical fluids, and gaseous atmospheres, which were inferred by the authority of analogy from physical theories of the earth. Here cometary theory serves as the exemplar of a change from theological to physical explanation whose basic form is ascertained from the specific example.

Lorraine Daston's concerns center on the classic statement of determinism made by P. S. Laplace in the *Essai philosophique sur les probabilités* that an intelligence "sufficiently vast" could submit data to analysis with the result that "nothing would be uncertain and the future, as the past, would be present to its eyes."[7] Daston explores the historical roots of this metaphysical viewpoint in seventeenth-century discussions of probability which, paradoxically, were part and parcel of a strictly deterministic epistemology. "Is it coincidence," she asks, "that Bernoulli and De Moivre came from Huguenot backgrounds, and that Pascal was a passionate convert to Jansenism?" Daston's claim is that theology, not natural philosophy, eliminated miracles from the world.

Joan Richards, too, explores the influences of theological presuppositions upon mathematical interpretations in England in the early nineteenth century. She contrasts French emphasis on rigor and certainty in establishing the foundations of the integral and differential calculus with English concern for arriving at empirically-based and conceptually clear statements, as defended by the English mathematicians John Herschel and William Whewell. In a novel interpretation, Richards argues for a shift in religious significance of the practice of English mathematics: from the idea that the practice of mathematics is an act of participation in divine knowledge to the different claim that the attainment of mathematical certainty is evidence of God's truth. The change, she argues, was significant to developments of both mathematics and theology later in the century.

Problems of Contingency, Coherence, and Truth

Beginning with the instance of nineteenth-century theologians' responses to Charles Darwin's theories, Frederick Gregory discusses the pervasiveness in the late nineteenth century of debates about the nature of truth and the relationship between scientific knowledge and transcendental truth. In particular, he argues the increasing influence upon some German theologians of a coherence theory of truth, in contrast to the traditional correspondence theory of truth.

This is a reversal of the chain of influence analyzed in Daston's and Richards' essays, since in Gregory's case study, pronouncements on the nature of truth are carried from the domain of science into theology, rather than the other way round. The result historically was the philosophical move among German theologians toward delineating stricter boundaries between religion and science. While scientists were to concern themselves with the physical world of natural phenomena, theologians were to focus on the moral domain of human ethics.[8]

Like the theologians studied by Frederick Gregory and the physical chemists analyzed by Diana Barkan, late nineteenth-century mathematicians turned to history in order to investigate and demonstrate the character of their knowledge. Skuli Sigurdsson uses Moritz Cantor's and Hieronymous Zeuthen's rival interpretations of the "discovery" of the calculus to illustrate conflicting views in the early twentieth century about whether there is a single reality "out there" which has been captured equivalently by diverse mathematical forms and notations.[9]

The problems which Sigurdsson addresses reflect the preoccupations with notions of transcendental truth and metaphysics characteristic of German and northern European culture in the late nineteenth and early twentieth centuries. Sigurdsson also provides a history complementary to Richards' discussion of the "limit," cautioning the reader against assuming the historical equivalence of similar mathematical statements worked out in very different contexts.

The Aims and Foundations of Physical Science at the Turn of the Century: The Cases of Electrical Physics, Psychophysics, and Physical Chemistry

By the mid-nineteenth century, "physics" was becoming a scientific discipline distinct from the eclectic natural philosophy which in the very early nineteenth century still included aspects of chemical philosophy and the "Humboldtian" sciences that investigated Nature on the grand scale and in the world-at-large.[10] The conceptual foundations and aims of the disciplinary physical sciences, their increasing dependence upon instrumentation and engineering, and the outline of their future progress were matters of considerable debate among scientists, philosophers, and scientists writing as philosophers. Among these we must rank most eminently Heinrich Hertz, Hermann von Helmholtz, Ernst Mach, and Pierre Duhem, all discussed in Part III of this volume.

In his account of the education of Hertz, Jed Buchwald gives a finely-etched account of the cultural milieu of a German youth oriented toward engineering and physics. Hertz, a student of mathematics, mechanics, and electrodynamics, was to become a superb experimentalist and masterly critic of the foundations of mechanics. Here we have a representative case of German physics at its finest, in a chapter which helps to answer the question of how young men came to be experimental and theoretical physicists in late nineteenth-century Germany.[11] Buchwald's chapter also brings out the extraordinary influence in late nineteenth-century science of Helmholtz, the "Helmholtzian" school, and the Helmholtzian "philosophy."[12]

This Helmholtzian theme is developed further in the chapter by Richard Kremer. Kremer, as much as any author in this volume, demonstrates the intersections in nineteenth-century physical science of ostensibly separate domains, analyzing how, as physical scientists, Helmholtz and Mach sought to unify the explanation of mind (psychology, psychiatry, biology) with the explanation of matter (chemistry, physics, biology). A familiar underlying motif is present in this essay: the ongoing debates over whether to base knowledge in sensation or in abstract and transcendental principles.[13]

If debates among Helmholtz, Mach and the physiologist Ewald Hering bring into view the varieties of empiricism practiced in the study of color vision, the debates analyzed by Diana Barkan in her chapter bring into focus fundamental differences of opinion among the founders of physical chemistry about the nature and aims of what they claimed to be a new physical science. In constructing different accounts of the history of physical chemistry, Wilhelm Ostwald, Pierre Duhem, Walther Nernst, and J. H. van 't Hoff not only claimed to have inaugurated a science with roots both in chemistry and physics, but they revealed startlingly different emphases in conceptions of physical chemistry. To some it was a bridging science; to others, it constituted a practical science; and to others still, its promise lay in its nature as a theoretical science reducing chemistry to absolutely fundamental, mathematical principles of physics.

Explanation and Discovery: The Claims of Chemistry, Physics, and Fortran

Mary Jo Nye takes up precisely this question of the historical and epistemological commensurability of chemistry and physics, or the reduction of chemistry to physics, by examining some of the relations between the respective aims and methods of chemistry and physics since the eighteenth century, as stated by their practitioners. While the use of quantification in chemistry was well-established by the end of the eighteenth century, the use of mathematical *explanation* remained one of the principal gulfs between physics and chemistry through most of the nineteenth century.

In examining the emergence by the mid-twentieth century of quantum chemistry and theoretical chemistry, Nye explores the issue of the reducibility of chemical processes to mechanical and mathematical physics. She concludes with a discussion of the priority and prestige of mathematical principles and mathematical reasoning in physical science, but argues the continued centrality to modern chemistry of the explanatory vocabulary and methodology typical of mid-nineteenth-century chemistry.

In the volume's concluding essay, Peter Galison analyses the development of two radically different strategies in modern particle physics for analysing data from bubble chambers. At Lawrence Radiation Laboratory in Berkeley (LBL), Luis Alvarez's physics group trained women to use reading machines to scan photographs which also were subject to computer analysis and to visual review by follow-up scanners and by physicists. In contrast, at the Center for European Nuclear Research in Geneva (CERN), Lew Kowarski backed development of a partially automated scanner which he envisioned as the first step in the elimination of human beings from the process of reading bubble-chamber photographs.

In his detailed study of the work regimen of the scanners, physicists, and engineers who worked in these laboratories, Galison demonstrates the evolution of a fundamental controversy about the nature of modern physics and the relationship of physicists to the instruments, data, and interpretation of data by machine computers that are not human and are not directly interacting with the physicist. The epistemological question at issue was whether bypassing the human roles of the scanner and the physicist, in the identification of particle events in the bubble chamber, eliminated a crucial epistemic role of the human for the practice of physics. A second problem was whether Kowarski's strategy devalued the completely unexpected event (the "golden event"), in favor of the statistically predictable event? As Galison puts it, there was some fear that Kowarski's method would exchange the physicist's laboratory for a library, i.e., an archive of magnetic tapes to be re-scanned periodically from a new point of view. Much as Nye shows how some chemists resisted the reduction of their work to mathematical physics, Galison analyzes how some physicists resisted the reduction of their work to FORTRAN.

In Galison's study, the role of mathematics indirectly highlighted lies in the computer language which provides the design and implementation of the instrumental hardware which physicists and other scientists employ in laboratory science. This laboratory science is physically removed from the natural world in which seventeenth-century natural philosophers began their experimental investigations and in which seventeenth-century theologians found evidence and experience of transcendental meaning and value.[14]

At the mid-nineteenth century, physicists supplanted both natural philosophers and natural theologians as (lay) divines in the study of the natural world. As Erwin Hiebert has discussed in his teaching as well as in published essays, physicists often wrote as philosophers. However, while some late nineteenth-century scientists claimed ethics and values as a domain of scientific inquiry, most scientists assented to twentieth-century theologians' assertion of the distinction between the domains of existential moral laws and phenomenological natural laws.[15]

A few, like the theoretical physicist and theoretical chemist Charles Coulson, whose work is discussed in Part IV by Nye, continued the time-honored tradition of serving as church preacher and practicing scientist. During the last decades of his life, Coulson, a lay preacher in the Methodist Church, spent so much time writing on religion, according to Erwin Hiebert, that London journalists began referring to him as the holder of the Chair of "Theological Physics."[16] Coulson delivered lectures in 1954 opposing the views that science and religion are the same or that they represent different domains of truth. In contrast, he argued that science is one part of the revelation of God, "consonant in its insistence on value and person with the traditional Christian conception, but adding certain elements which we could not otherwise ever know."[17]

In all of this, the high place of mathematics in the classification of the sciences and the hierarchy of knowledge remained practically untouched.[18] Indeed, in ordinary discussions about the sciences, mathematics sometimes has not even been characterized as a "science," presumably because the modern sciences are closely associated in common parlance with the material and the contingent, whereas mathematics is identified with the immaterial and the certain, i.e., with what is true.

Historians of science have been counselled to avoid the adoption of the scientist's characteristic attitude sanctioning the "transcendence" of science, one expression of which we find in Steven Weinberg's statement of 1974: "The laws of nature are as impersonal and free of human values as the rules of arithmetic. We didn't want it to come out that way, but it did."[19] Nonetheless, historical study demonstrates over and again the quest of a great number of mathematicians, theologians, and natural philosophers for a transcendent ideal, as they pursued and invented what has come to be called "physical science."

University of Oklahoma

NOTES

* Thanks to Joan Richards for substantive reworking of parts of this essay and to Skuli Sigurdsson for helpful comments.
[1] See Philip Kitcher, *The Nature of Mathematical Knowledge* (Oxford, 1983) and William Aspray and Philip Kitcher, eds., *Essays on the History and Philosophy of Modern Mathematics*, Volume XI, *Minnesota Studies in the Philosophy of Science* (1987).

2 Richard S. Westfall, *Science and Religion in Seventeenth-Century England* (New Haven, 1958), or more recently, Simon Schaffer, "Godly Men and Mechanical Philosophers: Souls and Spirits in Restoration Natural Philosophy," *Science in Context* 1 (1987), 55–85.
3 Nicholas Jardine, *The Birth of History and Philosophy of Science* (Cambridge, 1984), or, for a more specific case, Mary Jo Nye, "The Moral Freedom of Man and the Determinism of Nature: Catholic Science Viewed through the *Revue des Questions Scientifiques*," *British Journal for the History of Science* 9 (1976), 274–292.
4 As discussed in this volume, Hermann von Helmholtz is an important case in point. See his *Epistemological Writings*, ed. Paul Hertz and Moritz Schlick (transl. M. F. Lowe), Volume XXXVII, *Boston Studies in the Philosophy of Science* (Boston, 1977).
5 Quoted by Clark Kimberling, "Emmy Noether and Her Influence," in James W. Brewer and Martha K. Smith, eds. *Emmy Noether. A Tribute to Her Life and Work* (New York: Marcel Dekker, Inc. 1981), 3–61, on p. 11. I am indebted to Jayne Ann Arnold for this reference.
6 On relations between the "physical" and the "mathematical," see Section 4b in the chapter by Buchwald in this volume; also, David Rowe, "Klein, Hilbert and the Göttingen Mathematical Tradition," in Kathryn M. Olesko, ed., *Science in Germany. The Intersection of Institutional and Intellectual Issues*, Volume V, Osiris, 2nd series (1989), 186–213; and Erwin N. Hiebert, "The Transformation of Physics," in Mikulas Teich and Roy Porter, eds. *Fin de Siècle and Its Legacy* (Cambridge: Cambridge University Press, 1990), pp. 235–253.
7 P. S. de Laplace, *Essai philosophique sur les probabilitiés* (Paris, 1814) in *Oeuvres complètes*, ed. Académie des Sciences (Paris, 1878–1912), Volume VII, p. vi; English translation from F. W. Truscott and F. L. Emory, trans., *A Philosophical Essay on Probabilities* (New York: Dover, 1951), p. 3.
8 For a fuller treatment of themes in this chapter, see Frederick Gregory, *Nature Lost? Natural Science and the German Theological Traditions of the Nineteenth Century* (Cambridge, Mass.: Harvard University Press, 1992).
9 A classic and combative statement of the problem is found in Sabetai Unguru, "On the Need to Rewrite the History of Greek Mathematics," *Archives for History of Exact Sciences* 15 (1975), 67–114.
10 See Susan F. Cannon on "Humboldtian Science" in *Science in Culture: The Early Victorian Period* (New York: Dawson and Science History Publications, 1978). Also Keith Nier, "The Emergence of Physics in Nineteenth-Century Britain as a Socially Organized Category of Knowledge: Preliminary Studies" (Harvard University Ph.D. Thesis, 1975); and the articles by David Gooding and by Peter Galison and Alexi Assmus in David Gooding, Trevor Pinch and Simon Schaffer, eds., *The Uses of Experiment. Studies in the Natural Sciences* (Cambridge: Cambridge University Press, 1989).
11 The now classical work on the general subject is Christa Jungnickel and Russell McCormmach, *Intellectual Mastery of Nature*, 2 vols. (Chicago: University of Chicago Press, 1986).
12 An important new study of Helmholtz and the Helmholtzian school will soon be published under the editorship of David Cahan.
13 On the Helmholtz school and the pervasiveness of the empiricism/nativism debate in German and Austrian intellectual circles, see yet another example, in S. P. Fullinwider,

"Darwin Faces Kant: A Study in Nineteenth-Century Physiology," *British Journal for the History of Science* **24** (1991), 21–44.

[14] On laboratory science, see William Coleman and Frederic L. Holmes, eds., *The Investigative Enterprise: Experimental Physiology in Nineteenth-Century Medicine* (Berkeley: University of California Press, 1988); David Gooding *et al.*, eds., *The Uses of Experiment* (ref. 10); and Frank A. J. L. James, ed. *The Development of the Laboratory* (London: Macmillan, 1989).

[15] For example, see Erwin Hiebert, "The Integration of Revealed Religion and Scientific Materialism in the Thought of Joseph Priestley," in Lester Kieft and Bennett R. Willeford, eds., *Joseph Priestley: Scientist, Theologian and Metaphysician* (London: Associated University Presses, 1980), 27–61; "Einstein's Image of Himself as a Philosopher of Science," in Everett Mendelsohn, ed., *Transformation and Tradition in the Sciences. Essays in Honor of I. Bernard Cohen* (Cambridge, Mass.: Harvard University Press, 1984), 175–190; and "The Scientist as Philosopher," *Schriftenreihe für Geschichte der Naturwissenschaften, Technik und Medizin* **24** (1987), 7–17.

[16] Erwin Hiebert, "Modern Physics and Christian Faith," in Ronald Numbers and David Lindberg, eds., *God and Nature* (Berkeley: University of California Press, 1986), 423–447, on p. 441.

[17] Quoted in *ibid.*, from C. A. Coulson, *Science and Christian Belief* (Chapel Hill: University of Norton Carolina Press, 1955), pp. 2–3.

[18] On the history of classifications, see Nicholas Fisher, "The Classification of the Sciences," in R. C. Olby *et al.*, eds., *Companion to the History of Modern Science* (London: Routledge, 1990), 853–868.

[19] Paul Forman, "Independence, Not Transcendence, for the Historian of Science," *Isis* **82** (1991), 71–86; and Steven Weinberg, "Reflections of a Working Scientist," *Daedalus* (Summer, 1974), 33–46, on p. 43. This comment is given special attention by Evelyn Fox Keller, in *Reflections on Gender and Science* (New Haven: Yale University Press, 1985), p. 6.

PART I

NATURAL THEOLOGY, NATURAL PHILOSOPHY, AND THE CERTAINTY OF MATHEMATICS

SARA SCHECHNER GENUTH

DEVILS' HELLS AND ASTRONOMERS' HEAVENS: RELIGION, METHOD, AND POPULAR CULTURE IN SPECULATIONS ABOUT LIFE ON COMETS

According to eighteenth-century astronomical theory, comets were opaque, spherical, solid bodies shining by reflected light and revolving in elliptical orbits around the sun. The perceived similarities between them and the planets raised the question of their fitness for habitation. At the heart of this question was a tension among natural theology, scientific reasoning, and popular culture. Theology played a role in suggesting how the existence of extraterrestrial life was to God's credit, and its natural side sought to divine the Lord's purposes in populating the heavenly bodies. Scientific reasoning played a role in posing what that life might be like, given the material and biological conditions assumed to inhere on the heavenly bodies by analogy with the earth. And popular culture shaped the way that rules of reasoning and natural theology were applied to the case of comets. Before turning our attention to speculations about cometarians, let us examine the doctrine of the plurality of worlds with its religious and methodological underpinnings.

Though rooted in antiquity, the idea of the plurality of worlds was given new force in the seventeenth century by work in astronomy and natural history. Pointed skyward, telescopes revealed mountains standing on the moon and satellites orbiting around Jupiter and Saturn. The analogy between the earth and planets encouraged further comparison of the sun to the fixed stars, and stars soon became seen as other suns surrounded by their own planetary retinues. Pointed earthward, microscopes revealed that the smallest drops of water were crowded with minute creatures. This discovery led natural philosophers to ask why should the vast planets remain barren if nature was so fecund here. Were we not ascribing wastefulness to God if we thought he had squandered those opportunities to populate the celestial globes? Was it not more consistent with the doctrine of divine omnipotence and benevolence to maintain that God created innumerable worlds? These thoughts were expressed by Bernard le Bovier de Fontenelle in his lively and highly popular *Entretiens sur la pluralité des mondes* (1st edition, 1686). He endorsed the plurality of worlds for five reasons:

[1] The total resemblance of the planets to the earth which is inhabited; [2] the impossibility of imagining any other use for which they were made; [3] the fecundity and magnificence of nature; [4] the consideration she seems to have had for the needs of their inhabitants, as having given moons to planets remote from the sun, and more moons to those more remote; and [5] this which is very important, that everything is in its favor, and nothing is against it.[1]

Fontenelle's reasons were by no means original and to a great extent they hinged on what Arthur Lovejoy has termed the "principle of plenitude" – i.e., the doctrine that "no genuine potentiality of being can remain unfulfilled, that the extent and abundance of creation must be as great as the possibility of existence and commensurate with the productive capacity of a 'perfect' and inexhaustible Source, and that the world is the better, the more things it contains."[2] Tied to this principle of plenitude was a conception of teleology or final causes within the universe. All things were created for a purpose. Sacred texts such as Psalm 8 affirmed that the earth and its creatures existed for the sake of man, but were silent on the purpose of the stars, the planets, and their satellites. It was sheer hubris, however, to say that the dimmest of these, some visible only by means of a telescope, were created for man's benefit too, for that would imply that God did much in vain. If planets and stars were of no use to men, then, by parity of reason, they were created to serve their own inhabitants or those in their neighborhood.[3]

In making the case for pluralism, natural philosophers made much use of the methods that were codified by Isaac Newton in his "Rules of Reasoning in Philosophy." Newton's first rule was the principle of simplicity tinged with teleology: "We are to admit no more causes of natural things than such as are both true and sufficient to explain their appearances. To this purpose the philosophers say that Nature does nothing in vain, and more is in vain when less will serve." His second rule expressed the uniformity of nature: "To the same natural effects we must, as far as possible, assign the same causes." Newton applied this rule "to respiration in a man and in a beast; the descent of stones in *Europe* and in *America*; the light of our culinary fire and of the sun; the reflection of light in the earth, and in the planets." By way of justification for the principle of uniformity, Newton referred to the "analogy of nature, which is wont to be simple, and always consonant to itself." This explanatory statement was attached to Newton's third rule, which concerned analogical reasoning and universal qualities.[4] Following these premises of simplicity, uniformity, and analogy, and guided by empiricism whenever possible, the natural philosopher in theory was to feel secure in applying the physical laws governing terrestrial events to celestial phenomena.

Since the nature of cometary life was not open to direct inspection, the authors to be discussed in this paper often relied on these simple rules to guide them in comparing comets to the earth. Their speculations highlight the degree of latitude afforded by the rules and the pitfalls in trying to apply them in a case where there were few empirical controls. Of the principles of simplicity, uniformity, and analogy, analogy generated the most controversy in its practical application. First, in comparing comets to the earth, natural philosophers had to weigh the relative importance of likenesses and differences. Everyone agreed that the earth's nearly circular orbit was quite unlike those of the comets, which traveled in elongated ellipses, parabolas, and sometimes hyperbolas that were angled every which way to the ecliptic, but they disagreed on whether this made any difference in terms of their habitability. Ideally, it would have been prudent to establish a firm foundation on agreed physical or metaphysical preconditions before ascending the ladder of analogy, but as our case study of comets will show, there was little consensus on the selection of dependent and independent variables, or on whether an appropriate goal of analogy was to reason from form to function, or from function to form.

Among those who disregarded orbital characteristics, some were prompted by perceptions of similar physical forms (e.g., of comparable atmospheres) to advocate an analogy of function (e.g., to sustain life); conversely, others were led by their metaphysical belief in the uniformity of function (e.g., to sustain life) to posit the existence of analogous physical forms (e.g., atmospheres) on comets. Among those who factored in the stresses that orbits of different shape would introduce (e.g., by exposure to extremes of hot and cold), some concluded that though comets might be composed of matter similar to the earth, their functions could not be the same; others were led by their belief in analogous functions to conclude that the physical composition or life-forms had to be different in order to survive the stresses caused by the orbits. Here we have four possible arguments backed up by little or no tangible evidence. One consequence of such methodological ambiguity was the disagreement over whether comets were places of torment or delight. As this paper will suggest, the choice between hell and heaven was influenced by religious tenets and vestiges of folk beliefs.

WANDERING HELLS

The plurality-of-worlds view gained momentum during the course of the seventeenth century. After Fontenelle, the view was forthrightly advanced by Christiaan Huygens, quietly intimated by Isaac Newton, and propounded by Richard Bentley and John Ray.[5] In the *Astro-Theology* of William Derham, F.R.S., chaplain to the future king, George II, and Boyle Lecturer, the hypothesis was touted highly. Derham supported the "New Systeme" of the world, which "supposeth [that] there are many other Systemes of *Suns* and *Planets*, besides that in which we have our residence: namely, that every Fixt Star is a Sun, and encompassed with

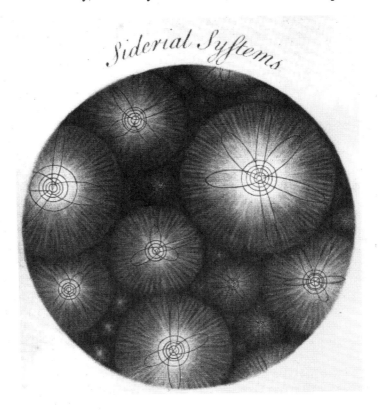

Fig. 1. Newtonian plurality of worlds. James Ferguson, *An Idea of the Material Universe* (London: 1754). [By permission of the Houghton Library, Harvard University.]

a Systeme of Planets, both Primary and Secondary, as well as ours." All planets, Derham judged, were "places as accommodated for Habitation, so stocked with proper Inhabitants" because myriads of systems befitted an infinite, glorious Creator.[6] It was perhaps due to the credentials of Derham that his pluralistic view became sanctioned and adopted by the majority of eighteenth-century natural philosophers.

Comets were a challenge to the pluralists. Once Newton and Edmond Halley had convinced natural philosophers that comets were members of the solar system, it was logical to compare their sizes, material composition, and atmospheres with the planets, and by extension, their habitability. But as Derham observed, their elliptical orbits differed greatly from those of the planets, indicating that their uses might differ as well.[7] Scorched and frozen in turns, they seemed unlikely abodes of happy creatures. Therefore, Derham thought that God appointed "such noxious Globes for the Executioners of his Justice, to affright and chastize sinful men at their approaches to the Earth, (and as some have imagined) to be the place of their Habitation and Torment after death."[8]

Here Derham, I suspect, alluded to the thought of Dr. George Cheyne. In a work on natural religion, Cheyne had noted that atmospheric conditions of comets created environments hostile to life as we know it on earth:

These *Comets* sometimes come so near the *Sun*, as to be heated to such a Degree that they cannot become cool again, in many Thousands of Years. This with its violent Motion in a *Curve*, which comes near to a streight Line, after it has pass'd its *Perihelium*, and the irregular Disposition of the confuss'd Mass of its *Atmosphere*, makes it an unfit Habitation for Animals, that are not in a state of Punishment, so far as we can conceive of the Nature of Animals.[9]

William Whiston concurred, writing:

The external Regions of Comets, which by passing through such immense Heat when nearest, and such prodigious Cold when farthest off the Sun; and by the confused and Chaotick State of their Atmospheres, do evidently appear incapable of affording convenient Habitations for any Beings that have Bodies, or Corporeal Vehicles, whether visible or invisible to us.[10]

With respect to habitability, comets were "design'd for very different Purposes from the Planets," Whiston and Cheyne concluded.[11] Unable or unwilling to imagine new forms of life suited to hostile cometary climates, they seized on comets as the sites of hell.

The theory had much to recommend it. Cheyne noted that it accommodated the philosophical rule of simplicity, for nothing in nature was wasted, nothing was in vain. From comets "we may learn that the *Divine Vengeance* may find a seat for the punishment of his Disobedient Creatures, without being put to the Expense of a new Creation."[12] It was economical and resourceful for God to use comets as hell. Whiston delighted in the theory's compatibility will both science and scripture:

> I observe, that the Sacred Accounts of *Hell*, or of the Place and State of Punishment for wicked Men after the general Resurrection, is agreeable not only to the Remains of ancient profane Tradition, but to the true System of the World also. This sad State is in Scripture describ'd as a State of *Darkness*, of *outward Darkness*, of *blackness of Darkness*, of *Torment* and *Punishment for Ages, or for Ages of Ages, by Flame*, or *by Fire*, or *by Fire and Brimstone, with Weeping and Gnashing of Teeth*; where *the Smoak of the* Ungodly's *Torment ascends up for ever and ever*; where they are Tormented *in the Presence of the Holy Angels, and in the Presence of the Lamb*; *when the Holy Angels shall have separated the Wicked from among the Just, and have cast them into a Furnace of Fire*. Now this Description does in every Circumstance, so exactly agree with the Nature of a Comet, ascending from the Hot Regions near the Sun, and going into the Cold Regions beyond *Saturn*, with its long smoking Tail arising up from it, through its several Ages or Periods of revolving, and this in the Sight of all the Inhabitants of our Air, and of the rest of the System; that I cannot but think the Surface or Atmosphere of such a Comet to be that *Place of Torment* so terribly described in Scripture.[13]

The blazing comet would be a "most useful Spectacle," a reminder to the blessed to preserve their obedience to the Lord.

Cotton Mather likewise advised "serious Christians" to take note of those hellish stars. In colonial New England, he preached that God directed noxious comet steams towards sinful planets capable of reform, but set incorrigible worlds ablaze and hurled these hells into cometary orbits.[14] Every comet, he declared, was a "wicked World *made a fiery Oven in the Time of the Anger of GOD!*"[15]

Mather's statement that comets were punished worlds raised the question of whether our earth could be transformed into a cometary hell. In an essay published as an appendix to Tobias Swinden's *Enquiry into the Nature and Place of Hell* (1714), William Wall lent indirect support to that idea. He suggested that there were thousands of planets beyond Saturn. When destroyed by divine decree, these fell towards the sun. At the Second Coming, once the righteous were gathered to meet the Lord in the air, the earth too would drop as a comet into the sun:

> in which descent ... not only the wicked Men, with all that is on the Surface, will be burnt up; but also the Heavens, i.e. the Sky about the Earth, will be dissolv'd, and the Elements (of Air, Water, &c.) be evaporated, or melt with fervent Heat, and the Body of the Earth burnt to a Coal.[16]

That was not the end of things, however. The scorched planet would "(like a Nut-shell let fall into a great Flame) be tossed out again, and carried to a new and better Place in the Firmament, and become a new Earth in a new Heaven or Sky, and there be the Scene of the millennial State."[17] Although Wall saw the sun as the ultimate prison for the damned, Whiston situated hell on the old earth, which would after the millennium be forced to travel through the heavens in a terrible blaze, leaving the blessed behind to watch the fearful spectacle from the vantage point of the earth's former orbit.[18] The cometary fate of the earth also impressed Mather Byles, a nephew of Cotton Mather. In vivid couplets inspired by the 1744 comet, he described the Day of Judgement when our guilty globe would be transformed into a fiery hell in cometary orbit.[19]

Implicit in these statements was an insistence on the uniformity and analogy of nature; or to put it more specifically, on the thesis that the material composition and life-forms of comets were essentially like the planets'. By presupposing the analogy and uniformity of material and biological structures, Derham, Whiston, Cheyne, and others were led to conclude that cometarians had a tortured existence. Were it not for the comets' eccentric orbits, these "Worlds in Confusion" would be in an orderly state fit for the domicile of happy creatures.[20] Whiston had first made this point in *The New Theory of the Earth* (1696), where he claimed that the earth had once been a comet and after the millennium would be transformed into a comet again.[21] Cheyne apparently subscribed to a similar view, describing comets as "the first *Rudiments* of *Planets*, not as yet brought into our *System*, or rather the *Ruins* of some banish'd thence."[22] The motif of comets as ruined, punished planets appears again and again in early eighteenth century texts. It was one basis for analogical arguments that established hell on comets.

The role of natural religion in the construction of these theories is worthy of comment, too. In hunting for a physical site of hell, the authors were motivated by a desire to provide a naturalistic interpretation of scriptural accounts of hell. In selecting comets, they adopted a view of divine power that emphasized both purpose and economy. God was magnificent in all his designs but frugal in the execution of them. Purposeful yet parsimonious, God was apparently unwilling to create life-forms happily suited to cometary climates.

The gruesome view that comets were penal worlds created to torment guilty creatures was widely adopted at the turn of the eighteenth century.

Nonetheless by the century's end, it had gone out of style, supplanted by the view that comets were populated not by sinners, but by the luckiest beings, blessed with the opportunity to explore the farthest reaches of the universe. In espousing this upbeat theory, natural philosophers emphasized different religious and physical premises, drawing inspiration from both the principle of plenitude and the latest theories of heat.

OTHER WORLDS

In the foregoing theories of comets as ruined planets, we may detect a bit of Cartesianism. René Descartes had kept silent on the possibility of other worlds and intellectual life among the stars, but his vortex cosmology was conducive to this opinion, for he had viewed both comets and planets to originate as darkened suns that no longer could sustain their own vortices and were consequently swept into neighboring ones.[23] Followers of Descartes extended his hypothesis to include as comets the old star's "lost" planets, which also left the ruined vortex and migrated into others.[24] The tails of these wayward planets were only optical illusions, and the cometary planets were otherwise the same as any other planet. If one assumed that normal planets were inhabited, then why not those that had lost their moorings? This was the point wittily considered by the Cartesian pluralist, Fontenelle, in his *Entretiens sur la pluralité des mondes*. In that work, the narrator chats with a Countess about cometarians, observing how sad it must be for them to see their own sun extinguished, but how diverting it must be to change vortices. Travelling through the heavens, cometarians would eagerly await the sighting of a new solar system. When one was spied, they would cry out, "*A new Sun, a new Sun*, as Sailors use to cry, *Land, Land*."[25]

Early Newtonians had a less sanguine view of cometary life, believing it to be a tortured existence at best, but by the mid-eighteenth century, we find natural philosophers embracing attitudes more in keeping with Fontenelle's pluralism. The first thing that distinguished these later scholars from the earlier hell-finders was the focus of their natural theology. In their opinion, the hellacious-comet hypothesis offered a limited and irreverent grasp of divine omnipotence and providence. God was a font of goodness, an inexhaustible source of life. It was more in keeping with his wisdom and benevolence that he should populate cometary worlds with creatures suited to their weather. These views were expressed in 1754 by James Ferguson, who wrote:

The extreme Heat, the dense Atmosphere, the gross Vapours, the chaotic-like State of Comets, seem to indicate them unfit for the Purposes of animal Life, and a most uncomfortable Habitation for rational Beings. Nevertheless, when we consider on the other Hand the infinite Power and Goodness of the Deity, the latter inclining, and the former enabling him to make Creatures suited to all States and Circumstances; that Matter exists only for the Sake of Intelligence, and wherever we find it, we find it always pregnant with Life, or necessarily subservient thereto; . . . when we reflect moreover, that some Centuries ago, 'till Experience undeceived us, a great Part of our Earth was judged uninhabitable; the torrid Zone by reason of excessive Heat, and the frigid Zones on account of exces-

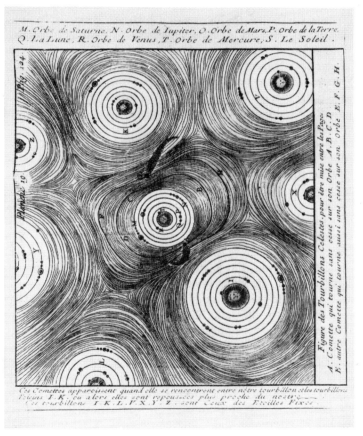

Fig. 2. Cartesian plurality of worlds showing comets within vortices. Nicolas Bion, *L'usage des globes celeste et terrestre, et des spheres suivant les differens systemes du monde* (Paris: 1744). [Courtesy of the Adler Planetarium, History of Astronomy Collection, Chicago.]

sive Cold: From these Considerations, it seems highly probable that such numerous and large Masses of durable Matter as the Comets, the most considerable Part of our System, however unlike they are to our Earth, are not destitute of Beings capable of contemplating with Wonder, and acknowledging with Gratitude, the Beauty, Wisdom, and Symmetry of the Creation; which is more plainly to be observed in their extensive Tour through the Heavens, than in our more confined Circuit. If farther Conjecture is permitted, may not one suppose they are peopled with guilty Creatures reclaimable by Sufferings, as we are on the Earth.[26]

Pierre Maupertuis, John Winthrop, Johann Lambert, and others joined Ferguson in finding extraterrestrial cometary life consistent with the inexhaustible variety of the works of God.[27] As Ferguson pointed out, God had created an "astonishing Diversity of Animals in Earth, Air, Water, and even on other Animals." Every blade of grass, every drop of water teemed with living things who enjoyed "such Gratifications as their Nature and State require[d]." If the earth teemed with life, then comets, like the planets and satellites revolving around every sun, could be commodious habitations for beings capable of knowing, obeying, and adoring God.[28]

In embracing the principle of plenitude, however, Lambert went one step further. Not only could comets be inhabited, but God had made them expressly for the purpose of packing the universe with as many creatures as possible. Angled every which way to the ecliptic, cometary orbits were suited to maximize habitability in every solar system. Lambert estimated that there were as many as five million comets around the sun alone, making them "much more necessary and useful for the habitability of the solar system than are the planets," he wrote, "and therefore contribut[ing] the most not only to the completeness but also to the perfection of the solar system."[29]

Believing it quite vain for men to suppose that a hundred or so comets had been created for the sake of punishing the solar system's sinners, Ferguson and Lambert emphasized the privileged state of cometarians who got to admire more of God's handiwork in their wide-ranging orbits. Some comets travelled in hyperbolic orbits, Lambert jealously noted, and therefore, were destined to visit one star after another. Far from peopling them with luckless sinners, Lambert "made astronomers of all of them, created for the purpose of viewing the edifice of the heavens, the position of each sun, the plane and course of their planets, satellites, and comets in their whole interconnectedness."[30]

Although the motif of ruined planets is frequently met within the works of those who thought comets to be hells, it seldom appears in the

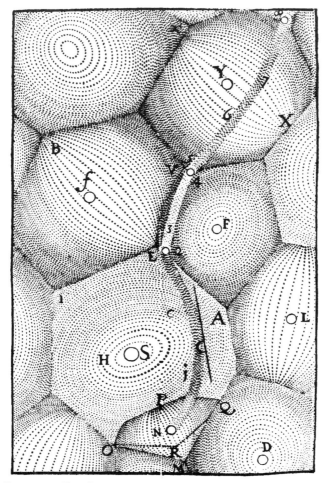

Fig. 3. Comet traveling from vortex to vortex. Descartes, *Principia philosophiae* (Amsterdam: 1644). [Courtesy of Department of Special Collections, University of Chicago Library.]

writings of those who thought them to be happy abodes. Those committed to cometary life tended to emphasize the permanence of celestial bodies. Judge Andrew Oliver, for example, rejected the possibility that comets and planets shared electrical atmospheres because this would be dangerous to world order. Planets with smaller atmospheres than comets would have less electrical fluid, and so would attract electrical tails. The discharge would result in an "instantaneous cataract of fire" between the two worlds, and an "explosion, which nothing could equal, short of the final voice of an archangel; and, if it were not sufficient to rouse the ashes of the dead, might reduce the living to their primitive dust." This was too frightening to contemplate, and Oliver hastened to reassure his readers:

> But such a catastrophe, we have not the least reason to dread, from the neighborhood of a Comet, unless we can suppose, that infinite wisdom and goodness would create one world, merely for the destruction of another; as we cannot conceive of any other ends, to which such huge electrical atmospheres could be adapted. Indeed the discharge would be equally fatal to both worlds; as it is certain from electrical experiments, that the effects of a stroke of lightning are the same, whether the flash proceeds from the cloud to the Earth, or is discharged from the Earth into the cloud, both of which have happened during the same thunder-gust.[31]

Serendipitous destruction of planetary worlds was not in keeping with God's benevolence or design. Whiston's theory, Oliver retorted, did not hold water.[32] Lambert thoroughly agreed, decrying theories in which planets became comets and comets became planets because, he said, "I do not at all think that a celestial body can make itself inhabited, or that, should the earth become a comet, the seed for industrious creatures could lie hidden in it and sprout up afterwards."[33] The region of comets was not a "storehouse of underdeveloped planets;"[34] every celestial body was to remain what it always had been. God saw that there was no need to transform one into another or repopulate it since "the preservation of celestial bodies and of their inhabitants is such an aim of the creation which allows no such exception that would include a complete destruction."[35]

The premises of permanence and preservation were fostered in part by the dedication of these natural philosophers to the principle of plenitude and their belief in the primacy of life, for it seemed unacceptable that God would go to the trouble of populating a comet only to destroy it. Again we see the role of natural religion in shaping scientific theory, and how the religious tenets at work differed from those held earlier by Whiston and others who had sought the site of hell. We may also here observe a distinction between the later pluralists and the earlier hell-finders in their applications of scientific method. Like the hell-finders, the pluralists agreed that nature was uniform, but they emphasized uniformity of function rather than of material and biological form. Or to put it more specifically, they accepted that the primary purpose of comets, like the planets, was to support life, and made their physical theories come in line with this assumption.

Late eighteenth-century physical speculations are noteworthy in showing how the commitment to the plurality of worlds was a driving force behind studies of the atmospheric and material conditions of comets that would make life possible. A common thread in these speculations was the view that the comas and tails of comets existed to protect the occupants. How this was accomplished varied from theory to theory. Lambert, for example, compared the tails and comas to the atmospheres of the planets, arguing that their distinctive appearance was an incidental by-product of the variation in solar heat experienced by a comet in its elongated orbit.[36] Composed of great vapory layers and mists, the comet's atmosphere expanded into a thick, enveloping fog as the comet approached perihelion and served to shield cometarians from the excessive heat of the sun.[37] The tail was formed from pieces of the atmosphere that were torn off when atmospheric heat rushed into the colder regions beyond the comet. Since tail matter was dispersed bit by bit through the aether, the comet lost a part of its atmosphere on each perihelion passage. In a twist on Newton's theory of comets' replenishing vital matter to the planets, Lambert proposed that comets were able to replenish their own stores of protective atmospheric matter by picking up fresh supplies as they travelled through the aether.[38]

Whereas Lambert had stressed the shielding nature of the comet's atmosphere, Dr. Hugh Williamson of Philadelphia turned his attention to temperature and pressure changes. Although many imagined that the surfaces of comets were alternately subjected to scalding heat and icy cold, Williamson thought that temperature fluctuations aboard a comet were less drastic and posed no obstacle to their habitation by beings of superior intelligence. He believed heat to be derived from a tremulous motion within bodies, and asserted that the heat of a comet depended both on its distance from the sun and its fitness to retain, propagate, and dissipate the vibrations that were communicated to its particles by light rays. The key was its atmosphere, which Williamson thought consisted of particles more subtle, elastic, tiny, and easily warmed than those in our air. He also reckoned the atmosphere on a comet to vary from eight to ten thousand miles in depth as compared to a depth of sixty to seventy miles on the earth. "Why should they have such a weight of atmosphere more than us? This is doubtless subservient to some very extraordinary purpose," he observed. Now where the earth's atmospheric pressure was greatest, as at sea level, the air was much heated by the sun; but where pressure was low, like that found on mountain tops, the air was cooler. The same should be true for comets, he claimed. At aphelion the comet's great atmosphere was highly compressed and responsive to the sun's feeble rays; the cometarians kept warm. Near perihelion the atmosphere was rarefied and rendered less fit for generating and retaining heat. The tail also helped to wick away excess heat. Thus, "we may easily see how they [cometarians] may be tolerably cool at noonday, on their nearest approach to the Sun."[39]

In Salem in 1772, Judge Oliver was so enthusiastic about Williamson's account of cometarians, that he penned a complete discourse on the origin and physical nature of comet atmospheres and tails, and their ameliorating effects on the abodes of cometarians. Oliver discussed meteorological conditions experienced by denizens of comets, and drawing on Halley's experiments with a diving bell, he considered how the cometarians' respiratory and optical organs might be adapted to an atmosphere whose density greatly fluctuated.[40] A distinguishing feature of Oliver's thesis was that all heavenly bodies including the sun had aerial atmospheres identical to the earth's. Formed from a highly elastic, transparent fluid, these atmospheres were mutually repellent. Their power of repulsion varied inversely as the distance, and Oliver believed that as celestial bodies gravitated towards each other, their respective atmospheres would endeavor to recede, while remaining gravitationally attracted to the bodies. As the solar atmosphere repelled a comet's, a tail was formed.[41]

Oliver illustrated this repulsion by means of electrical experiments using pith balls covered with threads, and an "artifical Comet, consisting of a small, gilt cork ball, with a tail of leaf-gold" suspended near a gilt sphere mounted on the stem of a Leyden jar. The electrical analogy had its limitations, however. Since electric sparks were self-luminous, Oliver was quick to assert that electric fluid was not responsible for comet tails, which shone by reflected light.[42] Electrified tails, moreover, would be dangerous to world order, and so were unacceptable on philosophical grounds. No matter how awful and portentous tails appeared, they were "designed for, and are wisely adapted to, the truly god-like purposes, of rendering habitable a vast variety of Worlds."[43] "Does it not redound more to his [God's] honour," Oliver asked, "to consider these bodies as

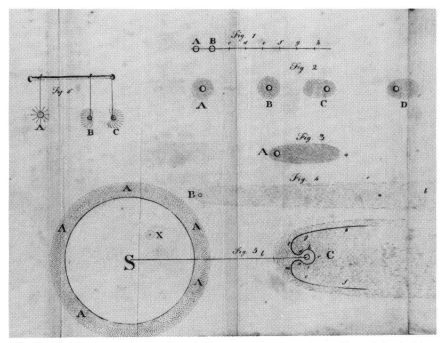

Fig. 4. Electric theory of mutually repellent cometary atmospheres. *Upper left*, electrified pith balls covered with strings that represent the atmospheres of comets at aphelion (*A*) and perihelion (*B*); *below*, the repulsion of a comet's atmosphere (*C*) by the sun (*S*). Andrew Oliver, An Essay on Comets (Salem: 1772). [Courtesy of the Adler Planetarium, History of Astronomy Collection, Chicago.]

so many inhabited Worlds, provided with every[thing] necessary for the comfortable subsistence of innumerable inhabitants, rational and irrational, like the Earth?"[44]

Oliver dedicated his essay to his former Harvard professor, Winthrop, who made the guarded comment that it was "really curious and uncommon."[45] Yet, it is a testimony to the temperament of the times that, like Williamson's paper, the tract was well received at home and abroad. Williamson was elected to a Dutch scientific society on the strength of his thesis, whereas Oliver's book was translated into French, reprinted in 1811, and treated seriously by Joseph Priestley. Priestley read the essay "with attention and pleasure," but urged Oliver to reconsider his rejection of any role for electrical fluids and to support Priestley's own theory that electrical atmospheres surrounded each heavenly body and interacted with aerial atmospheres at great distances.[46]

Oliver's, Lambert's, and Williamson's propositions concerning heat transfer within comet atmospheres were rather unsophisticated compared to the analysis of Pierre-Simon Laplace. In 1783 Laplace had made careful experiments with Antoine Laurent Lavoisier on latent and specific heats, and he incorporated this research into the treatment of the nebulosity of comets in his *Exposition du système du monde*.[47] Comet nebulosity was due to vapors raised from the comet's surface by the action of solar heat.[48] He suggested that near perihelion, the process of vaporization of cometary fluids would absorb heat, and so might temper the scorching temperatures a comet would otherwise experience near the sun, whereas near aphelion, the condensation of the same fluids might restore in part the loss of heat experienced as the comet traveled away from the sun into icy space. Therefore, the latent heat associated with changes of state would offset the temperature extremes encountered by comets, such that a thermometer placed on a comet would register little variation.[49]

Subscribers to Oliver's and Williamson's point of view were cheered by Laplace's work, which they thought reinforced the thesis of comet habitability.[50] It is not clear whether Laplace himself intended this discussion as a tentative preamble to statements of the possible habitability of comets. He embraced pluralism as a logical consequence of his nebular hypothesis. If the solar system evolved from the rotation and condensation of a primordial fluid, then planets and stars evolved together. When Laplace looked at the night sky, he saw "innumerable suns, which may be the foci of as many planetary systems,"[51] and he was disposed to populate these celestial bodies. He reasoned by analogy. Since terrestrial matter was so fertile, it was not natural for planetary matter to be sterile.

Man, formed for the temperature which he enjoys upon the Earth, could not, according to all appearance, live upon the other planets; but ought there not to be a diversity of organization suited to the various temperatures of the globes of this universe? If the difference of elements and climates alone, causes such variety in the productions of the Earth, how infinitely diversified must be the productions of the planets and their satellites? The most active imagination cannot form any just idea of them, but still their existence is extremely probable.[52]

By this logic, comets too might be inhabited by creatures suited to their climates. If Laplace believed this, however, he kept it to himself.

SECULARIZATION AND POPULAR CULTURE

In the example before us, we observe a chronological movement from abodes of the damned to realms of the blessed, and from scripturally dependent arguments to more naturalistic ones. It remains to ask what factors may have caused natural philosophers to give up their gloomy determination to locate hell in favor of their more optimistic search for an astronomer's heaven.

Part of the answer lies in the secularization trend of the eighteenth century. It is not the aim of this paper to examine the causes of the trend, but only to say that this case study exemplifies it. Whereas earlier astronomers had looked to comets in the heavens for astrological or divine messages pertaining to local affairs, those of the eighteenth century studied the skies primarily for practical or scientific ends. Insofar as religion was concerned, their aim was not to urge sinners to awake or repent, but to admire God's wise design of the cosmos. Their theories emphasized natural causes over the immediate hand of God.[53]

In the case before us, the secularization trend is reflected in the diverse role played by natural theology in speculations about cometary life. Those who endorsed comets as hells seemed motivated by a desire to find a physical site for hell that agreed with a rather literalist interpretation of scripture. Their God was an economizing master of vengeance. By contrast, those who populated comets with blessed individuals focused their attention on God as an inexhaustible source whose goodness, generosity, and providence inclined him to create life-forms suitable for cometary environments. Having said this, however, many of the latter pluralists concentrated on the physical and biological conditions that would be needed to sustain cometary life. In drawing on contemporary studies of heat and atmospheric science, they offered a view of the world-edifice in which God was a great architect, but otherwise seemed remote. The God of the earlier hell-finders, on the other hand, was an active executioner of justice who sent comets to chasten sinners and aggressively punished planets when their inhabitants refused to mend their ways.

Popular culture may provide another key to understanding the shift in outlook. On a folk level, comets were widely viewed with horror, as divine signs of impending judgement and harbingers of death, but by the end of the seventeenth century, the learned elite had dismissed the crudest folk beliefs as vulgar.[54] I say the crudest, because it can be shown that the elite never fully withdrew from the popular culture they had long shared with the masses. Although they rejected as vulgar the traditional view that comets imported famine, plague, and revolution, they continued to see them as seeds of destruction and renewal, as signs of the world's end or its reformation.[55] The authors discussed in this paper make this evident when they considered the viability of life on comets.

It is easy to see the connection in the thought of Derham, Cheyne, Whiston, and Mather. As hells, comets remained God's tools to punish the wicked and their apparitions served to remind sinners of future judgements. In the writings of Ferguson, Lambert, Williamson, and Oliver, the connection is more remote. As homes to happy beings, comets were assumed to be composed of life-sustaining materials. This assumption was related indirectly to a hypothesis of Newton that all life would cease were it not for comets circulating vital spirits and aqueous fluids throughout the heavens, including fuel to the stars.[56] As I have demonstrated elsewhere, the Newtonian view was derived from an older folk belief that comets influenced agriculture, health, and the weather.[57] We see that the views of early eighteenth-century hell-finders were closer in spirit to the old folk beliefs than views of the later pluralists were. But perhaps it comes as no surprise that vestiges of the old beliefs remained more clearly in the thought of Whiston, Cheyne and the others, for they grew up during the transition period when the learned were consciously withdrawing their support from the traditional belief system. By contrast, Ferguson, Lambert, and the other pluralists were children of the enlightenment. In so far as vestiges of folk beliefs remained in their philosophies, they were stripped of their religious veneer and highly naturalized.[58]

If the loosening grip of traditional beliefs and the secularization trend shed some light on the chronological transition from earlier to later conjectures, we may yet be startled by their antithetical nature. There was, after all, a world of difference between the hells of devils and the heavenly abodes of astronomically-minded creatures. The different conclusions reached about the fitness of comets to sustain life reveal how much latitude there was in the application of scientific method in cases such as this where empirical inspection was impossible. All the authors treated in this essay paid homage to the principles of analogy, uniformity, and simplicity, but it was up to each individual to decide where analogical reasoning was appropriate, when similarity of effects implied uniformity of causes, and how true causes could be distinguished from superfluous ones. In making these decisions, the natural philosophers were guided by their own metaphysical commitments to doctrines such as the principle of plenitude and purposefulness of nature. On the one hand, Derham, Cheyne, Wall, Whiston, and Mather saw God as purposeful and economizing in transforming planets into comets; hence, they chose to emphasize the similarity of material components and forms of life. Comets had atmospheres like the earth's, but their eccentric orbits caused the atmospheres to be "chaotic" and "confused," first scalding, then freezing. Insofar as cometarians resembled earthly creatures, they would find the conditions to be hell. Fontenelle, Ferguson, Lambert, Williamson, and Oliver, on the other hand, saw God as an inexhaustible creative power, and therefore emphasized analogy of purpose. Comets like the earth were designed for habitation, and their elliptical and hyperbolic orbits enabled cometarians to see more of the creation than earth-bound animals could. If comets were fit to be happy domiciles, it was likely that either their material conditions or forms of life must not be analogous to those on the earth. Williamson, for example, emphasized the different composition of the comets' protective atmospheres, whereas Ferguson and Oliver stressed the adaptation of cometarians to their environment. The historical question of life on comets is thus instructive for it reveals a dynamic interplay of natural theology, scientific method, and popular culture.

Department of Science, Technology, and Society
Sarah Lawrence College

NOTES

[1] Bernard le Bovier de Fontenelle, *Entretiens sur la pluralité des mondes*, critical edition, ed. Alexandre Calame (Paris: Librairie Marcel Didier, 1966), p. 161, my translation.
[2] Arthur O. Lovejoy, *The Great Chain of Being* (Cambridge: Harvard University Press, 1964), p. 52.
[3] Such arguments were made, for example, in the *Encyclopaedia Britannica*, 3 vols. (Edinburgh: 1771), 1: 444–445.
[4] Isaac Newton, *Philosophiae naturalis principia mathematica* (London: 1687), Book 3, 'Hypotheses,' which were renamed 'Regulae philosophandi' in subsequent editions.

Translation taken from Newton, *Mathematical Principles of Natural Philosophy*, Motte-Cajori edition (Berkeley: University of California Press, 1962), pp. 398–399. Similar statements appeared in Book 1, Part 1, Prop. 6, and Query 31 of the *Opticks*; see Newton, *Opticks*, 4th ed. (London: 1730; reprint, New York: Dover, 1952), pp. 76, 397.

5 Christiaan Huygens, ΚΟΣΜΟΘΕΩΡΟΣ [Cosmotheoros], sive de terris coelestibus earumque ornatu conjecturae (The Hague: 1698); *Idem, The Celestial Worlds Discover'd: or, Conjectures Concerning the Inhabitants, Plants and Productions of the Worlds in the Planets* (London: 1698). Newton, *Mathematical Principles* (ref. 4), General Scholium, p. 544; Conduitt Memorandum, March 1724/5, printed in Edmund Turnor, *Collections for the History of the Town and Soke of Grantham* (London: 1806), pp. 172–173. John Ray, *The Wisdom of God Manifested in the Works of the Creation* (London: 1691), pp. 1–2, 49. Richard Bentley's seventh and eighth Boyle Lectures, printed in Bentley, *A Confutation of Atheism from the Origin and Frame of the World* (London: 1693), part 2, pp. 14–15, 37, part 3, pp. 4–6; reprinted in I. Bernard Cohen, ed., *Isaac Newton's Papers and Letters on Natural Philosophy* (Cambridge: Harvard University Press, 1958), pp. 326–327, 349, 356–358. For detailed discussion of these and other authors, see Lovejoy, *Great Chain* (ref. 2), chapter 4; Michael J. Crowe, *The Extraterrestrial Life Debate, 1750–1900: The Idea of a Plurality of Worlds from Kant to Lowell* (Cambridge: Cambridge University Press, 1986), esp. pp. 9–26; and Steven J. Dick, *Plurality of Worlds: The Origins of the Extraterrestrial Life Debate from Democritus to Kant* (Cambridge: Cambridge University Press, 1982), esp. chapters 5 and 6.

6 William Derham, *Astro-Theology: Or a Demonstration of the Being and Attributes of God, from a Survey of the Heavens*, 1st ed. (London: 1715), 'A Preliminary Discourse Concerning the Systemes of the Heavens, the Habitability of the Planets, and a Plurality of Worlds,' pp. xl–lvi.

7 *Ibid.*, pp. 52–53; William Derham, *Astro-Theology*, 2nd ed., corrected (London: 1715), pp. 54–55. Cf. William Whiston, *Astronomical Principles of Religion, Natural and Reveal'd* (London: 1717), pp. 22–23.

8 Derham, *Astro-Theology*, 1st ed. (ref. 6), pp. 53, 218–219; Derham, *Astro-Theology*, 2nd ed. (ref. 7), pp. 55, 228–229.

9 George Cheyne, *Philosophical Principles of Natural Religion* (London: 1705), pp. 119–120, see also p. 122.

10 Whiston, *Astronomical Principles* (ref. 7), p. 92. Whiston wrote of "external regions" because he remained open to the possibility that the "internal regions" – i.e. cavities – within comets could be inhabited. See pp. 94–96.

11 *Ibid.*, p. 23.

12 Cheyne, *Philosophical Principles* (ref. 9), p. 151.

13 Whiston, *Astronomical Principles* (ref. 7), pp. 155–156.

14 [Cotton Mather], *A Voice from Heaven. An Account of a Late Uncommon Appearance in the Heavens* (Boston: 1719), pp. 8–11, 13–16, quotation on p. 11.

15 [Cotton Mather], *An Essay on Comets, their Nature, the Laws of their Motions, the Cause and Magnitude of their Atmosphere, and Tails; With a Conjecture of their Use and Design* (Boston: 1744), p. 7.

16 William Wall, 'Advertisement,' appended to Tobias Swinden, *An Enquiry into the Nature and Place of Hell* (London: 1714), pp. 287–292, see p. 291 for quotation.

17 *Ibid.*, pp. 289, 291–292. Final Judgement followed the millennium; the righteous would go to an even better place, while the damned would be imprisoned in the sun, which Swinden agreed was the site of hell.
18 William Whiston, *The Eternity of Hell-Torments Considered* [1740], 2nd ed. (London: 1752), p. 105; Whiston, *Astronomical Principles* (ref. 7), pp. 154–155; D. P. Walker, *The Decline of Hell* (Chicago: University of Chicago Press, 1964), pp. 100–101.
19 Mather Byles, *The Comet: A Poem* (Boston: 1744).
20 Whiston, *Astronomical Principles* (ref. 7), p. 23: "[Comets are] Worlds in Confusion, but capable of a change to Orbits nearer Circular, and then of settling into a State of Order, and of becoming fit for Habitation like the Planets." See also William Whiston, *A New Theory of the Earth* (London: 1696), pp. 48, and 74–75: "A *Planet* is a *Comet* form'd into a regular and lasting constitution, and plac'd at a proper distance from the Sun in a Circular Orbit, or one very little Eccentrical; and a *Comet* is a *Chaos*, i.e. a *Planet* unform'd, or in its primaeval state, plac'd in a very Eccentrical one."
21 Whiston, *New Theory of the Earth* (ref. 20) pp. 69, 378; Whiston, *Astronomical Principles* (ref. 7), pp. 153–154.
22 Cheyne, *Philosophical Principles* (ref. 9), p. 151.
23 René Descartes, *Principia philosophiae* (Amsterdam: 1644), part 3, sections 115, 119–120, 126–129, 133–139; recently published in translation as Descartes, *Principles of Philosophy*, trans. V. R. Miller and R. P. Miller (Dordrecht: D. Reidel Publishing Company, 1983), pp. 147, 150–151, 155–159, 163–168. Also see René Descartes, *Le monde, ou le traité de la lumière* [1st edition: 1664], French text of the 1677 edition, with translation by Michael S. Mahoney (New York: Abaris Books, 1979), pp. 44–45, 92–107.
24 See, for example, the remark made by John Flamsteed in a letter to Edmond Halley, 17 February 1680/1, that a "Comet ... may have beene some planet belonging formerly to another Vortex now ruined;" reprinted in *The Correspondence of Isaac Newton*, ed. H. W. Turnbull *et al.*, 7 vols (Cambridge: Cambridge University Press, 1959–1977), 2: 338.
25 Fontenelle, *Entretiens* (ref. 1), pp. 145–149; translation taken from [Fontenelle], *A Plurality of Worlds*, trans. J. Glanvill (London: 1702), 'Fifth Evening,' pp. 144–148, quotation, p. 148.
26 James Ferguson, *An Idea of the Material Universe, Deduced from a Survey of the Solar System* (London: 1754), pp. 25–27.
27 Maupertuis, *Lettre sur la comète* ([Amsterdam?]: 1742), p. 38; John Winthrop, *Two Lectures on Comets* (Boston: 1759), p. 40.; Bartholomew Burges, *A Short Account of the Solar System, and of Comets in General* (Boston: 1789), pp. 9, 15–16; J. H. Lambert, *Cosmologische Briefe über die Einrichtung des Weltbaues* (Augsburg: 1761); trans. and reprinted in *Idem, Cosmological Letters on the Arrangement of the World-Edifice*, trans. Stanley L. Jaki (New York: Science History Publications, 1976).
28 Ferguson, *Idea of the Material Universe* (ref. 26), pp. 9, 15–16, 23, 25–27, 29–30.
29 Lambert, *Cosmological Letters* (ref. 27), pp. 48–49, 55, 70–72, 87, 96, 103; quotation on p. 49. See also M. A. Hoskin, 'Lambert's Cosmology,' and 'Lambert and Herschel,' *Journal for the History of Astronomy* 9 (1978), 134–139 and 140–142.
30 Lambert, *Cosmological Letters* (ref. 27), pp. 83–84, and quotation, p. 73.
31 Andrew Oliver, *An Essay on Comets* (Salem: 1772), p. 48.
32 *Ibid.*, pp. 34–35.

33 Lambert, *Cosmological Letters* (ref. 27), p. 99.
34 *Ibid.*, pp. 48–49, 64–65, 84.
35 *Ibid.*, pp. 58–59, 62, 67–68, 99, 104–105, 129, quotation, p. 68.
36 *Ibid.*, pp. 55, 62, 84, 94, 103–104, 148. Like Kant, Lambert argued that the world of comets bordered on and fused with the world of planets. Cf. Immanuel Kant, *Allgemeine Naturgeschichte und Theorie des Himmels* [Leipzig: 1755].
37 Lambert, *Cosmological Letters* (ref. 27), pp. 83–84, 93, 100–101.
38 *Ibid.*, pp. 93–94.
39 Hugh Williamson, 'An Essay on the Use of Comets, and an Account of their Luminous Appearance; together with some Conjectures concerning the Origin of Heat,' *Transactions of the American Philosophical Society* 1 (1771), appendix, pp. 27–36; and in the second edition of the *Transactions* 1 (1789), 133–143.
40 Oliver, *Essay on Comets* (ref. 31), pp. 51–87.
41 *Ibid.*, pp. v–vi, 11–32.
42 *Ibid.*, pp. 40–49, quotation p. 46.
43 *Ibid.*, pp. 49–50.
44 *Ibid.*, p. iii.
45 John Langdon Sibley and Clifford Kenyon Shipton, *Sibley's Harvard Graduates*, 17 vols. (Cambridge and Boston: 1873–1975), **12**: 457.
46 Priestley to Oliver, 12 February 1775, *Proceedings of the Massachusetts Historical Society*, 2nd series, **3** (1886–1887), 13–14; letter reprinted in Robert E. Schofield, *A Scientific Autobiography of Joseph Priestley, 1733–1804* (Cambridge: M.I.T. Press, 1966), pp. 148–149. John Davis compared Oliver's theory to one of J. A. De Luc published in the *Journal de physique* (1802) and commented on its reception; see John Davis, ed., *Two Lectures on Comets, by Professor Winthrop, also An Essay on Comets, by A. Oliver, Jun. Esq. With Sketches of the Lives of Professor Winthrop and Mr. Oliver. Likewise, a Supplement, Relative to the Present Comet of 1811* (Boston: 1811), p. xxv. On Williamson's reputation, see Brooke Hindle, *The Pursuit of Science in Revolutionary America 1735–1789* (Chapel Hill: University of North Carolina Press, 1956), p. 172.
47 See Antoine Laurent Lavoisier and Pierre-Simon Laplace, *Mémoire sur la chaleur lû à l'Académie Royale des Sciences, le 28 juin 1783* (Paris: 1783); facsimile reprinted with translation in *Memoir on Heat Read to the Royal Academy of Sciences, 28 June 1783 by Messrs. Lavoisier & de la Place of the same Academy*, trans. Henry Guerlac (New York: Neale Watson Academic Publications, 1982).
48 Pierre-Simon Laplace, *Exposition du système du monde*, 1st ed., 2 vols. (Paris: 1796), **1**: 222.
49 This discourse on latent heat appeared in only the third and fourth editions. Pierre-Simon Laplace, *Exposition du système du monde*, 3rd ed., 2 vols. (Paris: 1808), **1**: 224–228; Laplace, *Exposition du système du monde*, 4th ed. (Paris: 1813), pp. 130–132.
50 Davis, *Two Lectures* (ref. 46), p. 187.
51 Laplace, *Système du monde*, 1st ed. (ref. 48), **2**: 304; translation taken from Laplace, *The System of the World*, trans. J. Pond [from the 2nd ed., 1799], 2 vols. (London: 1809), **2**: 366.
52 Laplace, *Système du monde*, 1st ed. (ref. 48), **2**: 294–295; translation taken from Laplace, *System of the World* (ref. 51), **2**: 355–356.
53 On the secularization of the study of comets, see Sara Schechner Genuth, 'From

Heaven's Alarm to Public Appeal: Comets and the Rise of Astronomy at Harvard,' in Clark A. Elliott and Margaret W. Rossiter, eds., *Science at Harvard University: Historical Perspectives* (Bethlehem: Lehigh University Press, 1992), pp. 28-54.

[54] On folk beliefs about comets, see Stith Thompson, *Motif-Index of Folk-Literature: A Classification of Narrative Elements in Folktales, Ballads, Myths, Fables, Mediaeval Romances, Exempla, Fabliaux, Jest-Books, and Local Legends*, rev. and enlarged ed., 6 vols. (Bloomington: Indiana University Press, 1955-1958), A124.5, A780, A786, A1002.2, A1050, D1812.5.1, F493.5, F797, F960-969, T525.2, V211.1.2.1; *Standard Dictionary of Folklore, Mythology and Legend*, ed. Maria Leach, 2 vols. (New York: Funk & Wagnalls, 1949-1950), s.v. 'comet.' More direct information on folk beliefs is obtained through the study of ephemeral texts such as ballads, broadsides, chapbooks, and almanacs, which often contained crude woodcuts depicting comets, through first-hand reports in diaries and letters, and through visual representations in folk art and artifacts. The texts and images that I have consulted are far too numerous to mention here, but many are discussed in my other works (see refs. 53, 55, 57, 58). Good illustrations of this genre can be found in J. Classen, *15 Kometenflugblätter des 17. und 18. Jahrhunderts*, Veröffentlichungen der Sternwarte Pulsnitz, 11 (Leipzig: Verlag Johann Ambrosius Barth, 1977); Gerhard Bott, ed., *Zeichen am Himmel: Flugblätter des 16. Jahrhunderts* (Nuremberg: Germanisches Nationalmuseum Nürnberg, 1982); Jean-Pierre Séguin, *L'information en France avant le périodique: 517 canards imprimés entre 1529 et 1631* (Paris: Éditions G.-P. Maisonneuve et Larose, 1964), pp. 95-100, and plates 3, 6, 18, 19; Roberta J. M. Olson, *Fire and Ice: A History of Comets in Art* (New York: Walker and Company for the National Air and Space Museum, Smithsonian Institution, 1985).

On the denigration of folk beliefs and customs by the elite in the 17th and 18th centuries, see Peter Burke, *Popular Culture in Early Modern Europe* (New York: New York University Press, 1978), chaps. 2, 8, 9; Roger Chartier, 'Culture as Appropriation: Popular Cultural Uses in Early Modern France,' and Jacques Revel, 'Forms of Expertise: Intellectuals and 'Popular' Culture in France (1650-1800),' both in Steven L. Kaplan, ed., *Understanding Popular Culture: Europe from the Middle Ages to the Nineteenth Century* (Berlin, New York, Amsterdam: Mouton Publishers, 1984), pp. 229-273; Harry C. Payne, 'Elite versus Popular Mentality in the Eighteenth Century,' *Studies in Eighteenth-Century Culture* **8** (1979): 3-32; Natalie Zemon Davis, *Society and Culture in Early Modern France* (Stanford: Stanford University Press, 1975), chaps. 7, 8; Keith Thomas, *Religion and the Decline of Magic* (New York: Charles Scribner's Sons, 1971); Patrick Curry, *Prophecy and Power: Astrology in Early Modern England* (Princeton: Princeton University Press, 1989); Robert W. Malcolmson, *Popular Recreations in English Society, 1700-1850* (Cambridge: Cambridge University Press, 1973).

[55] See Sara Schechner Genuth, 'Newton and the Ongoing Teleological Role of Comets,' in Norman J. W. Thrower, ed., *Standing on the Shoulders of Giants: A Longer View of Newton and Halley* (Berkeley: University of California Press, 1990), pp. 299-311.

[56] Newton, *Principia*, 1st ed. (ref. 4), p. 506; Isaac Newton, *Philosophiae naturalis principia mathematica*, 2nd ed. (Cambridge: 1713), pp. 480-481; for both passages in the Motte-Cajori edition (ref. 4), see pp. 529-530, 541.

[57] Sara Schechner Genuth, 'Comets, Teleology, and the Relationship of Chemistry to

Cosmology in Newton's Thought,' *Annali dell'Istituto e Museo di Storia della Scienza di Firenze* **10**, fascicolo 2 (1985), 31–65.

[58] The vestiges are better seen in theories about comets careening into planets than in those concerning comet habitability. See Genuth, 'Newton and the Ongoing Teleological Role of Comets' (ref. 55); and Sara Schechner Genuth, 'From Monstrous Signs to Natural Causes: The Assimilation of Comet Lore into Natural Philosophy' (Ph.D. Dissertation, Harvard University, 1988).

LORRAINE J. DASTON*

THE DOCTRINE OF CHANCES WITHOUT CHANCE: DETERMINISM, MATHEMATICAL PROBABILITY, AND QUANTIFICATION IN THE SEVENTEENTH CENTURY

INTRODUCTION

In 1814 the French mathematician Pierre Simon Laplace wrote:

All events, even those which on account of their insignificance do not seem to follow the great laws of nature, are a result of it just as necessarily as the revolutions of the sun ... Given for one instant an intelligence which could comprehend all the forces by which nature is animated and the respective situation of the beings who compose it – an intelligence sufficiently vast to submit these data to analysis – it would embrace in the same formula the movements of the greatest bodies of the universe and those of the lightest atom; for it, nothing would be uncertain and the future, as the past, would be present to its eyes.[1]

Despite the astronomical references, this classic statement of determinism occurs in Laplace's work on probabilities, not celestial mechanics.[2] Nor was this combination of probability theory and thoroughgoing determinism anomalous: Laplace is in fact echoing the words of Jakob Bernoulli, published a century earlier, right down to the astronomical standard of prediction. Between Bernoulli's *Ars conjectandi* (1713) and Laplace's *Théorie analytique des probabilités* (1812) a multitude of lesser probabilists insisted that chance was a mere sound signifying nothing; that every event in at least the natural world was a link in an unbroken chain of causes. They were not simply voicing a fashionable position obtained secondhand, for theirs were among the strongest and most influential proclamations of determinism. Nor was it a metaphysics free of consequences: their determinism committed them to an interpretation of probability as a degree of subjective certainty, rather than as a measure of objective variability. Far from signalling a new openness toward chance in the world, the advent of mathematical probability banished chance altogether.[3]

Why were the architects of the mathematical "doctrine of chances" so unanimously hostile toward chance itself? I shall argue that the paradoxical alliance between mathematical probability and determinism was the result of seventeenth- and eighteenth-century views about the preconditions for applying mathematics to phenomena. Although subsequent events ultimately revealed these views to be mistaken, at the time they not only made the initial link between probability theory and determinism conceivable; they made it almost a precondition for such a theory. I have divided my argument into four sections: I begin with evidence that the classical probabilists were in fact staunch determinists; I then briefly examine the various meanings of determinism in the seventeenth and eighteenth centuries, focusing more narrowly on the relationship between applied, or "mixed," mathematics and determinism during this period. The third section contrasts probability theory without and with determinism in the writings of Girolamo Cardano and Jakob Bernoulli, respectively. I conclude with some speculations on why determinism took hold when it did. Throughout, I shall be chiefly concerned with the early years of probability theory, ca. 1660–1720, although I believe that most of my claims would also apply to classical probabilists through Laplace and Augustus De Morgan.

DETERMINISTIC PROBABILISTS

Laplace's deterministic confession of faith was a cliché in the probabilistic literature by the time he uttered it. The Marquis de Condorcet had expressed nearly identical views in an open letter to Jean d'Alembert in 1768: "An intelligence that knew the state of all phenomena at any given moment, the laws governing matter and the effect of these laws after any given period of time, would have perfect knowledge of the system of the World."[4] And Condorcet was in turn ringing the changes on Jakob Bernoulli's introduction to Part IV of the *Ars conjectandi*, in which Bernoulli explained that events are contingent only relative to imperfect human knowledge. Even if we ignorant mortals call the fall of the die or next week's weather fortuitous, Bernoulli insisted that "these effects follow from their own proximate causes with no less necessity than the phenomena of eclipses follow from the movement of luminous bodies."[5] Abraham De Moivre, although he called his treatise on probability *The Doctrine of Chances* (1718; 2nd ed. 1738; 3rd ed. 1756), was equally emphatic that chance had no objective existence in the world; it could "neither be defined nor understood" and was but "a mere word."[6]

There were, to be sure, significant differences in the brands of determinism espoused by these probabilists. Bernoulli and De Moivre both came from Calvinist backgrounds, and there is more than a whiff of divine predetermination and providence in their tirades against chance. Bernoulli believed that even if future events are not brought about by "the inevitable necessity of some fate," they are nonetheless dictated with certainty by divine plan – anything less than this would be incompatible with the creator's "omniscience and omnipotence."[7] De Moivre was a proponent of Royal Society theology, which argued from design in the arrangement of the solar system, the ratio of male and female births, and other natural phenomena to the existence of a benevolent deity. He understood chance in the traditional sense of absence of purpose, here inferred from absence of design, and took Bernoulli's theorem to be a mathematical demonstration of how experience must in the end reveal "that Order which naturally results from Original Design."[8] In contrast, Condorcet's and Laplace's intelligence was distinctly secularized, a creature that calculates but does not plan the future. Condorcet and Laplace invoke only causes; Bernoulli and De Moivre could speak of reasons as well.

Whether we know both how and why the future must come about, or merely how, the classical probabilists agreed that it *must* come about, at least as far as natural events were concerned.[9] Whether the causes that brought it about were mechanical or occult, logically or merely *de facto* necessary, blind or purposeful – or even whether there were causes at all, or merely constant conjunctions – were of secondary importance to them. What mattered was that the laws of nature were inviolable, and that therefore the whole course of nature was in principle predictable. That we in fact cannot predict the future with the certainty of a deity or semi-deity was the common departure point for their interpretation of probability. Since Bernoulli's reflections in the *Ars conjectandi* were the locus classicus for this interpretation, I shall cite them at some length.

Having asserted that all events, past, present, and future, possess the "greatest objective certainty" in themselves, Bernoulli asks what meaning contingency can have in such a necessary world. The outcome of the cast of a die is as necessary as the next eclipse of the sun, both being subject to the laws of mechanics, and yet "only eclipses are calculated by necessities, but the cast of the die [is] calculated by contingencies." This distinction between necessities and contingencies relates solely to our state of knowledge, not to the state of the world. When we become more proficient in mechanics, we shall be able to calculate the casts of dice as accurately as we now calculate eclipses. Among peoples who know even less about mechanics than we do, eclipses may still be a matter of contingency rather than necessity. Probabilities therefore measure degrees of certainty, and change as our knowledge does: "Thus, it follows that what seems to one person at a given time to be a contingent event may seem to another person at another time (indeed, to the same person) to be a necessary event after its causes have been learned."[10] Probabilities were figments of human ignorance; a provisional way of taking stock of eventualities still imperfectly understood. The best-informed scientists would have no need of them. So did the classical probabilists reconcile the doctrine of chances to a world without chance.

MIXED MATHEMATICS AND THE PRECONDITIONS FOR QUANTIFICATION

Probability theory could thus be made compatible with determinism, but that is a long way from saying that probability theory *needed* determinism. In order to support this latter, stronger claim, I must first step back a bit from probability theory proper and look at seventeenth-century determinism more closely. I shall be particularly concerned with the relationship between a fortified, expanded form of determinism that emerged in the second half of the seventeenth century and the canons of mixed mathematics.

Determinism is a many-splendored thing. Some of its numerous forms have an ancient lineage. In a famous passage, Aristotle muses over whether tomorrow's battle at sea must necessarily occur or not, and concludes that particulars about the future are neither true nor false, but rather dwell in a limbo of possibility. Otherwise, there would be no need to deliberate, for nothing would really depend on our decisions.[11] Aristotle's concerns were typical of those that animated discussions of determinism until well into the nineteenth century. Until that time, as Ian Hacking has shown, the phrase "determinism" (or rather the "doctrine of fate or necessity," as it was then commonly known) was narrowly identified with questions of free will, predestination, and materialism.[12] There were nonetheless physical determinists *avant (ou sans) la lettre*. In fact, determinisms of both mind and matter boomed in the seventeenth and eighteenth centuries. On the side of mind, predestination underwent a revival in the theology of Jean Calvin and Jansenius;

Thomas Hobbes, Julien de La Mettrie, and Baron P. H. d'Holbach argued against free will as materialists; Benedict Spinoza's pantheism circumscribed even God's volition. On the side of matter, Francis Bacon believed that the "latent forms" of things necessarily determined their properties; René Descartes and his followers asserted an unbroken chain of mechanical causes; Gottfried Wilhelm Leibniz arrived at much the same conclusion on logical grounds; David Hume accepted a necessary order of events, even if he foreswore causal connections between them.

It is the determinism of matter that concerns us here. Although some of the classical probabilists wrote as if minds (e.g. the minds of jurors) were as predictable as matter, when they claimed that chance had no place in the world, they almost always meant the physical world. The flowering of seventeenth-century determinisms of matter was as diverse as it was vigorous. Within the space of about fifty years, two millennia's worth of commonplaces on the "habits" (as opposed to the "laws") of nature were overturned not by one clinching argument, but by a slew of plausible ones that did not always mesh with one another. Was the course of nature necessary because its constitution was corpuscular, or because it was mechanical, or because it was mathematical, or because it was logical? A survey of seventeenth-century texts turns up authors who answered one of the above, some of the above, or none of the above, and still professed nature's laws to be firm and fixed.[13] The situation is comparable to the nearly coeval emergence of the distinction between primary and secondary qualities: almost every major thinker from Galileo to John Locke subscribed to such a distinction, although hardly any two gave the same reasons for doing so.

This deterministic turn was a distinctly metaphysical one, and particular developments in natural philosophy had little to do with it. Although it is probably true that the triumph of Newton's new world system was an excellent advertisement for law and order in nature, the belief that nature was lawlike and orderly was firmly in place among the learned by the time the *Principia* appeared in 1687. Indeed, the *Principia* took this belief for granted, even though Newton himself subscribed to the most dilute version of the determinism of matter then available. According to Newton, the laws of nature, such as gravitation, just happened to be the case, and could well have been otherwise had God so freely chosen. But they were none the less lawlike, once ordained.[14] By the turn of the eighteenth century, the distinction between the exceptionless laws of nature and the more lax sort of the moral realm was clear. As Baron Charles de Montesquieu remarked, "the intelligent world is far from being so well governed as the physical."[15]

It is important to recognize the novelty and comprehensiveness of this determinism, even if it stopped short of the moral realm. Because Aristotelian natural philosophy was in principle demonstrative, a science of what must be the case, it was also, again in principle, deterministic. However, in practice the demonstrative standard was sometimes relaxed,[16] and it was moreover never intended to apply to all natural phenomena, but rather only those which happen "always or for the most part." Certain phenomena – the rare, the accidental, and the fortuitous – were excluded from scholastic natural philosophy as too irregular, unpredictable, and idiosyncratic to be studied scientifically. Such exceptions and anomalies were ascribed to chance, which was incompatible with the regularity requisite for natural philosophy: "But when an event takes place always or for the most part, it is not accidental or by chance. In natural products the sequence is invariable, if there is no impediment."[17] Even those few scholastic natural philosophers who did attempt natural explanations of marvels and oddities, such as Nicole Oresme, did not couch these in demonstrative form, and indeed despaired of untangling the web of causes responsible for each anomaly: "And for such things, who can give the reason why, other than the general one, namely that their causes are adequate, and no more and no less, for producing this? Therefore these things are not known point by point except by God alone, who knows unlimited things."[18]

In contrast, seventeenth-century determinism made it the business of natural philosophy to provide detailed causal explanations for all phenomena, however particular or even singular. Indeed, critics of scholastic natural philosophy often made such variable or anomalous events into a challenge, paying special attention to just those "chance" events that contradicted the rules of their rivals. Bacon called for a collection of such counterexamples, of all that is "new, rare, and unusual in nature," as a corrective to the alleged axioms of the Aristotelians, which were drawn solely from the observation of "nature in course." Moreover, Bacon exhorted natural philosophers to explain these apparent marvels: "For we are not to give up the investigation until the properties and qualities founded in such things as may be taken for the miracles of nature be reduced and comprehended under some form or fixed law, so that all the irregularity or singularity shall be found to depend on some common form . . . "[19] Another reformer, Descartes, proclaimed that there were "no properties so occult, no effects of sympathy and antipathy so marvelous and strange, nothing so rare in nature," that could not be handled by his mechanical philosophy.[20] Both Bacon and Descartes were determinists of the staunchest stripe, who believed that every natural phenomenon, no matter how bizarre or rare, could ultimately be subsumed under rigid natural laws. It was this fortified and expanded form of determinism in seventeenth-century natural philosophy that banished chance as a real cause in the material world.

The great expansion of mathematically-based science in the seventeenth century both promoted and profited from the subduing of variation and anomaly. For mathematically inclined natural philosophers of this period, determinism was a precondition for applying mathematics to nature, for they believed the connection between causes and effects must be as necessary as that between steps in a demonstration for the mathematical model to hold. They still worked within a framework of mixed mathematics, in which no sharp boundary separated the derivation of arithmetic and geometry (and later algebra and the calculus) from that of optics, mechanics, harmonics, and astronomy. According to the Aristotelian view, which remained standard well into the eighteenth century,[21] all of mathematics sprang from the same source, the abstraction of form from matter. The greater the simplicity of its objects, the more precise the science: hence the results of the "pure" mathematics of arithmetic and geometry were more precise than those of the "mixed" mathematics of astronomy and optics; and the mathematical sciences in general were more precise than all others.[22] This was understood by seventeenth-century interpreters to be a function of the proportion of form to matter: thus within mixed mathematics, optics might be rated more certain than harmonics, having less to do with physics.[23] A certain class of phenomena might belong to both mathematics and physics, depending on the perspective of the investigator – for example, optics considered *qua* lines was part of mixed mathematics; *qua* light rays, part of physics. Seventeenth-century natural philosophers differed from Aristotle not so much in their view as to what mixed mathematics was, and as to the necessity its conclusions implied, than as to the breadth of its domain. Once determinism had penetrated to the most obscure nooks and crannies of nature, all of nature (and much else besides) seemed potentially susceptible to mixed mathematical treatment.

This new and ebullient confidence in the power of mathematics to describe phenomena took both ontological and epistemological forms.

Galileo, Descartes, and Newton all believed that nature (or at least large parts of it) was mathematical in structure, and that ideally, knowledge of nature should be cast in terms of mathematical demonstration. These two precepts no longer go together: nature has never looked so mathematical as it does from the standpoint of today's scientist, but the Euclidian format of Newton's *Principia*, or of the Latin demonstrations appended to Galileo's *Discorsi* (1638) nonetheless strikes us as quaint. We no longer believe that the possibility of quantifying phenomena implies the possibility of demonstrating them: for us, numbers carry conviction, but not necessity. Matters were otherwise for the mixed mathematicians of the seventeenth century. The rising prestige of mathematics in the sixteenth and seventeenth centuries, vis-à-vis both scholastic logic and scholastic natural philosophy, actually gave the scholastic ideal of *scientia* as demonstrative knowledge a new, if brief, lease on life. Although the syllogism had been sharply criticized by humanists like Peter Ramus, mathematics seemed to offer an alternative way of drawing conclusions about nature (and much else besides) with certainty. Not all seventeenth-century natural philosophers of the reformed school embraced this pan-mathematicism: some, such as Robert Boyle, preached probabilism in epistemology and regarded mathematics in natural philosophy warily.[24] However, others were more ambitious in their hopes for a revived and enlarged *scientia*, now founded on mathematical rather than on syllogistic demonstrations.

Many seventeenth-century voices testify to this belief that mathematical knowledge was certain knowledge, and that its purview could be vastly extended. In the *Dialogo* (1632), Galileo's mouthpiece Salviati maintains that "with regard to those few (mathematical propositions) which the human intellect does understand, I believe that its knowledge equals the Divine in objective certainty, for here it succeeds in understanding necessity, beyond which there can be no greater sureness."[25] In the *Regulae ad directionem ingenii* (comp. 1619–28), Descartes dreamed of a universal mathematics that would "explain everything which can be asked concerning measure and order not predicated of any special subject matter," and we learn from Part I of the *Discours de la méthode* (1637) that he was early on impressed with "the certainty and self-evidence" of mathematical demonstrations.[26]

Leibniz had even grander plans for his Universal Characteristic, which would impart the certainty of mathematical proofs to discussions of nonmathematical matters by translating concepts into numbers and combining them according to the operations of arithmetic.[27] Marin Mersenne thought mathematics the best antidote to the extreme scepticism in vogue among Parisian intellectuals in the 1620s, who denied the possibility of genuine knowledge. In his dialogue between a skeptic, an alchemist, and a Christian philosopher, Mersenne subjects the skeptic (the alchemist appears to have wisely sneaked away after Book I) to a full course in mathematics running to over 600 pages, including whole tables of cube and square roots, until the exhausted skeptic admits that at least the mathematical sciences are "very certain and very true."[28] Mersenne's tour of contemporary mathematics embraced not only arithmetic, algebra, and geometry, but also pneumatics, mechanics, astronomy, geodesy, architecture, harmonics, geography, optics, and "automatopytique", or the science of making inanimate objects appear alive. For these writers, mathematics was not only the deepest form of knowledge; it was potentially the broadest as well.

But they did set some limits on the expanding domain of mathematics, and these limits reveal the criteria they believed a class of phenomena must meet in order to be mathematically tractable. Galileo doubted that there would ever be a true science – that is, a demonstrative, mathematical account – of air currents, because he believed them to be inherently variable.[29] Pascal foresaw similar problems in producing a "method of geometrical proofs in the art of persuasion": here it was the wellsprings of pleasure that "were variable in each particular, (and) with such diversity that there is no man who differs more than another than from himself at different times."[30] Bernard de Fontenelle, in his eulogy of Jakob Bernoulli, dwelt upon what he had heard about the applications of mathematical probability to civil life in Part IV of the as yet unpublished *Ars conjectandi*. His grounds for admiration shed light on what kinds of phenomena were thought amenable to mathematization and why: "It is not so glorious for the geometric spirit to reign over physics as over moral matters, [which are] so complicated, so contingent [*casuelle*], so variable; the more a subject opposes and rebels, the more honor there is in vanquishing it."[31]

These seventeenth-century writers believed that genuine science is mathematical science, and they also believed that far more of nature is "written in the language of mathematics" than their scholastic predecessors had. Yet they still acknowledged the existence of pockets of irreducible variability or contingency. In contrast to militants like Bacon and Descartes, their determinism was not yet all-encompassing. Where chance still intervened, there was no hope of a mathematical treatment: just as Aristotle had despaired of a demonstrative science of the accidental, so they despaired of a mathematical science of the variable and contingent.

We no longer share their pessimism, familiar as we are with the mathematics of randomness, variation, and even chaos. Nor do we still believe that the necessary connections between steps in a mathematical argument imply necessary connections between the events to which the mathematics is applied; that is, our science is mathematical without being demonstrative. Just because a class of phenomena can be described mathematically no longer implies for us the logical necessity of those phenomena. This is because we are no longer persuaded that any particular mathematical model *must* fit nature hand-in-glove. We can imagine that the world might be otherwise, and we can also imagine other axioms, postulates, and definitions. We are, therefore, content with the empirical adequacy of our mathematical models, and do not strive to demonstrate them to be the only possible ones. But for seventeenth-century thinkers, the mathematical and the demonstrative largely coincided. This is why they so often cast their results in terms of axioms, definitions, postulates, and theorems.

Thus seventeenth-century mixed mathematicians by and large believed that mathematics could only apply to those parts of experience where unexceptioned determinism reigned. The necessary connections that glued together the sequence of a mathematical demonstration must have their counterpart in the necessary connections that glued together the sequence of cause and effect. This belief did not always commit them to a thoroughgoing determinism: it was always possible to exempt whole classes of recalcitrant phenomena, as Galileo did with air currents. But that meant exempting them from any mathematical treatment whatsoever, which for Galileo was synonymous with abandoning any attempt to study them scientifically. Like Aristotle, Galileo and his like-minded contemporaries could not imagine a science of the accidental. Indeed, their mathematical bent pushed them further than Aristotle, and led them in principle to restrict science to what happens always, not just most of the time.

It is worth noting that this drive for demonstrative certainty within seventeenth-century mixed mathematics did not always go hand-in-hand with a demand for empirical accuracy. It was not so much that mathematical treatments of new classes of phenomena actually flew in the face of observations and measurements (although this was not wholly unknown), as it was that the measurements needed to test for empirical accuracy were practically or even in principle impossible. Consider Jakob Bernoulli's attempt to mathematize courtroom evidence in Part IV of the *Ars conjectandi*. The variables are cast in the language of counting, but a kind of counting that could hardly be imagined in practice: for example, the number of cases in which the pallor of a witness indicated guilt versus the number of cases in which it did not; or the number of cases in which a witness in fact turned pale under questioning (for whatever reason) versus the number of times he did not.[32]

Even if one were to accept Bernoulli's decomposition of courtroom evidence into the dimensions of pure versus mixed, existent versus nonexistent proof, the numbers needed to make calculations within this model were entirely illusory. It was not simply that such numbers had not yet been in fact gathered; even a bevy of legal statisticians armed with Bernoulli's theorem and logging decades of courtroom hours could hardly hope to exhaust the number of possible causes of (for example) witness-stand pallor, not to mention their relative frequencies. Although Bernoulli realized that the simplified coin or die analogy of a small number of equiprobable outcomes would not apply to the vast majority of real cases of uncertainty, and that these latter would have to be determined a posteriori, his treatment of the evaluation of evidence implicitly assumed just such a limited set of equiprobable outcomes. Examples of such quantification without numbers (or with invented numbers) might be multiplied within the annals of early probability theory, but the probabilists held no monopoly on mathematical models without measurement in the seventeenth century: Mersenne earnestly discussed how the Virgin Mary might have steadily augmented her degrees of grace in geometric progression to 64 terms;[33] Leibniz thought it would take a team of scholars less than five years to construct his Universal Characteristic.[34]

The point here is that although determinism was considered a precondition for mathematization in the seventeenth century, the possibility of making measurements was not. The realm of the quantitative could extend well beyond that of the measurable, so long as it did not overstep the bounds of the determinate. Leibniz made this distinction between the restricted domain of the measurable and the vast domain of the quantitative explicit in his 1677 musings on the Universal Characteristic:

> An ancient saying has it that God created everything according to weight, measure, and number. However, there are many things which cannot be weighed, namely, whatever is not affected by force or power; and anything which is not divisible into parts escapes measurement. On the other hand, there is nothing which is not subsumable under number. Number is therefore, so to speak, a fundamental metaphysical form, and arithmetic a sort of statics of the universe, in which the powers of things are revealed.[35]

It was this metaphysical conviction in the determinateness of the phenomena and their consequent tractability to mathematical treatment that spurred Leibniz and others to attempt the quantification of almost everything, from degrees of grace to the weight of evidence. They aimed as often at precision as they did at accuracy, which in part explains the insouciance with which they proposed mathematical models not tethered to measurement. Accuracy concerns the fit of numbers to some class of phenomena, and presupposes that the model can be anchored in measurement; precision concerns the clarity, distinctness, and intelligibility of concepts, and by itself stipulates nothing about how those concepts match the phenomena, only that the phenomena in question be, as it were, sharp-edged, or determinate. If the mathematization was to consist not only of a model but also of demonstrations, as in the case of Galileo's mechanics, seventeenth-century quantifiers further required that the phenomena be not only determinate, but determined. Early modern mathematicians cherished precision as a good in and of itself, even when not accompanied by accuracy.

PROBABILITY THEORY WITH AND WITHOUT DETERMINISM: BERNOULLI VS. CARDANO

We can now understand why Pascal described his newly invented "mathematics of chance (*géométrie du hasard*)" as paradoxical.[36] On the face of it, it seemed a contradiction in terms; a science of the accidental. In fact, however, it was both the outgrowth and the expression of the new determinism, a determinism so rigid and comprehensive that it reduced all chance and variability to an illusion. Given the assumptions of the mixed mathematicians, only the most watertight determinism could make the oxymoronic "mathematics of chance" conceivable. Mathematics dealt with the necessary, not the fortuitous and variable. Therefore, chance had to be made necessary; that is, it had to disappear in all but in name. This is why probability theory needed determinism, at least in the mid-seventeenth century.

As it happens, the historical record has here supplied us with that rarity in this line of research, the counterfactual: probability theory *without* determinism. Girolamo Cardano's curious little book *Liber du ludo aleae* (composed ca. 1520; published posthumously in 1663) reveals the difficulties in constructing a mathematics of real, as opposed to illusory chance during this period. Cardano, almost alone among the mathematicians who concerned themselves with probability theory, was an avid gambler, and his book is as much about unnerving your opponent and detecting a rigged game as it is about calculating chances. Like most gamblers then and now, Cardano believed fervently in good and bad luck, and for him chance (or "fortune") had an objective existence. Where the goddess Fortuna reigned, determinism was in exile (and vice versa). Cardano's stillborn attempt to subject the fickle Fortuna to calculation, particularly his own perplexities about how to do so, underscore the conflict between his precepts as a mathematician and his practice as a gambler.

There is much in Cardano to evoke a thrill of recognition in the latter day probabilist, devoté of the casinos or no. Cardano has a clear notion of the importance of equipossible outcomes – indeed, equality among players is his cardinal axiom – and he calculates odds of all manner of gambles using dice without difficulty. Yet, Cardano's explanation of the mathematical basis of these calculations turns upon a notion that even he admits, with some discomfiture, to be a blatant fiction. He bases his calculations of the chances of various casts of the dice on the idea of a "circuit," or number of throws necessary to realize all the possibilities. For a fair die, Cardano claims that a full circuit is completed in six throws; for two fair dice, in thirty-six throws, and so on. Of course, experience usually belies this claim, and in the next breath Cardano conceded that in practice more than six throws may be required to turn up all six faces: if a given face of the die turns up more or less often than it should in a given circuit, "that is a matter of luck." Luck is ubiquitous in Cardano's account, upsetting the determinism of his mathematical analysis, which latter, as he himself warns, contributes "hardly anything to practical play."

From our point of view, Cardano's problem was that he had conflated the probabilities with the observed frequencies. Subsequent probabilists, armed with Bernoulli's and Bayes' theorems and above all with a long term view of things, were untroubled by the discrepancy: in the very long run, the two will converge. Cardano shared their insight if not their theorems, and at one point claims that an infinite number of throws would render a calculated outcome "almost necessary [*proximé necesse*]". But this assurance did not assuage Cardano's own doubts about the matter. As a gambler, he was interested in the short run, or even the single instance; as a mathematician, he believed that his calculations must necessarily hold in each and every case, not just on average.

Cardano's own explanation for the discrepancy between calculation and actual outcome is "luck." Only in the case of many trials do the calculations come close to the fact of the matter, and even then they remain an "approximation" subject to the perturbations of fortune. Cardano did not use "luck" as later probabilists from Bernoulli to Laplace would, that is, as a shorthand for unknown causes and their equally inscrutable interactions. He believed firmly in its objective reality. In his book, neophyte gamblers are warned to avoid luckier opponents, for good fortune is as fixed a trait as skill. After an anecdote about a spectacular win of his own, Cardano argued that such protracted winning or losing streaks have to do with the favors of fortune rather than the fluctuations of mere chance.[37]

Cardano was no determinist, and his metaphysical conscience was not greatly troubled by the interventions of chance (small, temporary fluctuations) or luck (systematic trends or streaks). Indeed, since he believed in a daimon at his own side, they added more spice to the game. However, they all but vitiated the practical importance of his calculations, as he himself admitted. Given the contemporary view of mixed mathematics, this was only to be expected: mathematical calculations concern the necessary, but cannot be extended to the contingent. Inscrutable but real forces like luck would from time to time throw off the match between calculated and actual outcomes. Cardano's circuits represent the mathematician's vain attempt to impose necessity on the single contingent event; his admission of failure was also an admission of the reality of chance and luck. With that admission he effectively surrendered his claim to be the inventor of mathematical probability. Cardano believed mathematics applied strictly speaking only to the necessary, and that chance events are not necessary. Therefore, the mathe-matics of chance was doomed from the outset, at least at the level of the single event or game that mattered to the gambler. Even if Cardano had been more prompt in publishing, he could not have won credit for founding mathematical probability. Belief in luck – that is, that things sometimes occur without cause or reason – played havoc with the probabilities, and vitiated his mathematics.

However, later probabilists like Bernoulli and De Moivre were more ambitious for their calculations, and their unrelenting determinism licensed them to be so. They shared Cardano's understanding of mixed mathematics, but they were strangers to the gaming tables and branded luck a vulgar error. Their lack of hands-on gambling experience proved to be an advantage, for it allowed them to ignore the single event and the short term in favor of the long run. In fact, their world was as full of contingency and variability as Cardano's had been, but their metaphysics and their own inexperience with games of chance sunk that fact below the level of consciousness. Of course, observed frequencies still deviated from calculated probabilities, but the probabilists clung to their metaphysical belief in complete, watertight determinism: such contingencies were only apparent, and could be safely ignored, at least in the long run. Moreover, the theorems of Bernoulli and De Moivre reduced the length of the "long run" from Cardano's infinite number of trials to a finite, calculable number needed to guarantee that observed frequencies and calculated probabilities would most likely – to a moral certainty – match. It was the strength of their determinist convictions that allowed the late seventeenth-century probabilists to wave aside the deviations in the individual case that had so vexed Cardano in his attempt at a mathematics of gambling. Serene in the belief that chance was a figment of human ignorance, they could simply ignore short run variability. At least within the seventeenth-century context, determinism was a precondition for probability mathematics.

Jakob Bernoulli's approach to the problem of reconciling probabilities and frequencies throws this contrast between probability theory with and without determinism into relief. We have already heard Bernoulli's views on the "stark necessity" that governs all events, at least in nature. Good and bad fortune still had meaning for Bernoulli, but he redefined them as events less likely to have occurred, giving Aristotle's example of a man finding treasure while digging in his garden a quantitative twist: such windfalls happen less than one time in a thousand. (Characteristically enough, Aristotle's emphasis is not so much on the infrequency of the event as on the lack of intention: the digger was planting a tree, not excavating for buried treasure. In the original of Bernoulli's example, chance is the absence of purpose, not simply the rarity of an event.)

When Bernoulli turns to the theorem which bears his name, which we understand as a solution to Cardano's quandary of squaring calculated probabilities with observed frequencies, he instead concentrates on revealing the "hidden causes" of the weather or human mortality. These are peculiar sorts of causes, being neither microscopic mechanisms of the Cartesian sort or forces of the Newtonian sort, but rather "true ratios," or underlying probabilities. Bernoulli likens these ratios to those between black and white pebbles in an urn, and suggests that these ratios "cause" the frequencies of storms, deaths, etc. we actually observe. His theorem is meant to approximate these hidden causes a posteriori, with an accuracy that increases with the number of observations.[38]

Cardano was troubled by the discrepancy between calculated mathematical probabilities and observed frequencies in the short (or very short) run; Bernoulli pays the short run no heed. Rather, he is struck by how certainty increases with the number of observations, something which even "the stupidest man knows, by some instinct of nature, with no previous instruction" but which Bernoulli hopes to demonstrate "accurately and geometrically." Far from fretting over the individual case, Bernoulli worries about the pitfalls of the *too* long run for his theorem. In response to a query from Leibniz, he admits that the "true ratios" of, for example, deadly diseases might have changed since antediluvian times, and preaches suitable caution in circumscribing the observations to a period short enough to be homogeneous in this respect.[39]

In this context, gambling problems do not in the least trouble Bernoulli, for he assumes that the calculation of the probabilities follows straightforwardly from physical symmetry of the coins or dice. The short term discrepancies between these probabilities and the actual tosses, which persuaded Cardano of the existence of chance and luck, are not even mentioned. For Bernoulli the problem is no longer variability in the world, but gaps in our knowledge of the world – specifically knowledge of the hidden causes of "life and death in future generations," of next year's weather, of the outcome of a game of skill. It is not simply that Bernoulli's determinism solved Cardano's problem; it changed the problem almost beyond recognition.

Although Bernoulli conceived his theorem to be about the determinism of matter – weather, diseases, wine harvests, dice – his definition of probability as a "degree of certainty" seems at first glance to be about the indeterminism of mind. Contingencies that vary from culture to culture and from person to person hardly look like good candidates for necessity. Part of the problem is due to a genuine and much-remarked ambiguity in Bernoulli,[40] who applied the concept of probability indiscriminately to what we now distinguish as subjective and objective senses of the word; that is, both to states of mind and states of the world. Bernoulli's theorem exemplifies this problem in microcosm. In modern notation, it states that

$$\lim_{n \to \infty} P(|p - m/n| < \varepsilon) = 1,$$

where P is the (subjective) probability of our conjecture about the true, fixed (objective) probability p of a given event, actually observed to happen m out of n trials. But even Bernoulli's subjective probabilities were far steadier objects than Galileo's air currents or Pascal's rhetoric, and they were emphatically not subjective in the personalist sense now current among Bayesian probabilists of the Savage/De Finetti school. Insofar as Bernoulli's subjective probabilities have any twentieth-century analogues, they are the logical probabilities of John Maynard Keynes and Rudolf Carnap: that is, they represent the uniquely rational degree of certainty we should accord a given conjecture on the basis of the evidence available. If these probabilities were not objective in the hard-cornered sense of dice, they were at least intersubjective. Theirs was a kind of normative determinism, a coercion of reason which many seventeenth-century authors, including Hobbes and Leibniz, considered to be distinctive to mathematics. Bernoulli meant mathematical degrees of certainty to determine the intensity of belief, just as mathematical demonstrations command assent. Of course, free will makes it possible to act in defiance of both probabilities and demonstrations, but that was fool's freedom, as generations of classical probabilists were to claim again and again.[41] Hence, even Bernoulli's subjective probabilities were determined, albeit in a normative rather than in a descriptive fashion.

To sum up: Bernoulli and Cardano held similar views about the nature of mixed mathematics, but differed in their metaphysics. Cardano wrote in a period in which nature's singularities seemed more revealing of her essential structure than her regularities, and variability was studied rather than ignored or explained away. Cardano himself contributed to this way of thinking with his influential compendium of the strange and anomalous, *De subtilitate* (1550). His penchant for gambling riveted his attention on the individual event and the short run. Mathematics might deal with the necessary, but for Cardano, much of life and nature slipped through the meshes of necessity. Not so for Bernoulli. By the last quarter of the seventeenth century, the earlier fascination with the singular and the single case was fast disappearing from natural philosophy. Bernoulli chided Leibniz for suggesting that God could create anything "inherently uncertain and indeterminate," and his own *magnum opus* on mathematical probability is a ringing denial of the existence of anything like chance. He took the long view that eventually erased all the deviation and variability that had so preoccupied Cardano. Whereas Cardano's claim that a kind of necessity will triumph in the long run is incidental to his mathematics of chance, and hardly a compensation for its uselessness to the workaday gambler, it is the keystone of Bernoulli's theory. Belief in the longterm stability of statistical frequencies, despite ample evidence of wide variations, became the hallmark of applied eighteenth-century work in statistics: Johann Süssmilch's insistence upon a universal law of human fertility in the teeth of his own data to the contrary is a striking example of the lengths to which this faith could be carried.

Why these frequencies should turn out to be stable was a matter of theological debate for much of the eighteenth century. John Arbuthnot, De Moivre, Süssmilch, and others concluded from a variant on the argument from design that such stabilities were evidence of divine providence at work in maintaining monogamy and a constant population.[42] Although Nicholas Bernoulli objected to such claims when they were first advanced, it was not until the 1830s that Siméon-Denis Poisson could take the existence of such stabilities for granted, without invoking God's intelligent agency.[43] However, *that* frequencies were stable (for whatever reason) was already an article of faith by the mid-seventeenth century. At the conclusion of Bernoulli's demonstration of the "golden theorem" he wished to serve as his epitaph, he wrote: "Whence, finally, one thing seems to follow; if observations of all events were to be continued throughout all eternity ... everything in the world would be perceived to happen in fixed ratios and according to a constant law of alteration, so that even in the most accidental and fortuitous occurrences we would be bound to recognize, as it were, a certain necessity and, so to speak, a certain fate."[44] For mixed mathematicians of the late seventeenth century, determinism had made the world safe for probability theory. In the eighteenth and nineteenth centuries, probability theory was to return the favor.[45]

CONCLUSION

Where did this determinism come from, and how did it so quickly and decisively replace the natural philosophy of the anomalous that preceded it? Here, I can do little more than hint at an explanation to what surely was a convoluted historical episode. First, I believe we can eliminate some plausible candidates. Natural philosophy was more the beneficiary than the agent of this sea change in metaphysics. In the first instance, proponents of the new experimental philosophy were as likely to aid and abet the investigation of anomalies and variability as to rein it in. Conscientious empiricism lacked any criterion by which to reject the strange reports that later generations would dismiss out of hand: the moral of Locke's tale of the King of Siam and the Dutch ambassador was that there were more things in heaven and earth than had been heard of in our philosophy, and that we would do well to keep an open mind on the subject of marvels and even miracles.[46] Moreover, just because these unusual phenomena were then the focus of scholarly attention, they became the cases against which new systems of natural philosophy were tested, as we have seen in the case of Bacon and Descartes.

Nor did daily life come to be seen as markedly more tranquil during this period. Those changes of perception occurred in the mid-eighteenth, not the late seventeenth century – and even then, they were restricted by geography and class. Practitioners of risk such as gamblers and insurers largely ignored the theory of risk created by the mathematicians until late in the eighteenth century. Presumably, their attitudes mirrored their experience.[47] Gamblers agreed with Cardano that probability theory was disappointing in practice. The adventurer hero Ferdinand Count Fathom of Tobias Smollett's 1753 novel sizes up the oafish Sir Stentor as an easy mark, for the calculus of probabilities had been one of Ferdinand's "chief studies ... he could calculate all the chances with utmost exactness and certainty." Unfortunately, Ferdinand discovers that mathematics is of little use against Stentor's sharp dealing and (in all likelihood) loaded dice.[48] Ferdinand was not the only gambler to be converted from probability theory to cheating. Beau Hewit, a famous late seventeenth-century gambler, first studied Christiaan Huygens' treatise on probability theory, but

> ... finding his Rules of calculating chances most false and erroneous, he damn'd that Authour for as great a Blockhead as he was a Fool, in losing his Money upon such conceited whims; therefore learning the most profitable and surest way of tricking both at Cards and Dice, in which the Adversary could make no Calculation of Chances, he became so expert in the Dexterity of Slipping Cards, or Cogging a Dye, that in 4 years he was worth 6,000 Pounds.[49]

The rise of mathematics in the sciences certainly promoted the cause of determinism, although, as in the case of mathematical probability, the reverse was also sometimes the case. In addition to the difficulties of untangling cause and effect in the relationships between seventeenth-century mathematics and determinism, there is the problem of scope. The determinism of the late seventeenth century was sweeping, extending to all material phenomena, whereas the progress of the mathematical approach to nature halted at the frontiers of chemistry and natural history.

It is just this blanket character of the new determinism that implicates theology. Some of the most influential religious works of both Protestant and Catholic confessions during this period hammered home a message of predestination. Is it coincidence that Bernoulli and De Moivre came from Huguenot backgrounds, and that Pascal was a passionate convert to Jansenism? It is arguably theology rather than natural philosophy that eliminated miracles from the world of the learned during this period, and it was theologians who were the bitterest critics of "vulgar" beliefs in luck and fortune at this time.[50] Miracles and fortune were of course the greatest obstacles to the belief in inviolable laws of nature; it was largely theologians who removed them.[51] All of this is merely circumstantial evidence, but it is nonetheless suggestive. However, one must tread carefully here. Although the remarkable stability of certain demographic frequencies like the ratio of male-to-female births and mortality as a function of age were exploited to the hilt by natural theologians like William Derham, Süssmilch, and others, the belief of the mathematicians in such long term regularities antedates this particular variant of the argument from design. The so-called Royal Society theologians and their Dutch and German colleagues were the beneficiaries of this belief, not its originators.

It is more at the level of metaphysics than apologetics that seventeenth-century theology made its contribution to the determinism of the probabilists. The epistemological terms in which the determinism of the classical probabilists was invariably cast echo the contemporary discussions of God's powers and place in his creation. As in the case of the debates over predestination in both Calvinist and Jansenist circles, the emphasis is upon the predictability of the future; on the unbroken chain of cause and effect as grasped by (for Bernoulli and De Moivre) God or (for Laplace) a superhuman mind. It is relative to a divine standard that the imperfections of human knowledge are measured, just as probabilities are relative to a deterministic cosmos. As James Clerk Maxwell was to point out much later, this kind of determinism counts for little in the sciences or daily life, for we humans are destined to our mortal frailties, and therefore to dwell in the twilight of probabilities.[52] That is, for us, probabilities of at least some events, like turbulence, might as well be real. We shall never see the world from a God's eye point of view. But for the classical probabilists, the vision of a deterministic world order was just such a lofty one, even after it had been secularized. Only when they descended from these heights could chance become real again. Conversely, it may be that determinism became both thinkable and attractive, even though never attainable in its scientific details, because its proponents could imagine such a God's eye point of view, and, at least as important as omniscience, a God who planned ahead with unexceptioned consistency.

Since the mid-nineteenth century, it has been one of the glories of mathematical statistics that it can deal with events in which conditions weaker than causation obtain, and we now use probability theory to describe phenomena we believe to be genuinely random and incorrigibly variable. But this route was not open to the classical probabilists, who still associated the necessity of mathematics with the necessity of the phenomena it treated. Mixed mathematics demanded that match. In order to bring chance and variability into the mathematical fold, they required a broader and deeper brand of determinism than had previously been part of natural philosophy: broader, so as in principle to encompass all natural phenomena; and deeper, to penetrate to the stable underlying probabilities that ultimately fixed even the most unpredictable events. This second sense reveals why determinism also needed probability, insofar as it was to give a respectable account of variability. "Saving the phenomena" for determinism is a time-honored function for probability theory and statistics even today, at least in the social sciences. But in the beginning the debt was on the other side, when philosophers, theologians, and mathematicians joined forces to banish the harlequin goddess Fortuna, wheel and all.

Department of the History of Science
University of Göttingen

NOTES

* I would like to thank Lorenz Krüger for several very helpful discussions on the meanings of determinism, and Nancy Cartwright, Gérard Jorland, Ian Hacking, Theodore Porter, and Norton Wise for their comments on an earlier draft. Parts of this essay appeared in: 'Perché la teoria della probabilità aveva bisogno del determinismo: le origini,' *Intersezioni* 10 (1990), 541–567.

[1] Pierre Simon de Laplace, *Essai philosophique sur les probabilités* (1814), in Académie des Sciences de Paris, ed., *Oeuvres complètes* (Paris, 1878–1912), vol. 7, p. vi; English translation from F. W. Truscott and F. L. Truscott, *Philosophical Essay on Probabilities* (New York: Dover, 1951), p. 3.

[2] On the background to Laplace's determinism, see Roger Hahn, 'Laplace's First Formulation of Scientific Determinism in 1773,' *Actes XIe Congrès International d'Histoire des Sciences, 1965* (publ. 1968), 167–171; also R. Hahn, 'Determinism and Probability in Laplace's Philosophy,' *Actes XIIIe Congrès International d'Histoire des Sciences, 1971* (publ. 1974), 1, 170–176.

[3] See Maurice Kendall, 'The Beginnings of a Probability Calculus,' in E. S. Pearson and M. G. Kendall, eds., *Studies in the History of Statistics and Probability* (Darien, Conn.: Hafner, 1970), vol. 1, pp. 19–34, for the view that the emergence of mathematical probability was made possible by a new appreciation of chance; Ian Hacking, *The Emergence of Probability* (Cambridge: Cambridge University Press, 1975), pp. 2–3, makes the opposite claim.

[4] M. J. A. N. Condorcet, *Le Marquis de Condorcet à M. d'Alembert, sur le système du monde et sur le calcul intégral* (Paris, 1768), p. 5; quoted in Keith M. Baker, *Condorcet: From Natural Philosophy to Social Mathematics* (Chicago: University of Chicago Press, 1975), p. 167.

[5] Jakob Bernoulli, *Ars conjectandi* (1713), p. 212; in Naturforschende Gesellschaft in Basel, ed., *Die Werke von Jakob Bernoulli* (Basel: Birkhäuser, 1975), vol. 3, p. 240.

[6] Abraham De Moivre, *The Doctrine of Chances*, 3rd ed. (London, 1756), p. 253.

[7] Bernoulli, *Ars* (ref.6), p. 211; *Werke* (ref. 6), p. 239.

[8] De Moivre, *Doctrine* (ref. 7), p. 251.

[9] Laplace evidently was a determinist with respect to moral events as well, although he never published his views on this topic. The Dossier Laplace, Archives de l'Académie des Sciences, Paris, contains his manuscript reflections on free will and causation.

[10] Bernoulli, *Ars* (ref. 6), pp. 212–213; *Werke* (ref. 6) p. 240.

11 Aristotle, *De interpretatione*, 18a28–19a39, in Jonathan Barnes, ed., *The Complete Works of Aristotle* (Princeton: Princeton University Press, 1984), vol. 1, pp. 28–30.

12 Ian Hacking, 'Nineteenth-Century Cracks in the Concept of Determinism,' *Journal of the History of Ideas* **44** (1983), 455–475; also his *The Taming of Chance* (Cambridge: Cambridge University Press, 1990), pp. 150–159.

13 On the complexities of seventeenth-century conceptions of natural law, see Catherine Wilson, 'De Ipsa Natura: Sources of Leibniz's Doctrines of Force, Activity and Natural Law,' *Studia Leibnitiana* **19** (1987), 148–172.

14 See Samuel Clarke's half of H. G. Alexander, ed., *The Leibniz–Clarke Correspondence* (Manchester: Manchester University Press, 1976), written under Newton's tutelage.

15 Charles de Secondat de Montesquieu, *De l'esprit des lois* (1748), G. Truc, ed., 2 vols. (Paris: Garnier, 1961), vol. 1, p. 6.

16 Eileen Serene, 'Demonstrative Science,' in Norman Kretzmann, Anthony Kenny, and Jan Pinborg, eds., *The Cambridge History of Late Medieval Philosophy: From the Rediscovery of Aristotle to the Disintegration of Scholasticism, 1100–1600* (Cambridge: Cambridge University Press, 1982), pp. 496–517.

17 Aristotle, *Physics*, 199b25, in J. Barnes, ed., *Works* (ref. 12), vol. 1, pp. 341.

18 Bert Hansen, ed., *Nicole Oresme and the Marvels of Nature: The "De causis mirabilium"* (Toronto: Pontifical Institute of Mediaeval Studies, 1985), p. 279.

19 Francis Bacon, *The New Organon* (1620), Fulton H. Anderson, ed. (New York: Macmillan, 1985), II. 28–29, pp. 177–179.

20 René Descartes, *Principia philosophiae* (1644), IV., in Charles Adam/Paul Tannery, eds., *Oeuvres de Descartes* (Paris: Léopold Cerf, 1897–1913), vol. VIII, part IV, p. 187.

21 See Jean d'Alembert, 'Discours préliminaire', in *Encyclopédie, ou Dictionnaire raisonné des sciences, des arts et des métiers* (Paris, 1751) vol. 1, pp. I–XIV.

22 Aristotle, *Metaphysics*, 1078a10–16, in J. Barnes, ed., *Works* (ref. 12), vol. 2, p. 1704.

23 Marin Mersenne, *Les Questions theologiques, physiques, morales, et mathématiques* (1634), in *Questions inouyes* (Paris: Fayard, 1985), pp. 357–359; cp. Aristotle, *Metaphysics*, 1025b19–1026b32, in J. Barnes, ed., *Works* (ref. 12), vol. 2, p. 1619–1621.

24 Barbara J. Shapiro, *Probability and Certainty in Seventeenth-Century England* (Princeton: Princeton University Press, 1983), pp. 15–73; Steven Shapin, 'Robert Boyle and Mathematics: Reality, Representation, and Experimental Practice,' *Science in Context* **2** (1988), 23–58.

25 Galileo, *Dialogue Concerning the Two Chief World Systems* (1632) trans. Stillman Drake, 2nd ed. (Berkeley: University of California Press, 1967), p. 103; Antonio Favaro, ed., *Le Opere di Galileo Galilei* (Firenze: G. Barbera, 1965), vol. VII, p. 128.

26 Descartes, *Regulae* (comp. 1619–28), Rule IV, in Adam/Tannery, eds., *Oeuvres* (ref. 21), vol. X, p. 378; *Discours de la méthode* (1637), Part. I, in *Ibid.*, vol. VI, pp. 1–11.

27 Gottfried Wilhelm Leibniz, 'Preface to the Universal Science' (1677), in Philip P. Wiener, ed., *Leibniz, Selections* (New York: Charles Scribner's Sons, 1951), p. 15.

28 Marin Mersenne, *La vérité des sciences* (Paris, 1625), p. 226.

29 Galileo, *Discorsi* ... (1638), in Favaro, ed., *Opere* (ref. 26), vol. VIII, p. 277.

30 Blaise Pascal, 'De l'art de persuader,' in Léon Brunschvicg, Pierre Boutroux, and Félix Gazier, eds., *Oeuvres* (Paris: Librairie Hachette, 1914), vol. IX, pp. 276–77.

31 Bernard de Fontenelle, 'Eloge de M. Jacques Bernoulli,' *Histoire de l'Académie Royale des Sciences. Année 1705* (Paris, 1706), pp. 139–150.

32 Bernoulli, *Ars* (ref. 6), pp. 217–223; *Werke* (ref. 6), pp. 243–247.
33 Marin Mersenne, *Questions inouyes* (1634) (Paris: Fayard, 1985), pp. 45–46.
34 Gottfried Wilhelm Leibniz, 'Towards a Universal Characteristic' (1677), in Philip P. Wiener, ed., *Leibniz* (ref. 28), pp. 22–23.
35 *Ibid.*, p. 17.
36 Pascal, 'Celeberrimas mathesos Academiae Parisiensi' (1654) in Jean Mesnard, ed., *Oeuvres complètes de Pascal* (Paris: Bibliothèque-Européene/Desclès de Brouwer, 1970), vol. 1, Part 2, p. 1034.
37 Hieronymous Cardanus (Girolamo Cardano), *Liber de ludo aleae* (comp. ca. 1520; publ. 1663) in *Opere omnia* (Lyon, 1663), chs. 1, 9–11, 20.
38 Bernoulli, *Ars* (ref. 6), pp. 223–239; *Werke* (ref. 6), pp. 247–259.
39 For the Leibniz/Bernoulli correspondence (October, 1703–April, 1704), see G. J. Gerhardt, ed., *G. W. Leibniz mathematische Schriften* (Hildesheim: Georg Olms, 1962; reprint of 1855 edition), vol. 3, prt. 1, pp. 11–89.
40 Ian Hacking, "Jacques Bernoulli's Art of Conjecturing," *British Journal for the Philosophy of Science* **22** (1971), 339–355.
41 Lorraine Daston, *Classical Probability in the Enlightenment* (Princeton: Princeton University Press, 1988), pp. 49–111.
42 John Arbuthnot, 'An Argument for Divine Providence, Taken from the Regularity Observe'd in the Birth of Both Sexes,' *Philosophical Transactions of the Royal Society of London* **27** (1710–12), 186–190; Johann Peter Süssmilch, *Die göttliche Ordnung in den Veränderungen des menschlichen Geschlechts, besonders im Tode* (1756), 3rd rev. ed. (Berlin, 1775), 3 vols.; Karl Pearson, *The History of Statistics in the 17th and 18th Centuries*, E. S. Pearson, ed. (London and High Wycombe: Charles Griffin, 1978), pp. 279–347.
43 Siméon-Denis Poisson, *Recherches sur la probabilité des jugements en matière criminelle et en matière civile* (Paris, 1837), p. 118.
44 Bernoulli, *Ars* (ref. 6), pp. 228–239; *Werke* (ref. 6), pp. 250–259.
45 Theodore M. Porter, *The Rise of Statistical Thinking* (Princeton: Princeton University Press, 1986), pp. 151–192.
46 John Locke, *An Essay Concerning Human Understanding* (1690), Alexander Campbell Fraser, ed. (New York: Dover, 1959), vol. 2, Book IV, ch. 15, pp. 366–367.
47 Daston, *Classical Probability* (ref. 42), pp. 112–187.
48 Tobias Smollett, *The Adventures of Ferdinand Count Fathom* (Dublin, 1753), vol. 1, pp. 115–117.
49 Theophilus Lucas, *Memoirs of the Lives, Intrigues and Comical Adventures of the Most Famous Gamesters and Celebrated Sharpsters* (London, 1714), p. 285.
50 See for example the Protestant Jean Le Clerc, *Réflexions sur ce que l'on appelle bonheur et malheur en matière de loteries...* (Amsterdam, 1696).
51 Jean Calvin, "Epistre," *Institutions de la religion chrestienne* (n.p., 1541).
52 James Clerk Maxwell, 'Does the Progress of Physical Science Tend to Give any Advantage to the Opinion of Necessity (or Determinism) over That of the Contingency of Events and the Freedom of the Will?' (1873), in Lewis Campbell and William Garnett, *The Life of James Clerk Maxwell* (London, 1882), pp. 434–444.

JOAN L. RICHARDS*

GOD, TRUTH, AND MATHEMATICS IN NINETEENTH CENTURY ENGLAND

In the post-Newtonian tradition of English natural theology, the linkage between our understanding of mathematics and our understanding of God is strong and ubiquitous. A good starting point attesting to the connection is John Locke's *Essay Concerning Human Understanding*, which includes a chapter entitled "Of our Knowledge of the Existence of a God." This chapter appears in the context of a discussion of the three kinds, or tiers, of knowledge humans can claim. The first is immediate knowledge; following René Descartes, Locke argues that the only example of this kind is our knowledge of our own existence. The second kind comprises our knowledge of God, which Locke argues we can come to recognize from our experience of ourselves. The recognition is not immediate, however; it requires individual effort to attain it. Locke's third kind of knowledge comprises the existence of *"any other thing"* which, he asserts, "we can have only by *sensation*."[1] This kind of knowledge encompasses the vast majority of the things we know; defining its parameters focuses Locke's attention in much of the rest of his *Essay*.

The kind of knowledge we have of God is thus very specialized for Locke, neither immediate, like knowledge of ourselves nor mediated by the senses. Locke explained its intermediate position as follows:

But, though this [the existence of God] be the most obvious truth that reason discovers, and though its evidence be (if I mistake not) equal to mathematical certainty: yet it requires thought and attention; and the mind must apply itself to a regular deduction of it from some part of our intuitive knowledge, or else we shall be as uncertain and ignorant of this as of other propositions, which are in themselves capable of clear demonstration.[2]

Knowledge of God is thus available through a personal search; we arrive at it through reasoning on our direct experience of ourselves. For Locke, it is the roots of our religious convictions in this essentially personal quest that assures their ultimate certainty. It is this kind of certainty Locke meant when he wrote that our knowledge of God is a "certain and evident truth;" it is, he explains, known in a way "equal to mathematical certainty."[3]

The certainty Locke here attributes to knowledge of God is subtly different than the certainty available in scientific discourse, because it transcends the merely probable basis of scientific knowledge. There are other aspects of difference as well. Whereas in science arguments are fixed around repeatable experiences which occur in socially shared space, the path to mathematical and theological certainties cannot be shared in that way.[4] The certain knowledge attainable within these fields is profoundly individual, and cannot be convincingly communicated from one person to another. Thus, to those who do not agree with his theological argument, Locke can only reply with "that very rational and emphatical rebuke of Tully . . . 'What can be more sillily arrogant and misbecoming, than for a man to think that he has a mind and understanding in him, but yet in all the universe beside there is no such thing?'"[5]

In describing this kind of theological knowledge, Locke relies heavily on mathematical examples and language to clarify his meaning. This is not merely a superficial connection; Locke recognizes that the same kind of personal effort that he traces in theology is required to establish certainty in mathematics. Thus, in his later chapter, "Of Probability," Locke emphasises:

For example: in the demonstration of [a proposition] a man perceives the certain, immutable connexion there is of equality between the three angles of a triangle, and those intermediate ones which are made use of to show their equality to two right ones; and so, by an intuitive knowledge of the agreement or disagreement of the intermediate ideas in each step of the progress, the whole series is continued with an evidence, which clearly shows the agreement or disagreement of those three angles in equality to two right ones: and thus he has certain knowledge that it is so.[6]

Locke contrasts this man's certainty based on personal experience, with the merely probabilistic knowledge of one who accepted the same theorem on someone else's testimony.

In the course of the eighteenth century, many philosophers developed Locke's ideas in such a way that this second tier of knowledge was discredited. Ironically, many of the criticisms grew from Locke's own psychological theory in which all knowledge was built up from sensory experience. This approach led many of his followers to skepticism. Insisting that all of our knowledge comes from the senses, and recognizing that they provide neither direct nor infallible knowledge of the external world, these thinkers concluded that no knowledge was truly certain. This kind of eighteenth-century critique undermined the privileged position of both theology and mathematics. Thus in the tree of knowledge of the mid-century French *Encyclopédie* theology is banished to the far limits of our knowledge in one direction, abstract mathematics to the other.[7]

This was just the beginning. In subsequent decades, materialistic philosophers like Denis Diderot or Julien Offray de La Mettrie vigorously denied that our existence necessarily implies a creator. They insisted on Locke's impossibility: that all beings were the product of chance material interactions. Mathematics was under siege as well. Although in the eighteenth century it was universally agreed that all triangles have angle sums equal to two rights, the ontological significance of the theorem was hotly contested. Critics like Georges-Louis Leclerc, Comte de Buffon, found mathematics to be certain but empty, a mere tissue of deductive arguments with no substantial implications. This kind of approach to mathematics strengthened the hand of the materialists by undercutting not only the value of mathematical knowledge but of any other knowledge modeled upon it.

In France these kinds of considerations contributed to the dissolution of the Lockean equation. By the end of the Revolution, God had been wrested from a central place in the French state and relegated to the Catholic church. French mathematicians, for their part, were discouraged, and mourned the end of progress in their field. What Locke had presented as twin bulwarks of certain knowledge seemed to have crumbled in the face of eighteenth-century skepticism.[8]

This reading of the French case does not exhaust the history of the relationship between religion and mathematics, however. In the middle of the nineteenth century, more than one hundred years after Locke's *Essay* was first published, many English natural theologians similarly linked theological and mathematical knowledge. References relating mathematical to religious certainty can be found in sources as far ranging as William Whewell's *Philosophy of the Inductive Sciences* and Charles Kingsley's *Waterbabies*. An important part of this equation was the conviction, expressed in Locke, that these kinds of knowledge were neither empty nor contingent; they were uniquely compelling and central to human understanding and experience.

In the nineteenth-century context, though, defining the nature of this certain knowledge for both mathematics and theology was difficult. It was in an important sense experiential, in that its validity rested on convictions rooted in personal experience. Because it was ultimately personal, however, such experience was very different from that acknowledged in the scientific sphere. This meant, on the one hand, that it could not be communicated effectively from one person to another; it had to be experienced anew by each individual to be known. On a more positive note, though, its certainty was not bounded by the contingencies of sense experience. Both mathematics and theology moved beyond the reach of experience into realms of the perfect and the infinite. The certainty with which they were known itself transcended the boundaries of contingent scientific knowledge.

Thus, both natural theology and mathematics were intimately tied to the world of experimental science, but neither could be adequately understood or defined by those sciences. How they were to be defined or understood was not immediately obvious, however. This problem was not merely ignored nor was it shelved. In the natural theology of the 1830s and 1840s an attempt was made to discuss both areas at the same time, to unite them within a single view of the world.

I

The first decades of the nineteenth century in England was a very lively period, both in religion and in mathematics. Religiously it was characterized by the strength of the evangelicals; mathematically it opened with the works of the Cambridge Analytical Society followed by considerable developments in algebra. At first glance there might seem to be little to connect these two areas but in fact, much did. In both religion and mathematics Englishmen were struggling to justify their understanding in the face of those forces that would marginalize it.

The relationship between the fields was a somewhat lopsided one. Most English religious writers and thinkers were not in tune with developments in mathematics. English mathematicians, on the other hand, were not innocent of religion. To quote Joseph Altholz, "The most important thing to remember about religion in Victorian England is that there was an awful lot of it."[9] Whatever their individual religious affiliations, simply by virtue of living where and when they did English mathematicians were steeped in religion.

Looking just at those to be treated in this paper, the communality might seem to stop there. Whewell and George Peacock were ordained in the Church of England. Both Charles Babbage and John Herschel, on the other hand, considered entering the church when they finished

Cambridge, but decided on other directions instead. All four maintained strong religious convictions throughout their lives, though the contrasts between Peacock's institutional connections, Whewell's Tory orthodoxy, Herschel's steadfast piety, and Babbage's eccentric argumentativeness were marked in later life. Augustus De Morgan, on the other hand, was raised as a dissenter, and as an adult he was a fierce critic of all religious orthodoxies. This means that his own convictions are hard to know because, on principle, he kept them to himself. Judging from his wife's biography they were strongly held, however, and there is the direct evidence of his will where he commended himself "to God the Father of our Lord Jesus Christ, whom I believe in my heart to be the Son of God, but whom I have not confessed with my lips, because in my time such confession has always been the way up in the world."[10] Despite the deep differences which divided these men religiously, in essential ways their views of mathematics reflected a deep and common religious orientation they shared with their culture.

For lack of a better term, one might call the defining character of this early nineteenth-century English religiosity "evangelical." Three generations after John Wesley's revival, the movement he began had spread so far that its purview was no longer restricted to the confines of a single sect or even dogma. As Jane Austen wrote in 1814 to her friend, Fanny Knight, "I am by no means convinced that we ought not all to be Evangelicals."[11] With a movement so widespread the problem becomes defining it in such a way that the term has any meaning at all. Its identifying aspect, for the purposes of this study, is its essentially personal nature; in the final analysis, for the evangelical, true religion rested on the relation between the believer and God. Faith, the bedrock of evangelical life, was rooted in the human heart.

To many, this may seem to mean that there was no room for the rational in establishing religion; as a result evangelicals are often dismissed as "intellectually . . . narrow and naïvely reactionary."[12] This is inadequate to the breadth of the term as I am here trying to use it. Austen continued in her letter to say that those "who are so [evangelical] from Reason *and* Feeling, must be happiest and safest;"[13] for her the "feeling" element was best balanced by reason. Broadening the term evangelical to encompass Austen's use of it enables reason to work with feeling to support faith. At the same time, though, it requires examining the kind of reasoning that works in this context.

A brief episode reported in Charles Darwin's *Autobiography* may serve to illustrate some of the disparities between scientific and religious reasoning in this sense. There Darwin reports a statement by a Mrs Barlow, who was searching for an antidote to his father's scientifically-bred unbelief. "Doctor," she said, "I know that sugar is sweet in my mouth, and I know that my Redeemer liveth." This Darwin offered as an "unanswerable argument" but neither he nor his father was moved by it.[14]

The problem can be seen as one of private as opposed to public knowledge. Whereas Mrs Barlow placed the individual's private experience at the center of her argument, science dismisses it; sweetness has been classified as a secondary quality by virtually everyone after Descartes.[15] Scientific arguments are designed to stand beyond the individual's experience, experiments present "matters of fact" untrammeled by the confusions of individual perceptions. In science the emphasis on public knowledge was a very powerful strategy for establishing a neutral space for scientific discussion, but the record of the eighteenth century suggests that it was often not effective religiously. The emphasis on personal knowledge, on the other hand, might uphold religious faith but could be highly anti-social. Among early nineteenth-century evangelicals, this "excessive individualism" sometimes meant that "the corporate character of religious life ... was unappreciated;" in the larger sphere of public action it often bred an independence which could be infuriating.[16] Consensus, so highly valued in science, is not so important in the evangelical context.

These two kinds of knowledge were precariously balanced in natural theology. The typical eighteenth-century design argument can be seen as an attempt to establish religious verities in the new realm of public scientific truth, to establish the existence of God on the basis of evidence gleaned from the realm of natural "matters of fact." William Paley's *Evidences of the Creation*, a capstone of the tradition, explores and justifies every wrinkle of the thought processes leading to the conclusion of a designer from designed objects. Most of the work is devoted to multiplying examples of properly established "matters of fact" which evidence design, thereby establishing the extra-personal scientific basis for the conclusions.

By 1802, when Paley was writing, however, this kind of evidentiary argument for the existence of God was beset with problems. The question of what and how much could be proved scientifically by the design analogy had been hotly contested, notably in David Hume's *Dialogues*

Concerning Natural Religion of 1779. There were religious problems with it as well. Approached as a proof of the existence of God it was basically irrelevant to evangelical interests which focused not on whether God existed but rather on how one related to God. Nonetheless, the argument from design thrived in the early nineteenth century. That Paley's *Evidences* remained a standard text at Cambridge well into the century suggests that seeing this kind of argument as merely a proof of God's existence is too narrow a reading of the genre. Its longevity in the face of considerable scientific and religious obstacles invites a closer look.[17]

The most obvious place to consider the argument from design in the early nineteenth century is in the Bridgewater Treatises. These celebrated tomes were commissioned by the Royal Society under the terms of the will of the Reverend Francis Henry, Earl of Bridgewater who died in 1829. Bridgewater stipulated that works be written "On the Power, Wisdom, and Goodness of God, as manifested in the Creation; illustrating such work by all reasonable arguments . . . "[18] Much of the eight volumes written under these stipulations is taken up with embellishments on the argument from design. This was not the full extent of their message, however. In addition to this evidentiary argument, building upon scientifically established facts, there is another one which reflects more clearly the evangelical emphasis on establishing a personal relation to God. This second kind of argument, which I will call the participatory argument, emphasizes the value of the pursuit of science for one's personal religious health. It is exemplified by the approach outlined in the final chapter of Thomas Chalmers' treatise entitled *The Adaptation of External Nature to the Moral and Intellectual Constitution of Man*.

In this chapter Chalmers struggles with the problem of atheists, people who fail to see God's presence either in the natural world or in their lives. Certainly, Chalmers acknowledges, one cannot demonstrate this presence to them definitively. Neither, however, can any one disprove it. There is then a middle point, what Chalmers calls a "neutral atheism" at which one just stands unconvinced and uncommitted. What, he asks, can lead someone from this position to a recognition of God? Chalmers' answer to this question is a personal one; the issue he asserts lies in the individual will. If such a neutral atheist were to pursue the subject assiduously and actively, he or she would succeed in finding God; "for it is by the exercise of a strong and continuous will, upholding or perpetuating the attention, that what at the outset were the probabilities of a subject are at length brightened into its proofs, and the verisimilitudes of our regardful notice become the verities of our confirmed faith."[19]

From the perspective of evidentiary science, this kind of knowledge shored up and strengthened by an individual's attitudes may seem weak indeed; for Chalmers, however, it was the basis of the "verities" of "confirmed faith." He is here not concerned with proving God's existence in the world, but rather with showing how that existence can be recognized. The process of recognition is a personal one which entails active participation and will. Chalmers is describing a process of knowing very similar to that Locke acknowledged for his second, theological tier of knowledge.

Chalmers was a Scottish divine whose influence on science in England was indirect at best. However, his emphasis on the personal value of pursuing science as a way of knowing God is not hard to find within the scientific community as well. A clear and central example is in Herschel's *Preliminary Discourse on the Study of Natural Philosophy*, first published in 1830. In his first part, entitled "Of the General Nature and Advantages of the Study of the Physical Sciences," Herschel expounds on the world which opens before the gaze of the searching scientist. The external world is so multifarious, Herschel asserts: "that as the study of one [subject] prepares him [the scientist] to understand and appreciate another, refinement follows on refinement, wonder on wonder, till his faculties become bewildered in admiration, and his intellect falls back on itself in utter helplessness of arriving at an end."

Being thus overwhelmed by the enormity of the natural world is a positive first step in the scientists' pilgrimage. It turns his gaze inward, where again he

feels himself capable of entering only very imperfectly into these recesses of his own bosom, and analysing the operations of his mind, – in this as in all other things, in short, *'a being darkly wise'*; seeing that all the longest life and most vigorous intellect can give him power to discover ... serves only to place him on the very frontier of knowledge, and afford a distant glimpse of boundless realms beyond, ... [20]

Herschel's conclusion from these musings on the limitations of human knowledge in relation to the natural world was not despairing. On the contrary, his picture of the scientific quest gives further form to the kind of assurance Chalmers had claimed for the religious seeker. Herschel moves seamlessly from his recognition of the limits of human knowledge to an acknowledgement of divine omnipotence to a powerful statement about the afterlife:

> Is it wonderful that a being so constituted [the awed human scientist] should first encourage a hope, and by degrees acknowledge an assurance, that his intellectual existence will not terminate with the dissolution of his corporeal frame, but rather that in a future state of being ... he shall drink deep at that fountain of beneficent wisdom for which the slightest taste obtained on earth has given him so keen a relish?[21]

For Herschel, as for Chalmers, religious hopes grow into assurances in the pursuit of science.

The focus in this view of knowledge is on the individual's experience of it, rather than on its external fixity. Herschel's "being darkly wise" is from Alexander Pope; the resonances are to the Paul of I Corinthians 13:12: "For now we see through a glass, darkly; but then face to face: now I know in part; but then shall I know even as also I am known."[22] The sermonic overtones are particularly evident in the close of Herschel's chapter:

> There is something in the contemplation of general laws which powerfully persuades us to merge individual feeling, and to commit ourselves unreservedly to their disposal; while the observation of the calm, energetic regularity of nature, the immense scale of her operations, and the certainty with which her ends are attained, tends, irresistibly, to tranquillize and re-assure the mind, and render it less accessible to repining, selfish, and turbulent emotions. And this it does, not by debasing our nature into weak compliances and abject submission to circumstances, but by filling us, as from an inward spring, ... ; by showing us our strength and innate dignity, and by calling upon us for the exercise of those powers and faculties by which we are susceptible of the comprehension of so much greatness, and which form, as it were, a link between ourselves and the best and noblest benefactors of our species, with whom we hold communion in thoughts and participate in discoveries which have raised them above their fellow mortals, and brought them nearer to their Creator.[23]

Clearly, for Herschel as for Chalmers, scientific work is a process leading to a right relation with God. This is not the same as establishing or proving God's existence. Quite the contrary, it is in the willing personal encounter with the impossibility of the task thus conceived that knowledge of God appears.

Herschel continued to emphasise the value of science as a process rather than as a demonstration throughout his book. He takes an empirical view of science and is well aware of the skeptical arguments that all experimentally-based conclusions must be, in any ultimate sense, tentative. He does not react to this as troubling, however. In a characteristic passage dealing with our ability to know the essence of force Herschel notes:

> This one instance of the obscurity which hangs about the only act of direct *causation* of which we have an immediate consciousness, will suffice to show how little prospect there is that, in our investigation of nature, we shall ever be able to arrive at a knowledge of ultimate causes, and will teach us to limit our views to that of *laws* ... Nor let any one complain of this as a limitation of his faculties. We have here 'ample room and verge enough' for the full exercise of all the powers we possess ... [24]

It was not in the absolute truth of its results that science was valuable. It was a "way," rather than a finished journey.

Herschel's *Preliminary Discourse* was primarily devoted to the physical sciences; although he was earlier very interested in mathematics he here treated it merely as an aid to their study. This makes it somewhat hard to know his views on the relation between mathematical and religious knowledge. A clearer and more focused treatment of this question can be gleaned from the work of his contemporary, Whewell, who in 1833 published a Bridgewater Treatise entitled *Astronomy and General Physics considered with Reference to Natural Theology*. This work includes both evidentiary arguments for the existence of God, and participatory ones attesting to the value of science as a way to know God.

Whewell's evidentiary arguments deal with a somewhat different subject matter than did Paley, but follow the same general pattern. The coincidence of the cycles of botanical growth with the twelve-month astronomical cycle is an example of the kind of design which he finds in astronomy. He is equally struck by the particular physics of the atmosphere which seems so beautifully and specifically adapted to support life on earth. The permanence of the atmosphere behind the myriad changes of weather was, for him, another testament to the care of the designer.

It is not always clear that these arguments are focused primarily on establishing God's existence, however. At least equally important are the personal responses the designer's creations elicit. To quote just one example, Whewell wrote:

> The purpose for which the world was made could be answered only by its being preserved. But ... [this] is a permanence of a state of things adapted by the most remarkable and multiplied combinations to the well-being of man, of animals, of vegetables. The adjustments and conditions therefore, ... by which its permanence is secured, must be conceived as fitted to add ... to the admiration which the several manifestations of Intelligent Beneficence are calculated to excite.[25]

Here, again, design does not prove or establish anything. Instead it excites admiration and disposes those who recognize it to a positive attitude towards the God who lies behind it.

In the latter half of his treatise, Whewell elaborated the participatory argument for the spiritual value of science more directly. Although the point is essentially a personal one, in keeping with the distancing erudition which characterizes his style in general, Whewell argued this position on the basis of historical rather than personal evidence. He used the example of past discoverers – Nicholas Copernicus, Johannes Kepler, Galileo Galilei, Robert Boyle and Isaac Newton – to argue that science strengthened the religious convictions of its practitioners. This approach enabled Whewell to be more distant and erudite in tone than Herschel had been. At the same time, however, it caused him considerable difficulty because he found equal numbers of scientific thinkers, particularly among his closer contemporaries, whose religious convictions were not strengthened by their scientific work. Clearly, men like Jean d'Alembert, Alexis Clairault, Leonhard Euler, Pierre-Simon, Marquis de Laplace, and Joseph-Louis Lagrange were not religiously inspired by their science.

These illustrious counterexamples to the claim that scientific pursuits deepened religious faith led Whewell to develop a distinction between truly great discoverers, members of the first group, and merely good scientists. The discoverers, he claimed, were those truly engaged in inductive science, who came to recognize unexpected new truths in the world around them. They were moving into areas where they had no predecessors, charting new scientific territory. This experience was, Whewell emphasized, an essentially humbling one, since the discoverer

> is compelled to look beyond the present state of his knowledge, and to turn his thoughts to the existence of principles higher than those which he yet possesses .... The effort and struggle by which he endeavours to extend his view, makes him feel that there is a region of truth not included in his present physical knowledge; the very imperfection of the light in which he works his way, suggests to him that there must be a source of clearer illumination at a distance from him.[26]

For Whewell, as for Herschel, it is the experience and recognition of a truth beyond what is now known that strengthens the religious convictions of true scientific pioneers.

Having made this point, Whewell goes on to argue that the scientific enterprise is different for the discoverers' posterity. This second tier of thinkers, which inherits the insights of the great innovators, proceeds deductively, working out the details of already established truths. "Such persons," Whewell explained,

> are not led by their pursuits to anything beyond the general principles, which form the basis of their explanations and applications . . . . [T]hey make these their ultimate grounds of truth; and they are entirely employed in unfolding the particular truths which are involved in such general truths. Their thoughts dwell little upon the possibility of the laws of nature being other than we find them to be, . . . still less on those facts and phenomena which philosophers have not yet reduced to any rule . . .[27]

Satisfied with wholly human constructions, these are the kinds of thinkers who are likely to pursue science without recognizing a larger reality which stands outside of the human mind. They are apt to agree with Laplace, who stands for Whewell as their epitome, that as for God, they have no need of that hypothesis.

Whewell did not stop at admitting that merely good scientific work did not unambiguously lead to God; the example of irreligion among continental scientists was powerful enough that he was led to write:

> We may thus, with the greatest propriety, deny to the mechanical philosophers and mathematicians of recent times any authority with regard to their views of the administration of the universe; we have no reason whatever to expect from their speculations any help, when we attempt to ascend to the first cause and supreme ruler of the universe. But we might perhaps go further and assert that they are in some respects less likely than men employed in other pursuits, to make any clear advance towards such a subject of speculation.[28]

This statement by Whewell seems almost as far as possible from the linkage of mathematics and theology found, for example in Locke. However, his strictures are aimed quite specifically at "mechanical philosophers and mathematicians of recent times." Whewell is being critical just of a specific view of mathematics being developed, for the most part, on the continent. Mathematics in England was developing rather differently. The early decades of the nineteenth-century have long been viewed as a time when English mathematicians merged their field with their continental counterparts. A closer look, however, reveals that significant differences continued to separate them and that these differences were closely connected to the issues Whewell raised against the irreligious continental group.

II

The nineteenth-century development of English mathematics can be seen as starting with the work of the Cambridge Analytical Society in the second decade of the century. Intrigued by the explosion of mathematical activity across the channel in Napoleonic France, the Cambridge Analytical Society's mission was to bring it to England.[29] In 1816, Babbage, Herschel, and Peacock published a translation of the abridged French edition of Sylvester-François Lacroix's calculus text.

The translation is interesting not just as evidence of the attempt but also of its failure. Lacroix's translators retained his Leibnizian symbology, but in other respects his work suffered a considerable seachange as it crossed the channel. The English version is studded with revisionary footnotes and endnotes correcting Lacroix's presentation of the calculus, reinterpreting its foundations, recasting his proofs. The issues at stake were not minor ones; the young analytics disagreed with Lacroix at the most fundamental level. Whereas Lacroix adopted the concept of the limit as the basis of the calculus, his English translators insisted on retaining the algebraic foundation put forward by Lagrange. The reasons behind their revisions were philosophical rather than mathematical; they did not find inconsistencies or other errors in Lacroix's work. Rather Babbage, Herschel and Peacock disagreed with the approach to the subject which led him to favor the limit.

The English rejected the limit because of the difficulty of understanding the concept. As Peacock wrote in a note to the translation:

Our notion, indeed, of a ratio, whose terms are evanescent [the limit] is necessarily obscure, however rigorously its existence and magnitude may be demonstrated; and its introduction into all our reasonings in the establishment of this Calculus, is calculated to throw a mystery over all its operations, which can only be removed by our knowledge of its more simple and natural origin.[30]

It was the personal standard of clear conception which guaranteed the necessity of mathematical truth; foundational constructions which did not generate this kind of conceptual clarity were by definition inadequate.

The disagreement over the limit is a specific instance of the larger issue about the characteristics of legitimate mathematical knowledge, which divided French and English mathematics. One of the effects of rigorizing mathematics is to devalue personal intuition in favor of impersonal standards. In this, it is similar to basing scientific knowledge on communally legitimated matters of fact. The search for rigor was justified in terms of the rigid certainty it would guarantee to mathematical demonstration. In the theologically saturated context of early Victorian England, however, this kind of extra-personally established, fixed certainty was only of peripheral value. Just as the religious question was not whether God existed but rather how God was recognized, so the mathematical issue was not whether the limit existed but rather how it was understood. Peacock's adherence to the value of conceptual clarity can be seen as an insistence on the preeminent value of personally apprehended truth over that of absolutely established rigor.

Peacock's orientation is characteristic of early nineteenth-century English mathematicians, and the distance between English and French analysis grew rather than lessened in the subsequent decades. Lacroix's text was revised as it was translated, but Augustin Cauchy's seminal *Cours d'Analyse* of 1821 was never translated at all. In this work Cauchy defined and delineated the limit and its uses with an unambiguous precision far greater than Lacroix's. On the continent, the *Cours d'Analyse* heralded the tradition of rigor which characterized much of nineteenth-century mathematics.

The French interest in formal rigor was not shared by the English in the early decades of the nineteenth century, however. In the 1830s, they loosened their insistence on Lagrange's interpretation of the calculus, but even as they embraced the limit it was in a conceptual as opposed to a rigorous formal context. Whewell endorsed the centrality of the limit to calculus in his 1835 *Thoughts on the Study of Mathematics as Part of a Liberal Education* and in his 1838 textbook, *The Doctrine of Limits*. De Morgan did the same in his *Differential and Integral Calculus*, which appeared serially, between 1836 and 1842, and was then published as a single volume.[31] However, neither man accepted Cauchy's mathematical definition as foundational. In his 1842 *Philosophy of the Inductive Sciences*, Whewell identified the limit as the "Fundamental Idea" which underlay the calculus; De Morgan's work opens with a 27-page introductory chapter explaining the concept. Even as they conceded the centrality of limiting processes to the calculus, Whewell and De Morgan were not joining Cauchy's search for a rigorous foundation for mathematics. For both of them, as for Peacock twenty-five years before, personally grounded conceptual clarity rather than externally established mathematical rigor was the touchstone of well-grounded mathematics.

In the meantime English mathematics continued to develop in the algebraic direction marked out by Lagrange; in the 1820s and 1830s their major contribution was to symbolical algebras. While this approach allowed the English to avoid the specific problem of clearly conceiving the limit, it led to its own set of epistemological challenges. Powerful results could be generated through symbolical manipulations but it was not always clear how they were to be understood.

The same emphasis on personal clarity of conception which made the limit such a problematic basis for the calculus brought this problem to the fore. The symbols of algebra drew their validity from arithmetic and the concept of number, but in the 1820s De Morgan and others interpreted this concept so tightly that even negative quantities were technically illegitimate.[32] This meant that even the simplest algebraic forms like: $a^2 - b^2 = (a + b)(a - b)$ posed serious philosophical problems. Beyond these lurked the whole field of complex numbers as well as the variety of strange results which fell out of infinite series.

In 1831, Peacock proposed a solution to this problem. Instead of defining symbolical algebra as a universal arithmetic, Peacock regarded it as a separate subject which did not rely on arithmetic concepts for its validity. Arithmetic served merely as the "science of suggestion" for symbolical algebra; arithmetic forms like $4^2 - 2^2 = (4 + 2)(4 - 2)$ pointed the way to symbolic forms like $(a^2 - b^2) = (a - b)(a + b)$. Beyond this suggestive function, Peacock argued, arithmetic bore no foundational relationship to symbolical algebra. This study focused on the symbolical forms themselves and on their interrelationships.

Central to Peacock's view was the somewhat elusive principle of equivalent forms. He had trouble encapsulating this principle in a single consistent phrase, but a representative statement of it reads: "Whatever equivalent form is discoverable in arithmetical algebra considered as the science of suggestion, when the symbols are ... specific in their value, will continue to be an equivalent form when the symbols are general in their nature."[33] This principle guaranteed the truth of symbolic forms even if their conceptual interpretations were, for the moment, unclear.

Peacock's approach was very effective in freeing algebra from the constraints imposed by interpreting it strictly as universal arithmetic. At the same time, it is not hard to see why De Morgan was perplexed by it: "at first sight it seemed to us something like symbols bewitched, and running about the world in search of meaning."[34] He was, however, mollified by the principle's success, notably in the effectiveness of the geometrical model for negatives and imaginaries being developed in France.[35] By the mid-30s, De Morgan essentially endorsed Peacock's principle.

The striking thing about the principle of equivalent forms, as De Morgan construed it, was its open-endedness; it attested to the validity of algebraic forms, which meant that they had conceptual interpretations. It did not mean, however, that these interpretations were known. The individual mathematician might not ever attain them but that was no reason for despair; just as in physics Herschel was undaunted by the kinds of limits so clearly demarcated by the skeptics, in mathematics De Morgan was undaunted by those laid out by French analysts. His position in this regard can be seen, for example in his argument with the French mathematician, Siméon-Denis Poisson, over the use of divergent series.

One of the most innovative aspects of Cauchy's program of rigor was his rejection of divergent series. These had been widely used in the eighteenth century, before Cauchy declared that they were unacceptably ill-defined, and produced ambiguous or even erroneous results. Picking up on this point in several papers of the early 1830s, Poisson tried to come to a clearer understanding of these series and the boundaries of their legitimacy. In his 1844 paper, "On Divergent Series and Various Points of Analysis Connected with Them," De Morgan blasted not only Poisson's ideas and Cauchy's definition of the integral on which they were based, but the whole preoccupation with certainty which valorized the search for rigor. "Divergent series, at the time Poisson wrote, had been nearly universally adopted for more than a century, and it was only here and there that a difficulty occurred in using them," he fumed.[36] The knowledgeable mathematician, De Morgan pointed out, could easily detect and correct such problems when they arose. To artifically control their use just in order to guarantee rigorous exactitude was at best unnecessary and ridiculous. At worst it could stand in the way of deeper understanding of the truth embodied in these series, which was as yet still poorly comprehended. As De Morgan wrote:

> We must admit that many series are such as we cannot at present safely use, except as means of discovery, the results of which are to be subsequently verified. But to say that what we cannot use no others ever can, to refuse that faith in the future prospects of algebra which has already realised so brilliant a harvest ... seems to me a departure from all rules of prudence.

For De Morgan, to draw back from poorly defined or understood mathematical conclusions was a grievous error. "The motto which I should adopt against a course which seems to me calculated to stop the progress of discovery would be contained in a world and a symbol – remember $\sqrt{-1}$."[37]

De Morgan never abandoned this motto and searched all of his life for interpretations of algebraic forms. As late as 1865 he wrote to J. S. Mill: "With regard to the acceptance of . . . [Peacock's] system, the time is not yet come. The algebraists almost all make algebra obey their preconceived notions . . . . So long as an algebraist has preconceptions which his science must obey, so long is he incapable of true generalisation."[38] Reining in algebra to conform to clearly established axioms or preconceptions might guarantee a kind of certainty to its results; throughout his life, however, De Morgan never felt that this kind of certainty was worth the safety it afforded. Like Herschel he was willing to remain on the frontiers of knowledge with its "distant glimpses of boundless realms beyond."[39]

III

Upon closer examination, then, England's mathematicians were developing their subject differently than were their French counterparts. The differences were such that they might circumvent the strictures which Whewell had leveled against the strictly defined, deductively limited subject exemplified by Cauchy's rigorous calculus. The distinction is a subtle one, though, and it was neither always conscious nor obvious. Whewell's remarks about mathematicians in his 1833 *Astronomy* were directed against a continental tradition which simple empirical evidence indicated was not religiously uplifting. By condemning what he called deductive science, Whewell was trying to defend the pursuit of science from the bad reputation threatened by those who were religiously wayward. Although his definition of the parameters of such a subject would not technically encompass the work of his compatriots, it is not surprising that some feared his allegations would give the study of mathematics a bad name. Notable among these was the maturing francophile, Babbage, who was so incensed that he wrote an uncommissioned *Ninth Bridgewater Treatise* to rebut Whewell.

Babbage's defense of the religious value of mathematics rested on a different characterization of the nature of the subject than that proposed either by Whewell in his *Bridgewater Treatise* or by Herschel in his *Preliminary Discourse*. Whereas in these texts mathematics was presented essentially as a subsidiary to physics, Babbage considered it as a separate subject in its own right; whereas they valued it, if at all, in participatory terms, Babbage found its value to lie in the nature of its results.

In the preface to the second edition of his *Treatise* Babbage formulated a scale of knowledge ranging from the most to the least certain. Mathematics held pride of place on this scale because "The truths of pure mathematics are necessary truths; they are of such a nature, that to suppose the reverse involves a contradiction." On this scale, our knowledge of the laws of nature represented a step down in certainty because "although some of them are considered as necessary truths, [they] depend, in many instances, on the testimony of our senses. These derive their highest confirmation from the aid of pure mathematics, by which innumerable consequences, previously unobserved, are proved to result from them."[40] Below these two categories, Babbage placed two others, natural religion and revealed religion. Of these, natural religion had the brighter future, because its reliance on knowledge supplied by the first two classes promised progress. Revealed religion, the fourth and final rung, was considerably weaker. Following Hume, Babbage recognized the flimsiness of any knowledge which depended solely on the credibility of witnesses. He did not completely abandon such knowledge, however; he accepted it until that time "when, by the progress of knowledge, internal evidence of the truth of revelation may start into existence with all the force that can be derived from the testimony of the senses."[41] His celebrated defense of miracles modeled on a computer programmed to do something exceptional in a long series of uniform events, represents an attempt to take such a progressive step.

Thus, for Babbage, mathematics held essentially the same epistemological position that Locke had claimed for it more than a century before; it was the exemplification of the highest kind of knowledge of which humans were capable. As such it was neither peripheral nor dangerous for natural theological discussion; it was central. The goal of the religious search remained to know God as well as one could know mathematics.

In making this argument, Babbage was not merely defending mathematics, however. He was shifting the ground under it. He focused on mathematical truth itself, not the process by which it was attained. The certainty of the subject he described is not the glimpsed certainty of the awed seeker; it is an absolute certainty of clearly grasped knowledge.

In Whewell's 1833 *Treatise* it had been the "effort and struggle" of pursuing that truth that was crucial to the great discoverer's religious health. It was the "very imperfection of the light in which he works his way" that bolstered his religious convictions. What is more, it is the solidity of their grasp of truth that makes Whewell's "mechanical philosophers and mathematicians" so smug about their own powers that they can ignore God's presence.[42] Thus, in Whewell's early work, clearly established and recognized scientific truth is highly suspect. This is, however, just the kind of truth Babbage is proposing for mathematics as well as religion.

In thus construing mathematics, Babbage was constructing a middle ground between the absolutely established but possibly meaningless certainties of French rigor and the meaningful glimpses of English algebra. His emphasis on clearly grasped mathematical truth combined the concreteness of the English approach with the definitiveness of the French. Theologically it enabled him to hope that, in time, knowledge of God would also move from the shaky to the secure. Pauline glimpses were of little value to Babbage.

Babbage's brief and typically eccentric treatise did not develop his alternative, truth-emphasizing theme at length. Nor, apparently did it persuade Whewell, who in 1837 published an "Open Letter to Charles Babbage" firmly defending his initial position.[43] In the long run, though, it seems that Babbage's treatise might have had an effect. In the late thirties and into the forties, Whewell elaborated a comprehensive philosophy focused largely on the issue of truth. A critical part of his structure entailed classifying truth in a way that paralleled the first two rungs of Babbage's truth hierarchy. Whewell asserted that there was a "fundamental antithesis" in philosophy between necessary and contingent truths. Contingent truths were merely empirically based; they were observed facts about phenomena which "for aught we can see ... might have been otherwise." Necessary truths, on the other hand, were those whose opposites were inconceivable.[44] They were known certainly.

Whewell used the example of mathematics at once to establish the reality of necessary truth, and to demonstrate that human minds could attain it. In geometry he pointed out, one not only knew that the sum of the angles of a triangle was 180°, it was impossible to imagine that it could be otherwise. When a theorem like this one was fully understood, it was impossible to clearly conceive of its being contradicted: such a conception would inevitably violate some other indubitable piece of knowledge like the nature of the straight line, the right angle or the triangle. Here then was certainty grasped; there was nothing open-ended or limited about it.

This truth-oriented view succeeded, as Babbage had hoped it would, in redeeming mathematics from the kinds of strictures Whewell had leveled against it in 1833. In fact, in the hands of the more mature Whewell it served as the philosophical bulwark which protected mathematics' pride of place in the Cambridge education. Thus, in a typical defense of the Cambridge curriculum Whewell wrote:

The peculiar character of mathematical truth is that it is necessarily and inevitably true; and one of the most important lessons which we learn from our mathematical studies is a knowledge that there are such truths, and a familiarity with their form and character.[45]

The value of knowing this kind of truth in mathematics lay in the assurance it provided that the same kind of knowledge was possible in all areas, including those where it may seem remote. This assurance included centrally the knowledge of God, which thereby changed from being a mere glimmer of recognition lighting the way of committed seekers into a firmly grasped truth which none could gainsay.

Whewell's move away from the participatory emphasis of his *Astronomy* to a fixed, truth-oriented one was not a solitary one. In 1841 when Herschel reviewed Whewell's philosophy he took the same step and repudiated his earlier agnosticism about human beings' ability to attain real knowledge. Herschel maintained his basically empirical stance but found a way to allow absolute knowledge into it. As he put it:

among the infinite analogies which may exist among natural things, it may very well be admitted that those only are designed ... to become elaborated into general propositions, and finally to work their way to universal reception, and attain to all the recognizable characters of truth, which are really dependent on the intimate nature of things as that nature is known to their Creator, and which have relation to their essential qualities and conditions as impressed on them by Him;...

Herschel's admission of divine aid in the process by which one generalized from experiment enabled him to get around the sceptical roadblock to finding certain truth. His earlier participatory emphasis had enabled him to skirt this problem; in 1841 when he believed it solved, however, he could join Whewell in abandoning that approach in favor of a truth-oriented one. His admission of the possibility of real knowledge extended beyond the natural world to the God who had made it. Herschel declared himself:

satisfied that ... the mind of man is represented as in harmony with universal nature; that we are consequently capable of attaining to real knowledge; and that the design and intelligence which we trace throughout creation is no visionary conception, but a truth as certain as the existence of that creation itself.[46]

This emphasis on the attainment of truth, in particular of necessary truth, as the religiously validating aspect of scientific study marked a significant change from the earlier participatory account. Whereas in their earlier works it was when standing at the limits of human knowledge that one could catch glimpses of God, in the later view such knowledge was knowledge attained.

Within mathematics, the growing strength of this view of truth created a hierarchy of value which can be seen in decisions about what was central and what peripheral for the Cambridge curriculum. Even though he embraced Peacock's principle of equivalent forms as the Fundamental Idea on which algebra was based, it bore too many marks of incompleteness to be satisfactory. Whewell's interpretation of truth led to an insistence that clearly understood and unambiguously established Euclidean geometry and Newtonian dynamics be the central focus of the Cambridge curriculum. Open-ended, progressive studies, including the whole of rapidly developing analysis and mathematical physics, were relegated to the sidelines because they could not so well model necessary truth.[47] Thus in the hands of the maturing Whewell, truth became a very conservative force. His approach was never acceptable to De Morgan, for example, who defended and pursued open-ended mathematics until the end of his life. By the 1850s, however, this set De Morgan off as something of an eccentric in mathematical circles.[48]

Within natural theology, the clearest indication of the importance of this change is an indirect one pointing not to its establishment, but rather to its collapse. Moving knowledge of God from the shifting sands of individual experience to the firm pedestal of necessary truth proved to be merely a temporary way of strengthening it. Later in the century one might even say it served to weaken religion; the emphasis on absolutely known geometrical truth as the exemplar of religious certainty exposed religious conviction to a serious challenge with the introduction of non-Euclidean geometry.

It was in the decades immediately following the publication of Darwin's *On the Origin of Species* that the absolute truth of Euclidean geometry, the seemingly impregnable mathematical wall against agnosticism and doubt, was challenged. The most knowledgeable and powerful voice raised against the absolute truth of mathematics, specifically of geometry, was that of the mathematician William Clifford. Geometry, Clifford argued, was a physical science limited like all physical sciences to those things which we can directly experience. It cannot lay any claims to infinite or perfect knowledge; its truths are not known certainly or necessarily. It is not final knowledge of an absolute or transcendental realm lying behind the world as we know it.

Non-Euclidean geometry played a key role in Clifford's attack on necessary truth. Clifford not only translated Bernhard Riemann's non-Euclidean "Habilitationsvortrag" into English and published it in *Nature*, but in a series of popular addresses carefully spelled out its devastating implications for the notion of absolute certainty in mathematics. He wrote:

I shall be told no doubt, that we do possess a great deal of knowledge of this kind, in the form of geometry ... If this had been said to me in the last century, I should not have known what to reply. But it happens that about the beginning of the present century the foundations of geometry were criticized ... And the conclusion to which these investigations lead is that, although the assumptions which were very properly made by the ancient geometers are practically exact ... for such finite things as we have to deal with, and such positions of space as we can reach; yet the truth of them for very much larger things, or very much smaller things, or parts of space which are at present beyond our reach, is a matter to be decided by experiment.[49]

From this mathematical base, Clifford moved out to challenge the larger scheme of truth detailed by Whewell and his posterity. The titles of the articles he published in the 1870s – "The Ethics of Belief," "On the Scientific Basis of Morals," "The Influence upon Morality of a Decline of Religious Belief," – give a sense of the range of issues Clifford considered. All of these areas and more had to be re-examined, he maintained, in light of the new recognition of the inadequacy of the Whewellian view of mathematical truth. Clifford's entré into all of these areas was ultimately the equation between mathematical and religious knowledge, interpreted in the truth-emphasizing tradition of Babbage and the later Whewell and Herschel. The sum of the angles of a triangle was *not* equal to 180°, Clifford emphasized again and again. The major exemplar of necessary truth was itself not true, and the central epistemological structure within which English religious ideas had been interpreted began to crumble.

IV

It was apparently in 1869 that Thomas Huxley first used the term "agnosticism" to describe his philosophical and religious orientation.[50] A crucial element of Huxley's concern in coining this term was epistemological. It entailed a profession of essential ignorance, and emphasis on the limits of human knowledge. These limits applied equally to science and to religion; in both areas Huxley and his fellow agnostics would deny the possibility of moving beyond contingency to certain or absolute knowledge. Although Huxley, himself, vehemently denied that agnosticism was tantamount to atheism, it appeared that way to his many religious critics. It seemed to them that to limit knowledge as Huxley had done, removed the essential epistemological underpinnings for religious knowledge. With historical hindsight they seem to have been right. The sea of faith was drawing back from Matthew Arnold's "Dover Beach," leaving dry the "naked shingles of the world."[51] The progress of science seems to have inexorably drained away the religion which had been so confidently joined to it earlier in the century.

What has not always been so clearly recognized was that this phenomenon was not an inevitable result of the theories being developed in science and mathematics, that the tidal action in religion was not a necessary effect of the scientific and mathematical composition of the gravitating moon. Equally if not more important to the historical development, was the specific relationship maintained between the scientific and religious spheres; the gravitational law pulling on the religious sea.

In physics the gravitational law does not change; following the analogy into history, however, it may, and in nineteenth-century England it did. In the twenties and thirties, the evangelical tradition supported a view of mathematical and religious knowledge which was in important ways open-ended and flexible. Certainty was an individual rather than absolute issue; knowledge tended to be construed in a Pauline mode of flickering glimpses vouchsafed on a personal journey. It was not clearly established and forever fixed, able to be directly passed on from generation to generation. This view had its clearest expression in evangelical religion. It can also be found, though less obviously, in mathematics where the principle of equivalent forms and similarly open-ended constructs were accepted as the underpinnings of mathematical understanding.

The absolutist view of mathematical and relatedly religious knowledge, here found in Babbage and later Whewell, was significantly different than the evangelical one. The certainties the evangelicals merely glimpsed were grasped and established. Truths were not "visionary conceptions" but rather certainly known; "the mind of man" was "in harmony with universal nature," "capable of attaining to real knowledge."[52] The experience of peering into unknown infinities was still acknowledged but more as a mark of inadequacy than of grace. The absolutely known immutable truths of Euclidean geometry became the norm for both mathematical and religious certainty.

This is the view which set the stage for the religious crises of the late nineteenth century. The notion of absolute certainty was undercut from within science by the great innovations of the period, most directly by developments in non-Euclidean geometry. To the young Clifford, this change bore, not the marks of defeat but rather those of new vision. He once admitted "that I personally have often found relief from the dreary infinities of homaloidal space in the consoling hope that, after all, this other [non-Euclidean, non-homaloidal space] may be the true state of things."[53]

Clifford's excitement at the possibilities opened by the collapse of geometrical certitude might have appealed to the young Herschel who could still rejoice that "'there was ample room and verge enough' for the full exercise of all the powers we possess."[54] Religion in the second half of the century was less malleable, however; it was more attuned to the certainties Herschel and his friends had grasped for in their middle age. These were certainties marked not by their personally transformative power but by their absolute fixity. Ultimately this fixed notion of certainty and truth proved to be the rock against which English natural religion foundered late in the nineteenth century.

Department of History
Brown University

NOTES

* I am grateful to Lorraine Daston, Menachem Fisch, Thomas Jenkins, and Mary Jo Nye for their remarks on earlier versions of this essay. I have tried to respond to their suggestions but take full responsibility for any weaknesses it still contains.

[1] John Locke, *An Essay Concerning Human Understanding*. 2 vols., collated and annotated by Alexander Campbell Fraser (New York: Dover Publications, 1959), **2**: 325.

[2] *Ibid.*, p. 306.

[3] *Ibid.*, pp. 309, 306.

[4] For a more detailed analysis of the kinds of assumptions surrounding the construction of experimental "matters of fact," see Steven Shapin and Simon Schaffer, *Leviathan and the Air-Pump* (Princeton: Princeton University Press, 1985).

[5] Locke, *Essay* (ref. 2), p. 310.

[6] *Ibid.*, p. 363.

[7] For a discussion of mid-eighteenth century classifications of knowledge see Robert Darnton, 'Philosophers Trim the Tree of Knowledge: The Epistemological Strategy of the *Encyclopédie*' in *The Great Cat Massacre* (New York: Vintage Books, 1984), pp. 191–213. The marginal position of mathematics and theology is not as clear in the chart Darnton considers for this paper as it is in the real tree pictured in the first volume of the Index to the *Encyclopédie*.

[8] The anti-Christian aspects of the French enlightenment have long been recognized. They are a major theme in Peter Gay, *The Enlightenment: An Interpretation*, 2 vols. (New York: W. W. Norton and Co., 1969). More recently in *The Cultural Origins of the French Revolution*, trans. by Lydia G. Cochrane (London: Duke University Press, 1991), Roger Chartier has recognized the same theme of de-Christianization which he has located in the larger social context. The history of mathematics is less extensively worked, but the *fin de siècle* malaise has long been recognized. Its most poignant expression is perhaps in Joseph-Louis Lagrange's letter of 21 September, 1791, to Jean D'Alembert. "Il me semble aussi qui la mine [of mathematics] est presque déjà trop profonde, et qu'à moins qu'on découvre de nouveaux fillons il faudra tôt ou tard l'abondonner. La Physique et la Chimie offrent maintenant les richesses plus brillantes et d'une exploitation plus facile; aussi le goût du siècle paraît-il entièrement tourné de ce côté-là, et il n'est pas impossible que les places de Géométrie dans les Académies ne deviennent un jour ce que sont actuellement les chaires d'arabe dans les Universités." *Oeuvres de Lagrange*, ed. M. J.-A. Serret, 14 vols. (Paris: Gauthier-Villars, 1882) **13**: 368. Other, similar sentiments are quoted in Morris Kline, *Mathematical Thought from Ancient to Modern Times* (New York: Oxford University Press, 1972), p. 623.

[9] J. L. Altholz, 'The Warfare of Conscience with Theology,' in Gerald Parsons ed., *Religion in Victorian Britain, Vol IV, Interpretations* (New York: Manchester University Press in association with the Open University, 1988), p. 150.

[10] Sophia Elizabeth De Morgan, *Memoir of Augustus De Morgan* (London: Longmans, Green and Co. 1882) p. 368. About Whewell see Menachem Fisch and Simon Schaffer, *William Whewell: A Composite Portrait* (Oxford: Clarendon Press, 1991); Mrs. Stair Douglas, *The Life and Selections from the Correspondence of William Whewell* (London: Kegan Paul, Trench, 1882). On Herschel see G. Buttman, *The Shadow of the Telescope: A Biography of John Herschel*, trans by B. E. J. Pagel, edited with introduction by David

S. Evans (New York: C. Scribner's Sons, 1971); S. S. Schweber, 'John F. W. Herschel: A Prefatory Essay' in S. S. Schweber, ed., *Aspects of the Life and Thought of Sir John Frederick Herschel* (New York: Arno Press, 1981). On Babbage see Anthony Hyman, *Charles Babbage: Pioneer of the Computer* (Princeton: Princeton University Press, 1982).

11 Quoted in Boyd Hilton, *The Age of Atonement: The Influence of Evangelicalism on Social and Economic Thought 1795–1865* (Oxford: Clarendon Press, 1988), p. 29. This book is an excellent starting point for considering evangelicals in the broad sense I am using here.

12 Bernard M. G. Reardon, *From Coleridge to Gore* (London: Longman, 1971), p. 29.

13 Hilton, *Age* (ref. 11), p. 29.

14 *The Autobiography of Charles Darwin, 1809–1882*, edited with appendix and notes by Nora Barlow (New York: W. W. Norton & Company, 1958), p. 96.

15 I owe this observation to Lorraine Daston.

16 Reardon, *Coleridge to Gore* (ref. 12), p. 30. In the public sphere, it is hard not to sympathize with the frustration of those like Evelyn Baring, a British agent in Egypt, who complained: "A man who habitually consults the prophet Isaiah when he is in a difficulty is not apt to obey the orders of anyone." Quoted in Geoffrey Best, 'Evangelicalism and the Victorians,' in Anthony Symondson, ed., *The Victorian Crisis of Faith* (London: The Camelot Press Ltd., 1970), p. 55.

17 A similar point is made by John Hedley Brooke, *Science and Religion: Some Historical Perspectives* (Cambridge: Cambridge University Press, 1991), p. 210 ff. It is considerably elaborated in the essay 'Indications of a Creator: Whewell as Apologist and Priest' in M. Fisch and S. Schaffer, *Whewell* (ref. 10), pp. 149–73; see also Hilton, p. 21ff.

18 W. H. Brock, 'The Selection of the Authors of the Bridgewater Treatises,' *Notes and Records of the Royal Society* 21 (1966), 164.

19 Thomas Chalmers, *On the Power, Wisdom and Goodness of God as Manifested in the Adaptation of External Nature to the Moral and Intellectual Constitution of Man*, 2 vols. (London: William Pickering, 1835) 2: 259.

20 John R. W. Herschel, *A Preliminary Discourse on the Study of Natural Philosophy* (Chicago: The University of Chicago Press, 1987), pp. 4, 5, 6.

21 *Ibid.*, p. 7.

22 Alexander Pope, *An Essay on Man*, epistle 2, li. 4. in *Poems of Alexander Pope*, ed. Maynard Mack, vol 3, pt. 1 (London: Methuen, 1950), p. 53.

23 Herschel, *Discourse* (ref. 20), pp. 16–17.

24 *Ibid.*, pp. 87–88.

25 William Whewell, *Astronomy and General Physics Considered with Reference to Natural Theology* (London: William Pickering, 1839), pp. 109–110.

26 *Ibid.*, pp. 330–31.

27 *Ibid.*, p. 331.

28 *Ibid.*, p. 334.

29 For more information on this society, see Philip C. Enros, 'The Analytical Society (1812–1813): Precursor of the Renewal of Cambridge Mathematics,' *Historia Mathematica* 10 (1983), 24–27.

30 S. F. Lacroix, *An Elementary Treatise on the Differential and Integral Calculus*, translation and supplementary notes by C. Babbage, J. Herschel and G. Peacock (Cambridge, 1816), p. 612. The issues surrounding this translation are discussed in more detail in

Joan L. Richards, 'Rigor and Clarity: Foundations of Mathematics in France and England, 1800–1840,' *Science in Context* **4** (1991), 297–319. There, however, I erroneously attributed the notes to Babbage (who translated the text to which they are appended) whereas they were written by Peacock. I am grateful to Menachem Fisch for pointing out this error.

[31] William Whewell, *Thoughts on the Study of Mathematics as a part of a Liberal Education* (Cambridge, 1835): William Whewell, *The Doctrine of Limits* (Cambridge, 1838): Augustus De Morgan, *The Differential and Integral Calculus* (London, 1842).

[32] See further Helena M. Pycior, 'George Peacock and the British Origins of Symbolical Algebra,' *Historia Mathematica* **8** (1981), 23–45; Helena M. Pycior 'Early Criticism of the Symbolical Approach to Algebra,' *Historia Mathematica* **9** (1982), 392–412.

[33] George Peacock, 'Report on the Recent Progress and Present State of Certain Branches of Analysis,' *Report of the Third Meeting of the British Association for the Advancement of Science*, 1833, p. 199.

[34] Augustus De Morgan, review of George Peacock, *A Treatise on Algebra*, in *Quarterly Journal of Education* **9** (1835), 311.

[35] The works generally cited were John Warren, *A Treatise on the Geometrical Representation of the Square Roots of Negative Quanties* (Cambridge, 1828); and Adrien Quentin Buée, 'Mémoire sur les quantités imaginaires,' *Philosophical Transactions of the Royal Society of London*, 1806, pt. 1., pp. 23–88.

[36] Augustus De Morgan, 'On Divergent Series and Various Points of Analysis Connected with Them,' *Transactions of the Cambridge Philosophical Society* **8** (1844), 184.

[37] *Ibid.*, pp. 183.

[38] Quoted in Sophia De Morgan, *Memoir* (ref. 10), p. 328.

[39] Herschel, *Discourse* (ref. 19).

[40] Charles Babbage, *The Ninth Bridgewater Treatise. A Fragment*, 2d ed. (Philadelphia: Lea & Blanchard, 1841), pp. v–vi.

[41] *Ibid.*

[42] Whewell, *Astronomy* (ref. 25), pp. 330–31, 334.

[43] I know of this letter from Anthony Hyman, *Charles Babbage: Pioneer of the Computer* (Princeton: Princeton University Press, 1982), p. 139, but have not been able to find it in the United States. I am currently pursuing it in the British Library.

[44] William Whewell, 'On the Fundamental Antithesis of Philosophy,' *Transactions of the Cambridge Philosophical Society* **8** (1849), 170.

[45] William Whewell, *Of a Liberal Education in General* (London, 1845), p. 163.

[46] [John Herschel], 'Review of Whewell's *History of the Inductive Sciences* (1837) and *Philosophy of the Inductive Sciences* (1840),' *The Quarterly Review* **68** (1841), 182. For a more detailed treatment of these discussions see the first chapter of Joan L. Richards, *Mathematical Visions* (Boston: Academic Press, Inc., 1988).

[47] Harvey Becher, 'William Whewell and Cambridge Mathematics,' *Historical Studies in the Physical Sciences* **11** (1980): 1–48.

[48] For a fuller treatment of De Morgan's views see Joan L. Richards, 'Augustus De Morgan, the History of Mathematics, and the Foundations of Algerbra,' *Isis* **78** (1987), 7–30; Helena M. Pycior, 'The Three Stages of Augustus De Morgan's Algebraic Work,' *Isis* **74** (1983), 211–226.

49 William Clifford, 'On the Aims and Instruments of Scientific Thought,' *Macmillan's Magazine* **26** (1872), 504.
50 For a discussion of this term and its origins see Bernard Lightman, *The Origins of Agnosticism: Victorian Unbelief and the Limits of Knowledge* (Baltimore: Johns Hopkins University Press, 1987).
51 *Poems of Matthew Arnold*, ed. Kenneth Alcott (London: Longmans, 1965), p. 242.
52 Herschel, 'Review' (ref. 46), p. 182.
53 William Clifford, 'The Philosophy of the Pure Sciences: The Postulates of the Science of Space,' *The Contemporary Review* **25** (1874–5), 376.
54 Herschel, *Discourse* (ref. 20), p. 88.

PART II

PROBLEMS OF CONTINGENCY, COHERENCE, AND TRUTH

FREDERICK GREGORY

THEOLOGIANS, SCIENCE, AND THEORIES OF TRUTH IN NINETEENTH-CENTURY GERMANY

The query made famous by Pontius Pilate in the first century A.D. has reverberated down through the centuries in many forms and in a variety of settings. But in seeking to answer the question, "What is truth?", one quickly discovers that the result depends above all on what is allowed to count as true. There is no necessary agreement from one cultural context to another and from one time to another about how truth is to be established. The problem is especially acute where the historian is concerned. The noble dream of providing a true, unbiased, objective account of the past "as it really happened" has been both defended as necessary and attacked as impossible virtually ever since Leopold Ranke set forth the challenge in the nineteenth century.

American historians, as Peter Novick has recently shown so impressively, have wrestled with the "objectivity question" over the entire course of the twentieth century.[1] Indeed, Novick's treatment has itself become part of the historical debate he describes, for his account has attempted to make clear the background to and the contours of present arguments about what constitutes proper history. These issues can be brought to bear on the specific historical problem under consideration here, namely, the account of the reception of the growth of natural science by nineteenth-century German theologians.

Most of us are familiar with the differences in approach between traditional intellectual historians and so-called contextualists. Since the mid-sixties social historians have insisted that the setting of an event is more determinative of proper historical understanding than is the intellectual framework of either an individual or a community. Now, however, both of these historiographical approaches are opposed by post-structrualists – historians who are so inherently suspicious of the misleading effect of structure that they prefer fragmentation, not synthesis, as the highest value of historical scholarship.[2]

Historians of science have certainly been active players in the development of contextualism. Because it was concerned with natural science, which was seen as a last reliable and relatively fixed repository of knowledge, Thomas Kuhn's pathbreaking focus on the *structure* of scientific revolutions helped as much as any other single historical work to move the center of cultural history away from the cognitive content of beliefs about nature toward the social factors conditioning those beliefs.[3] A generation of historians has felt the influence of Kuhn's achievement. At present, however, a new question facing historians of science is the same one facing their counterparts in the wider mainstream: must we go beyond structuralism altogether?[4]

The challenge implicit in these historiographical issues becomes real when faced with a specific historical problem. How, for example, could one investigate the understanding theologians possessed of natural science in nineteenth-century Germany? In England and America the relationship between theologians and scientists traditionally has been described in terms of a warfare that allegedly existed between the forces of the scientific and religious communities. As a result of the insights of contextualist historians, however, the center of attention moved away from the cognitive issues that separated theologians and scientists and came to rest on the question of social authority; i.e., historians described the significance of the interaction of science and religion in terms of a transfer of social authority from the representative of the church to the new "high priests" of science. In these accounts the notion that one side or the other was "right" about, say, human origins was irrelevant. The development of science was of interest because it served as a reflection of cultural change. For pure contextualists, incidentally, that was all it was, and any cognitive dimension that remained was made possible by the culture's structure and was associated with the synthesizing function of the cultural context.

Because of their rejection of the very possibility of synthesis, poststructuralist historians cannot be impressed with the contextualist account of nineteenth-century science and religion. It relies, after all, thoroughly on the alleged existence of a social historical context. Poststructuralists are convinced that one can have no more knowledge of "the social context" than one can cognitively apprehend "objective reality." To them the contextualist who thinks it useful to employ a notion of "the historical context" commits an error similar to the one which contextualists themselves identify in the traditional intellectual historian's reference to the touchstone of "objective reality." In both cases the historian has relied on a construct whose reality, while it imparts synthesis to the historical account, is inaccessible to us. Consequently neither a notion of objective reality nor of social context can serve as a foundation for writing about the history of science.

One should not expect a resolution of the differences between traditional intellectual, contextual, and post-structural approaches to history of science. While proponents of the first two perspectives share the conviction that synthesis and integration *is* a worthy goal for the historiography of science, they disagree how the integration is to be achieved. And if there has been sufficient common ground between traditionalists and contextualists to have permitted a fruitful dialogue between them for some time, no meeting of the minds is possible between either of these groups and the post-structural historians who declare that "all calls for synthesis are attempts to impose an interpretation."[5] Here the situation is a classic clash between mutually exclusive dogmas.[6]

In the treatment of science and religion in nineteenth-century Germany found below there is no attempt to resolve the differences described above nor to choose among the three historiographical approaches. On the contrary, aspects of each perspective are present. By choosing, for example, to focus on the concept of truth where science and religion are concerned, a link is forged to an older, more traditional historiographical style in which standards of truth and falsehood are assumed to exist. It would be difficult to argue that truth claims were not of major importance to the protagonists of the lively exchanges that characterized nineteenth-century treatments of the relationship between science and religion. Consequently one cannot as a historian ignore the content of these arguments, however differently one might view them today.

Yet the account depicted here is also contextualist enough to insist that it is meaningful to try to assess the various historically conditioned cognitive claims on their own terms. To refuse the attempt either because one holds that it is impossible to recreate the terms of the nineteenth-century debates or because one believes that intellectual development since the nineteenth century has exposed all cognition to be problematical represents to me a haughty close-mindedness.

To insist that one cannot abandon the cognitive dimension of the historical encounter of theologians with science is not, however, to say that one must judge the nineteenth-century debates in terms of a "proper" cognitive framework, or even in terms of one's own understanding of cognition. There is advantage in adopting the post-structuralist's healthy suspicion of synthesis and integration. Could not one identify the various ways in which the subject of science and theology was treated as a cognitive issue in the nineteenth century without privileging one way over another? Could not one suspend the synthesis that a quest for truth normally presumes necessary and deliberately view concepts of truth and cognition as *heuristic devices* of historical investigation? Admittedly there were numerous stances possible, and many different approaches were presented. Some decision, therefore, has to be made to delimit the historical account to make it manageable.

One avenue of approach to the German theological scene is to choose criteria that would regulate the historical investigation. First, one could insist that in order for a particular theological position on the relationship between natural science and religion to be included in the account it would have to be representative of the beliefs of a large theological group recognized at the time. It would of course be foolish for any author to claim that he or she could include all significant theological perspectives; hence use of this criterion cannot remove all hint of arbitrary choice.

If an additional criterion of organization is employed, the arbitrariness can perhaps be somewhat lessened. Where cognition is the focus, the question of truth is never far behind. But one must not presume that the understanding of what constitutes the nature of truth is the same for everyone. It is here that the philosopher can lend a hand to the historian who seeks a way to employ the concept of truth as a heuristic tool, for philosophers have distinguished three distinct theories of truth which have been dominant in the modern period. Although there are numerous possible expressions of these three perceptions of truth, those figures from the past who have been interested in truth have invariably espoused one or another of the three.

Where the German theologians of the nineteenth century are concerned, the combination of these criteria can be of considerable help. Anyone who has studied German thought in the nineteenth century knows how difficult it can be. It is especially challenging to try to acquire an overview of the century where an issue like the relationship between science and religion is in view. The proposal being pursued, here, then, is to employ recognizable divisions within German theological circles in conjunction with philosophically categorized theories of truth to enable us to compare the ideas of representative theologians who address this subject.

In what follows four different schools of theological thought, which can be linked to four distinct groups in nineteenth-century German culture, are identified. There follows a brief examination of how the treatment natural science receives in each is connected to the theory of truth each preferred. This approach permits us to draw together how the German story illuminates the historical interaction of theology and natural science.[7]

First, however, some description of the ways truth has been understood is in order. How have philosophers recently characterized theories of truth? The first, the correspondence theory, suggests that truth consists of some form of correspondence between what we think and what is "out there." In this schema nature is thought of as wholly independent of us, and the challenge is to devise a rational system that in some sense matches up with an indifferent nature. That the world is rational is not open to question in this perspective; it is assumed that, although it may be difficult to establish the correspondence sought, it is theoretically possible. Isaiah Berlin, recalling his youth at Oxford, reveals how enticing this attitude could be when he described the Platonic ideal to which he then found himself committed. The ideal implied

> ... that, as in the sciences, all genuine questions must have one true answer, and one only, all the rest being necessarily errors; in the second place, that there must be a dependable path towards the discovery of these truths; in the third place, that the true answers, when found, must necessarily be compatible with one another and form a single whole, for one truth cannot be incompatible with another.[8]

In this understanding of truth one appreciates that humans are not free simply to make up what is to count as true; on the contrary, one is here impressed with the element of constraint encountered as one tries to understand the natural world. As the Dutch physiologist Jakob Moleschott put it in 1867: "The natural scientist does not give into the belief that he has created the law. He feels in his innermost being that the facts impose it on him."[9] The correspondence theory of truth is the choice of those who in our own day identify themselves as scientific realists.[10]

The correspondence theory of truth is by far the oldest of the theories before us. It has appeared in one form or another at least since Plato's day, and it does seem to capture the common understanding of what the investigator of nature is up to. Certainly something like establishing a correspondence between our ideas of nature and nature itself was what many in the nineteenth century judged the scientist's task to be. The emphasis here was on discovery of the unknown, on the need to conform to nature's independent authority, on the determining role of fact, and on the incredible power of reason.

In the coherence theory, on the other hand, truth is seen to be affected by the assumptions and presuppositions which we *bring to* our attempts to know nature. Since these can and do vary from individual to individual, from culture to culture, and from age to age, our knowledge of nature is to some degree relative. Here a statement is true or false if it coheres or fails to cohere with a system of other statements.[11] There is no less desire to master the unknown here than in the correspondence theory, but the emphasis now is on invention over discovery, on human assumptions over nature's autonomy, on value over fact, and on recognizing the limits of reason over a naive trust in its power.

According to the philosopher A. N. Prior, the coherence theory of truth is a comparatively late invention, most probably, he says, owing its origin to the stimulus to thought that Kant's work left in its wake.[12] Often Georg Wilhelm Friedrich Hegel is cited as the premier example of one who defines the truth of a proposition through its coherent relation with other propositions and ultimately with the axioms of the system.[13] Hegel carries this sentiment beyond the formal science of logic into metaphysics. Since the required inter-connection of propositions can only be confirmed when they are all known,[14] Hegel's claim regarding the truth of his metaphysical system brings with it an inflexible and dogmatic tone.

Hegel's coherence theory has been called pure coherence in contrast to the so-called impure version of Immanuel Kant and his followers, on the grounds that Hegel did not restrict his attention to the level of phenomena as did the Kantians. The Romantic Kantian Jakob Fries demonstrated that truth need not be described in the metaphysical terms Hegel had adopted when he defined what he called "inner or empirical truth" in opposition to "transcendental truth" in his *Wissen, Glaube, und Ahndung* of 1805.[15] Drawing on Fries and his student Ernst Apelt, Hermann Lotze later in the century spoke of the strict distinction "between truths which are valid and things which exist."[16] In this coherence view, which casts doubt on all facile apprehension of truth, there is never doubt about the need to assume that the phenomenal world can be rationally explained. Those who rely on this understanding of truth willingly acknowledge that the rationality of nature's appearances is being assumed as an hypothesis. About nature *in itself* Kant of course will not speak. The Kantian tradition of impure coherence is, therefore, anti-metaphysical, and is the choice of the contemporary scientific anti-realist.[17]

At the end of the nineteenth century a new theory of truth emerged which would make its presence felt especially in the twentieth century. This theory of truth, the pragmatic theory, contrasted with both its predecessors. Unlike the correspondence theory, truth in the pragmatic theory did not depend on a metaphysical realism in which truth was established by bringing thought and things into correspondence. Truth was determined by what worked. Whatever permitted one to make successful predictions was deemed to be true. Consequently there was no metaphysical assumption made about nature's rationality, nor was there any demand that scientific explanations be wholly coherent. Self-contradictory concepts could be allowed and were said to give a "true" scientific account if the results that were obtained confirmed the predictions made.

None of the German theologians of the nineteenth century embraced the pragmatic theory of truth that was contained in the work of the American mathematician and philosopher Charles Sanders Peirce; indeed, there is no evidence that they were even aware of it. Some did read William James, although they did not find his work to be of central importance to them. As a result, the two older theories of truth represent the two varieties of understanding that characterized the thinking of the German theologians, and they will be the theories of truth employed in this investigation.

It must be said that no theologians from the four German traditions of the nineteenth century here described seemed eager to pay detailed attention to developments in natural science. Those in the first group, called supernaturalists by their fellow theologians, remained preoccupied long into the century with the confessionalist controversies surrounding the forced union in 1817 of the Lutheran and Reformed denominations in Prussia. Members of the second school, speculative theologians who followed the philosopher Hegel, became embroiled in the radical issues of the higher criticism of the Bible. The so-called theologians of mediation, who took their cue from Friedrich Schleiermacher, were content to spend the bulk of their time with esoteric questions of Christology. And last, though hardly least, neo-Kantian theologians, members of the school of Albrecht Ritschl, felt that religion had to do with ethics, and argued that natural scientific questions should be kept separated from the important questions of religion.

When we turn our attention to the supernaturalists, it should perhaps come as no surprise that people of a conservative mentality tended to see scientific debates the same way they viewed theological issues. It was a straightforward matter of right versus wrong, true versus false. The key here was that what was true, in science or in theology, was not something dependent on human achievement. Truth, to use the theological term, transcended human experience. This notion blended very easily with the traditional correspondence theory. To ascertain truth one had to match human conceptions to what God had predetermined. Reality was one, reflecting the unity of its creator, and humankind was obligated to submit to it.

Those who were sympathetic to the supernaturalist position included the many Germans who wanted to hold to the traditions of their faith in spite of assertions from some quarters of the scientific community that they would have to change. Such individuals found the new ideas about evolution particularly unsettling because they agreed with the majority of scientists about the nature of truth. They concurred that there was a correspondence between thought and things, and because of this the argument came down to a disagreement about whose thought matched reality. When their understanding of nature was challenged by alternative propositions about the age of the earth or the origin of human beings, the issue took on aspects of a warfare.

The champion of the conservative position in Germany was Otto Zöckler, a theologian at Greifswald who wrote more on the relationship between science and theology than *any* theologian in the nineteenth century, German, English, or American. Zöckler began his theological career with an essay on Martin Luther's (and therefore Lutherans') understanding of nature, followed a year later by an 800-page study entitled *Theologia Naturalis: A Sketch of a Systematic Theology of Nature from the Standpoint of the Believer* (1860). Both works betrayed Zöckler's interest in nature, and both made clear their defense of the conservative Lutheran perspective.

But it was in Zöckler's 1861 review of what he referred to as the species question, which is among the earliest extended reviews of Darwin's work by a theologian, where the tenor of Zöckler's argument was made clear. Zöckler compared the work of Charles Darwin and Louis Agassiz. While offering minor criticism of Agassiz, Zöckler preferred the work of the Swiss naturalist to Darwin's. He endorsed what he identified as the leading idea of Agassiz's system, namely, that an "objective or natural" system embodied the thoughts of the Creator. "In this rigorous claim of the *objectivity* of natural historical systematics and of the fixed character of species lies essentially and especially the genuinely theistic element of Agassiz's world view."[18]

In a host of essays Zöckler continued for the remainder of the century to examine the conclusions of scientists regarding the evolutionary history of life on earth. He did not shy away from the difficult challenges facing a nineteenth-century creationist; on the contrary, he repeatedly demonstrated his thorough mastery of the issues as the scientists saw them. He believed, of course, that his position, in which the human race had appeared approximately 6000 years ago as the direct result of God's creation, was the one that would ultimately prevail. He believed this because of his primary conviction that truth as humans came to it had to conform or correspond to God's truth. "Belief," he wrote in 1869, "(is) the immediate apprehension of truth through divinely illuminated reason, (and it) gives rise to knowledge, (which is) the mediated apprehension *of the same truth* through reason struggling for light."[19]

If Zöckler was the first protestant theologian to oppose Darwin in a substantial fashion, David Friedrich Strauss was the first to claim him. "To Strauss," wrote Adolph Kohut in 1908, "goes the renown of being the first one of the leading protestant theologians ... who boldly and openly, even enthusiastically made Darwin's theory his own."[20] Unlike Zöckler, however, Strauss's encounter with Darwin came at the end, not at the beginning of his theological career.

Some have suggested that in his book from 1872, called *The Old Faith and the New*, Strauss had abandoned the Hegelian speculative theology he had championed earlier in his career. Because in the book he completely and thoroughly embraced Darwin's *Origin of Species*, they argue, Strauss must have given up his Hegelian idealism for crass materialism, or else he had simply misunderstood Darwin. When one realizes, however, that another infamous Young Hegelian, Ludwig Feuerbach, was viewed by his contemporaries as the intellectual mentor of the German scientific materialists of the 1850s,[21] and that the program of these materialists retained the very idealistic flavor its adherents so frequently tried to denounce,[22] one must take seriously the claim, which was urged by none other than Friedrich Nietzsche, that in *The Old Faith and the New* one does encounter a Hegelian theological response to Darwin.

Throughout his career, from his epoch-making book on the life of Jesus in 1835 on, Strauss drew on the Hegelian premise that there was a complete correspondence between what was rational and what was real. Along with Feuerbach he understood his service to be the restoration of a proper respect for the real. In this endeavor he embraced what he assumed were the values of the natural scientist; i.e., Strauss believed that one could only demonstrate the correspondence if one began without presuppositions. Even in *The Life of Jesus* he noted that his work differed from that of his opponents because his was without presuppositions. He was simply proclaiming the truth and calling on all people to face up to it.

And what was this truth? In *The Old Faith and the New* Strauss described a universe devoid of the traditional God and one which got along quite well without Him. Darwin was portrayed by Strauss as the first scientist to replace a superintending agency with natural forces to explain the development of life. Strauss felt that the match between Darwin's reasoning and the way things really are was so clear that it was simply irrational to deny it. In his book of 1872 he declared that he was speaking for an "innumerable multitude" who no longer were willing to deceive themselves by the old faith, but who insisted on thinking logically about religion regardless how hallowed the doctrine.[23]

It is important to note the determining role that presumptions about a correspondence between thought and things played in the two positions described so far. These are both extreme responses to the challenge natural science presented to religion in nineteenth-century Germany, although they were representative of the views of large numbers of people. As is frequently the case, extreme positions share something fundamental. In this case both sides assumed that there was one truth, that it was possible to know it, and that the task was to get it right. Because each side accused the other of doing violence to the truth, each felt justified in denouncing the other in the name of truth.

A third position was also widely held in Germany in the latter half of the century. If Zöckler spoke for those who felt that the results of Darwinian science had to be rejected, and Strauss for those who believed that traditional religion must be abandoned, then Rudolf Schmid, who wrote two books and several essays about science and religion in the age of Darwin, represented the attempt to reconcile science and religion.

Schmid's thought reminds one most of the nineteenth-century theological tradition known as the theology of mediation, or *Vermittlungstheologie*.

In his treatment of the problems raised by Darwin, Schmid argued that it would be necessary to give up some of the pronouncements about nature found in historic religion. To him questions like the age of the earth or the evolution of life were not central to religion. Much of Schmid's effort was given to defending the right of scientists to investigate nature freely, without interference from theologians.

Where Schmid balked was at the suggestion that Darwin had successfully eliminated God and purpose from the universe. He insisted that the mechanical descriptions of nature given by scientists and the teleological accounts provided by theologians were not contradictory, but complementary. In his justification of this stance he revealed that he too was assuming a correspondence theory of truth, for it never occurred to him that there might be something arbitrary about either the mechanical or the teleological perspective. They were so necessary that every phenomenon in the world "is in fact regarded from either one or the other of these standpoints."[24]

Something cannot be true from a natural scientific viewpoint and false from a religious one, nor, conversely, (can it be) true from a religious viewpoint and false from a natural scientific one. Further, I cannot, to appease myself, be a Christian with my heart and an atheist with my intellect.[25]

One can see from the analysis so far that a diversity of positions is possible on the assumption that truth consists of establishing a correspondence between reality and the mind. The requirement common to all three views is that reason must be employed properly. If one begins with false or biased presuppositions, then there cannot be a match between the real and the rational. But, because truth must be one and cannot be fractured into parts that are at odds with one another, human beings can be confident that the truth of nature they uncover will stand.

It is precisely this attitude toward the conclusions of science that motivated the last group of theologians to be considered here. Neo-Kantian theologians in the late nineteenth century took their cue from the writings of Albrecht Ritschl, whose emphasis on value over cognition found resonance among those who found it difficult to oppose the alleged victory of unbiased scientific truth over dogmatic theology in the waning decades of the century. These theologians did not at all see the relationship of natural science and theology to hinge on one's ability to establish a correspondence between the real and the rational.

Writing in the 1870s the theologian Wilhelm Herrmann severely criticized all attempts to make metaphysical claims about the world, especially those which materialists like Strauss wanted to justify through the results of natural science. Metaphysical claims, Herrmann asserted, arose because of the human need to know and manipulate nature. Such a need had not always been part of theology or the church. One could not, said Herrmann, find it in ancient Christianity. Only as the church itself succumbed to the wish to dominate all social life did it make use of the sciences as a means of enhancing its domination.[26] The metaphysical idea of the world as a unified whole which can be known is functionally related to the assumption of the natural scientist that nature is rational. This Herrmann called the "hypothesis of the comprehensibility of the world."[27]

But according to Herrmann it was the character of metaphysics to take accepted hypotheses about the world of phenomena and to absolutize them. It was not given to humans, however, to be able to penetrate nature to such depths. Herrmann protested vehemently against the notion that an explanation of nature could ever be final, a notion he associated with the natural science of his day. He complained bitterly about the dogmatism of the idea of a final or completed explanation of the world.[28] Natural science was equated with *Naturbeherrschung*, the mastery of nature. It was an essentially utilitarian enterprise whose results enabled us to dominate nature. He did not say it in so many terms, but his claims about the necessary assumption of the rationality of nature by the scientist's position amounted to asserting that in natural science a theory could be called true if it cohered together in a consistent system in a useful way. But sometimes, he wrote, namely in metaphysical claims, the conclusions of science were mistakenly elevated to forms of the being of the world.[29]

Herrmann felt that he could rescue theology from unnecessary conflict with natural science by making sure that theology removed from its purview all contact with metaphysics. Theology must restrict itself to the openly subjective realm of the ethical; it must forgo all desire to speak of nature. The price for such a step was extremely high. Science and religion were cleanly separated from one another. They could not conflict because they did not intersect. But neither did they reinforce one another. Gone was the possibility of referring to God as the creator or the preserver of the world. Theology and science represented separate realms of truth which were to be judged in separate terms. What was necessary was that we not mix or confuse the one with the other.

The reader will not be surprised to learn that both Karl Barth and Rudolf Bultmann, whose works helped to carry theology into existential directions quite compatible with those described here, were Herrmann's students. This sharp separation of science and religion could not, of course, please the nineteenth-century scientists, philosophers, and theologians who were committed to truth as correspondence. But the theological direction opened up by Ritschl and Herrmann, and the coherence conception of scientific truth with which it was consistent, have both enjoyed widespread appeal in the twentieth century.

The question of truth was, as far as most people in the nineteenth century were concerned, at the heart of the issue of the relationship between science and theology. What has become evident from this investigation is that the various claims regarding the relationship between theology and science, when examined with respect to how the question of truth was framed, emerge in a very different fashion from the alignment they received in nineteenth-century authors like John Draper and Andrew Dickson White.[30] Because the various theological treatments of natural science can be grouped according to at least two different understandings of truth, positions usually thought of as at war emerge not only as belonging to the same group, but as viewpoints that possessed a common desire to arrive at final truth. Supernaturalists, speculative Hegelians, and mediation theologians shared a commitment to an understanding of truth as correspondence with reality, and as such they all characterized what might loosely be called a nineteenth-century German interpretation of the relationship between theology and science. That there was fundamental questioning of this approach in the last decades of the nineteenth century becomes clear from the emergence of the alternative understanding of truth embraced by theologians in the Kantian tradition. Indications of unease were not, of course, confined to theology. In several fields there were those who were uncomfortable with the alleged triumph of the Darwinian worldview and even with the occasionally expressed attitude that physics was nearing the completion of its task, that it had gotten beyond the prospect of revolutionary new ideas. Many scientists avoided such declarations, while others, like Ernst Mach and the philosopher Hans Vaihinger, offered alternative interpretations of the same mechanical science that inspired many to boldness.

Thus developments in German theology serve as a mirror of a broader cultural pattern; i.e., they reflected the contrast between those who retained confidence that truth would someday be captured and those who began to subordinate the acquisition of knowledge and truth to a greater enterprise. In particular, those theological responses which rested on a correspondence theory of truth reflected the self-assured temperament consistent with the widespread spirit of progress of the nineteen century, a theological temperament that is not without its representatives in our own time. But a theological tradition that took its cue from a coherence theory of truth provided a hint that the discomfort and dissatisfaction with that same confident air portended for some a rearrangement of how the relationship between theology and science was to be understood.

Department of History
University of Florida

NOTES

[1] Peter Novick, *That Noble Dream: The "Objectivity Question" and the American Historical Profession* (Cambridge: Cambridge University Press, 1988).
[2] On fragmentation as a higher value than synthesis, see Allan Megill, 'Fragmentation and the Future of Historiography,' *American Historical Review* **96** (1991), 694. For a lively discussion of Novick's book see comments by J. H. Hexter, Linda Gordon, David A. Hollinger, and Allan Megill, followed by a response from Novick and an Afterword by Dorothy Ross, in the 1991 Forum of the *American Historical Review* **96** (1991), 675–708.
[3] This is not to say that Kuhn himself endorsed without reservation the trend he helped to produce. Reflecting on the move away from the older intellectual approach of his teachers to contextual history of science Kuhn suggests some important questions were left behind, namely questions about how science works and of what scientific progress consists. See Thomas S. Kuhn, 'The Histories of Science: Diverse Worlds for Diverse Audiences,' *Organon* **22–23** (1986–1987), 15. In the course of reviewing Kuhn's influence on the historical profession, Novick notes that although Kuhn opened the door to contextualism (externalism) Kuhn himself never left the camp of the internalists. Cf. Novick, *That Noble Dream*, pp. 524ff, especially pp. 533–35.
[4] Cf. my essay, 'Is There an Integrative History of Science?' Proceedings of the Conference *Critical Problems in the History of Science*, Madison, Wisconsin (October, 1991), forthcoming.
[5] Megill, 'Fragmentation and the Future of Historiography' (ref. 2), p. 694.
[6] Without doubt many will object to this characterization of the current debate. For an extended defense of this claim, see my 'Is There an Integrative History of Science?' (ref. 4).

[7] For a fuller examination of the treatment of natural science among German theologians see my *Nature Lost? Natural Science and the German Theological Traditions of the Nineteenth Century* (Cambridge: Harvard University Press, 1992).

[8] Isaiah Berlin, *The Crooked Timber of Humanity: Chapters in the History of Ideas*, ed. Henry Hardy (New York: Knopf, 1991), pp. 5–6. Berlin's remarks pertain to the 1920s. Peter Novick's ample and informed discussion of the three classic theories of truth also has reference to the twentieth century. See Novick, *That Noble Dream* (ref. 1), *passim*.

[9] Jakob Moleschott, *Ursache und Wirkung in der Lehre vom Leben* (Giessen, 1867), p. 8.

[10] Hilary Putnam, 'What is Realism?', in Jarrett Leplin, ed., *Scientific Realism* (Berkeley: University of California Press, 1984), p. 140.

[11] Alan R. White, 'Coherence Theory of Truth,' *The Encyclopedia of Philosophy*, 4 vols. (New York: Macmillan Publishing Company, Inc. and The Free Press, 1967), II, p. 130.

[12] A. N. Prior, 'The Correspondence Theory of Truth,' *The Encyclopedia of Philosophy*, II, p. 224. Prior endorses this sentiment from an article on truth by G. E. Moore. For an analysis of the relation of Kant's thought to the emergence of the coherence theory of truth, see Gregory, *Nature Lost?* (ref. 7), pp. 20–22; 285, n. 38; 286, n. 40.

[13] Herbert Keuth, *Realität und Wahrheit: Zur Kritik des kritischen Rationalismus* (Tübingen: Mohr, 1978), p. 7.

[14] *Ibid.*, p. 38.

[15] See J. F. Fries, *Knowledge, Belief, and Aesthetic Sense*, trans. Kent Richter (Köln: Dinter Verlag, 1989), p. 31, where he says: "We cannot, as is usually done, speak of truth as opposed to error by saying that truth is the correspondence of a representation with its object. We can only say that the truth of a judgment is its correspondence with the immediate cognition of reason in which it is grounded. ... I call this truth the *inner* or *empirical truth* of cognition, because we concern ourselves in this idea only with the unification of all our cognitions into a system of reason."

[16] Quoted from the third edition of Lotze's *Microcosmus*, Bk. 8, ch. 1, 'Truth and Science', in William Woodward, *From Mechanism to Value: Hermann Lotze and Nineteenth-Century German Thought*, ch. 11, forthcoming.

[17] Ralph C. S. Walker, *The Coherence Theory of Truth: Realism, Anti-Realism, Idealism* (New York: Routledge, 1989), pp. 19, 35, 38.

[18] 'Über die Speziesfrage nach ihrer theologischen Bedeutung mit besonderer Rücksicht auf die Ansichten von Agassiz und Darwin', *Jahrbücher für Deutsche Theologie* VI (1861), p. 708. Emphasis mine.

[19] *Über Schöpfungsgeschichte und Naturwissenschaft* (Gotha: Perthes, 1869), p. 7.

[20] Adolph Kohut, *David Friedrich Strauss als Denker und Erzieher* (Leipzig, 1908), p. 95. Strauss's recent biographer agrees. Cf. Horton Harris, *David Friedrich Strauss and His Theology* (Cambridge: Cambridge University Press, 1973), p. 248.

[21] Frederick Gregory, *Scientific Materialism in Nineteenth Century Germany* (Dordrecht: Reidel, 1977), chapter 1.

[22] *Ibid.*, pp. 156, 213.

[23] *Der Alte und der Neue Glaube. Ein Bekenntniss* (Leipzig: Hirzel, 1872), pp. 5–6.

[24] *Das Naturwissenschaftliche Glaubensbekenntnis eines Theologen. Ein Wort zur Verständigung zwischen Naturforschung und Christentum*, 2nd ed. (Stuttgart: Kilemann, 1906), p. 9. Cf. also pp. 12–13.

[25] *Ibid.*, p. 2.
[26] Cf. Ueli Hasler, *Beherrschte Natur. Die Anpassung der Theologie an die bürgerliche Naturauffassung im 19. Jahrhundert* (Bern: Peter Lang, 1982), p. 256.
[27] *Die Religion im Verhältnis zum Welterkennen und zur Sittlichkeit. Eine Grundlegung des systematischen Theologie* (Halle, 1879), p. 35. Herrmann borrowed the expression from A. Stadler. Cf. Hasler, *Beherrschte Natur* (ref. 26), pp. 271–72.
[28] *Religion im Verhältnis zum Welterkennen und zur Sittlichkeit* (ref. 27), pp. 25f.
[29] *Ibid.*, pp. 69f.
[30] John Draper, *History of the Conflict between Religion and Science* (New York: Appleton, 1875); Andrew Dickson White, *A History of the Warfare of Science with Theology in Christendom*, 2 vols (New York: Dover, 1896).

SKULI SIGURDSSON*

EQUIVALENCE, PRAGMATIC PLATONISM, AND DISCOVERY OF THE CALCULUS

> Thus, just as pragmatism faces forward to the future, so does rationalism here again face backward to a past eternity. True to her inveterate habit, rationalism reverts to "principles," and thinks that when an abstraction once is named, we own an oracular solution.
> William James[1]

The Newton–Leibniz discovery of the differential and integral calculus in the latter half of the seventeenth century is a classic example of a multiple scientific discovery. It is a Mertonian doublet.[2] The meaning of the simultaneous construction of Newtonian fluxions and Leibnizian infinitesimals continues to haunt the historical consciousness – not the least because with the complacent wisdom of hindsight these two formalisms can be regarded as "equivalent" versions of the calculus. The central aim of this paper is to discuss how logically equivalent formalisms are non-equivalent in practice, and to sketch how this can have surprising consequences for the history of mathematics.

In the first section I examine how two historians of mathematics viewed the discovery of the calculus. They were Moritz Cantor (1829–1920) and Hieronymous Zeuthen (1839–1920). Their interpretations were shaped by their professional experiences: Cantor was a historian of mathematics but Zeuthen a mathematician who worked in the history of mathematics as well. The former wrote an encyclopedic history of mathematics, the latter emphasized conceptual analysis at the cost of historical comprehensiveness. In spite of their different styles as historians both recognized the superior power of Leibniz's formalism, Cantor unequivocally but Zeuthen with reservations. In the second section I discuss how Zeuthen's ambivalence toward Leibniz's achievement stemmed from his dislike of mass-produced knowledge. Scientific mathematics was produced by paying close attention to foundations and not by relying on ready-made algorithms. He believed furthermore that a study of old mathematics texts with modern tools might open up unexplored avenues for mathematics. Equivalence enabled Zeuthen to journey into the past and set out anew in a direction determined by modern research.

The history of the discovery of the calculus and the subsequent priority dispute between Newton and Leibniz is roughly as follows.[3] Isaac Newton (1642–1727) made his discoveries of the method of fluxions in the years 1664–1666. He made them known in manuscript form to a small circle of initiates in *On Analysis by Means of Equations Having an Infinite Number of Terms* (1669) and *Method of Series and Fluxions* (1671).[4] From the early 1670s Gottfried Wilhelm Leibniz (1646–1716) was in contact with the English through Henry Oldenburg, secretary of the Royal Society. Leibniz had begun an intensive study of mathematics, and in 1675 he made his fundamental discovery of the basic ideas of the infinitesimal calculus; some of them he published in the Leipzig journal *Acta Eruditorum* in 1684. Newton responded in two letters in 1676 to Leibniz's inquiries concerning his mathematical results without divulging the underlying theoretical framework. Leibniz was in London for the second time in 1676 and hastily studied Newton's *On Analysis*. Possessing the basic ideas of the calculus, he was only interested in Newton's results concerning infinite series and the binomial theorem. Leibniz later asserted that he owed nothing to Newton, whereas Newton asserted the opposite. The priority dispute that flared into the open at the end of the century was bitter, nasty and lasted a number of years beyond Leibniz's death. That Leibniz had seen some of Newton's writings made it easier to accuse him of plagiarism. It is now accepted by scholars that Newton and Leibniz discovered their respective formalisms independently, but there has been less agreement whether they discovered the same thing or not.[5]

Steven Shapin has remarked in a review of A. Rupert Hall's *Philosophers at War* (1980): "As much as has been written about the Newton–Leibniz disputes, more doubtless would have been written had the protagonists' conduct not offended historians' sensibilities or done such massive violence to idealized conceptions of how revered men of science ought to behave." According to Shapin the dispute has been approached from a variety of perspectives: the hagiographic (David Brewster), the metaphysical in the Leibniz–Clarke exchanges (Alexandre Koyré), the political (J. R. and M. C. Jacob), the psychological (Frank Manuel), and the sociological/normative (Robert K. Merton).[6]

My approach differs from these. I wish to explore Cantor's and Zeuthen's account of the discovery of the calculus in preparation for interpreting Zeuthen's historiographic position. That I do in the second section in light of recent studies that emphasize the role of scientific practice.[7] My intentions are sociological but not Mertonian; I want to understand the implications of the Leibnizian version of the calculus for the production of new mathematical knowledge, the hierarchical structure of the scientific community and the question of equivalence. That the last issue lies at the heart of the Newton–Leibniz dispute is strikingly demonstrated at the end of Hall's book, where he writes:

> I end this story with a question: Did Newton and Leibniz discover the same thing? Obviously, in a straightforward mathematical sense they did: Calculus and fluxions are not identical, but they are certainly equivalent.

After further discussion he adds:

> Yet one wonders whether some more subtle element may not remain, concealed, for example, in that word "equivalent." I hazard the guess that unless we obliterate the distinction between "identity" and "equivalence," then if two sets of propositions are logically equivalent, but not identical, there must be some distinction between them of a more than trivial symbolic character. Perhaps this is in some sense metaphysical rather than operational, of psychological interest rather than cash value.[8]

I would take issue with Hall's claim that Newton and Leibniz discovered the same thing "in a straightforward mathematical sense." That position may be tenable on Platonic grounds where competing mathematical formalisms merely map a single unchangeable reality in different ways. The English mathematician G. H. Hardy expressed this viewpoint clearly which he ascribed to Plato when he wrote: "I believe that mathematical reality lies outside us, that our function is to discover or *observe* it, and that the theorems which we prove, and which we describe grandiloquently as our 'creations,' are simply our notes of our observations."[9]

Psychologically pleasing formalisms can have high cash value by suggesting new directions for research. Therefore it is questionable that the metaphysical and psychological can be laid so neatly aside as Hall suggests. Yet he has identified the question to which I will return later: namely, if logical "equivalence" does not guarantee identity and if "equivalence" is not the same as "identity," what consequences does the difference have? Couched in the language of William James one might say that logic is static, conservative and backward looking, whereas practice is dynamic, open-ended and forward looking. The working mathematician, although subscribing to some version of Platonism, knows that one can construct very different road maps of "reality" and that some are vastly superior to others for the trek into the unknown.

CANTOR, ZEUTHEN AND THE DISCOVERY OF THE CALCULUS[10]

Moritz Benedikt Cantor, who was born in Mannheim in 1829, started lecturing on elementary mathematics at the University of Heidelberg in 1853 and lectured on the history of mathematics from about 1860 onwards. He became "Extraordinarius" in 1863 and "ordentlicher Honorarprofessor" in 1875. He died in Heidelberg in 1920. Besides writing many books on the history of mathematics, he edited the *Abhandlungen zur Geschichte der Mathematik*, and he wrote countless reviews on a variety of works: historical, technical, textbooks.[11]

Cantor was faced with a difficult problem in his monumental *Vorlesungen über Geschichte der Mathematik* which appeared in four volumes from 1880 to 1908. In light of the exponential growth of mathematical knowledge and rise of academic history since the Scientific Revolution, how was it possible to write a total history of mathematics? Could one follow in the spirit of the Enlightenment Jean Étienne Montucla's *Histoire des mathématiques*, where both pure and mixed mathematics (astronomy, mechanics, optics, music) had been treated? Cantor now had to restrict his treatment to the former in order to be able to incorporate the result of recent historical research, and even so he never managed to bring the discussion up to the present. Part of the price to be paid, as Dirk J. Struik points out, is to "miss the agreeable style of the eighteenth century friend of the philosophers." We must "be satisfied with the serious style of the persevering nineteenth century Herr Professor."[12]

One cannot help agreeing with Struik's assessment. Pedantic thoroughness pervades Cantor's work. What is more, a certain ambiguity underlies his writings. The third volume of the *Vorlesungen* dealt with the period from 1668 to 1758. Ample space was devoted to the rise of the calculus and the priority dispute between Newton and Leibniz.[13] His sympathies in the Newton–Leibniz controversy were clearly on the side of Leibniz since the German philosopher-scientist had discovered a more flexible and fruitful formalism. Yet, in writing his history of mathematics Cantor paid scant attention to the choice of notation.

Cantor saw Leibniz's introduction in 1675 of the symbols for summation and difference, ∫ and d, as being the significant breakthrough that marked the discovery of the calculus. This approach enabled Cantor to sidestep questions of Leibniz's conceptual debt to others. Whatever Leibniz might owe his predecessors and contemporaries, the development of this symbolism had "risen from Leibniz's own mind." Newton was in the possession of something similar from 1671, but "as it is immaterial who clears the new paths for science, it was fortunate that Leibniz discovered the sign language uninfluenced." Cantor continued in the same vein, employing mixed metaphors of geographical conquest and organic growth to account for the successes of Leibniz's sign language:

It has that in common with other languages, that first it developed gradually from imperfect beginnings to ever higher forms, that is spread continually wider, conquering one region, one country after the other, while the spirit of the language remained the same unchanged, while the concepts summation and difference were always revealed with the same clarity.[14]

In Cantor's view, infinitesimal considerations had progressed so far before the arrival of Newton and Leibniz that further progress depended crucially on the discovery of an effective notation. And that was what had happened, as "a notation was now available which had indisputably been discovered independently by Leibniz."[15]

Cantor, however, did not take his own teaching to heart. He did not discuss the appropriate notation to use while writing the history of mathematics, and he apologized rarely for using a modern way of expression.[16] Even though he made clear that Newton's dot notation was first made public in the 1690s, that did not affect his discussion of Newton's work on fluxions.[17] It seems fair to assert that these concerns did not play an important role in his historiography.[18] Cantor's unself-conscious use of a modern notation was nurtured by his progressivist view of the history of mathematics: in this view different formalisms and systems of notation might either speed up or slow down mathematical growth, but they would not change its course significantly.

This viewpoint found a clear expression in late nineteenth-century puzzlement over the mathematical employed in Newton's *Philosophiae Naturalis Principia Mathematica* (1687). The physicist and philosophical critic Ernst Mach (1838–1916) admitted in *Die Mechanik* (1883), that he found Newton's geometrical procedure "frequently so artificial that, as Laplace remarked, it is unlikely that the propositions were discovered in that way."[19] Some years later, in a newly added section on Newton's achievements, Mach noted:

> In his papers and books on optics Newton showed the paths which led to his discoveries quite frankly and without any restraint. Apparently the unpleasant controversies in which these first publications of his involved him had an influence on his exposition in the *Principia*. In the *Principia* he gave the proofs of the theorems that he had discovered in a synthetic form, and did not disclose the methods which had led him to these discoveries.[20]

Mach's reading of the *Principia* illustrates the strength of an enduring tradition in the history of mathematics. When a cumbersome notation from a latter-day viewpoint is employed in mathematics it may be explained away by an appeal to extraneous reasons. Notation is a packaging device that does not shape the actual content of mathematics any more than wrapping paper affects the contents of commercial products. Competing mathematical formalisms are merely alternative ways of describing a single "outer reality."

Newton lent his authority to this tradition. During the priority dispute with Leibniz he advanced anonymously the idea that the geometrical mode of presentation in the *Principia* did not reflect the way the results had been found, that is, with the aid of fluxions. This position may have served Newton's purposes in his dispute with Leibniz, but it clearly was untrue. The mathematics used in the *Principia* is neither traditional geometry nor fluxions but something else, although the use of geometrical limit-increment procedures abounds.[21] The mathematics has been characterized as the geometry of the ultimate, where the diagrams used for demonstrations are assumed to move and change their shapes.[22] But a calculus-inspired reading of the *Principia* has an intuitive appeal and continues to enjoy great popularity – so much that D. T. Whiteside recently proclaimed in exasperation: "How often am I still asked: '*Did* Newton use calculus to obtain the theorems in his *Principia*?' How, without seeming to patronize, do you lay the groundwork on which you can reply that the question is ill-formed and therefore meaningless?"[23]

Hieronymous Georg Zeuthen, who was born in West Jutland in 1839, lectured in mathematics at the University of Copenhagen from 1871 and was a professor from 1883; he lectured at the Polytechnic Institute as well. He died in Copenhagen in 1920. His research field was enumerative geometry, but in addition he began to work in the history of mathematics from the mid-1870s onwards. He published widely in mathematical journals and was the editor of *Tidsskrift for Mathematik*. He wrote the article on enumerative methods in the *Encyklopädie der mathematischen Wissenschaften* (1906).[24]

Among Zeuthen's numerous historical writings were lectures on the history of mathematics, *Forelæsning over Mathematikens Historie* that appeared in two volumes in 1893 and 1903 and were aimed at mathematicians and mathematics teachers. He emphasized that his audience need not concern itself with isolated historical details, nor know who first discovered this or that truth (*Sandhed*) or procedure (*Fremgangsmaade*). What mattered was to recognize the form in which these truths and procedures had emerged and the use to which they had been put.[25]

Because of the didactic character of the lectures – they were meant to teach mathematical thinking and not primarily history – Zeuthen did not list all the references to other historians, and he apologized for that. He referred to Cantor's unusually complete *Vorlesungen* for details. He freely acknowledged his debt to other historians: Allmann, Brettschneider, Braunmühl (*Geschichte der Trigonometrie*), Chasles, Colebrooke, Curtze, Enestrøm, Hankel, Heiberg, Hultsch, Loria, Mach (*Die Mechanik*), Ritter, Tannery, Wertheim, Woepcke, and the philosopher Høffding (*Filosofiens Historie*).[26]

The focus of the second volume of the lectures was the history of mathematics in the sixteenth and seventeenth centuries. It was in two parts: a chronological overview and a technical exegesis. In his discussion of Newton's indebtedness to Isaac Barrow (1630–1677), Zeuthen introduced the recurrent theme of oppositions (*Modsætningsforhold*) which was represented here by the inverse nature of differentiation and integration of which Leibniz became aware in his work on characteristic triangles.[27] He stayed in Paris in the 1670s and met Christiaan Huygens (1629–1695). The latter's emphasis on specialized methods provoked an opposite reaction in Leibniz: he desired to search for general procedures, an approach which resonated with his general philosophical outlook.[28]

Zeuthen stressed the social nature of Leibniz's reaction to Huygen's working style. The explicit rules of Leibniz's calculus enabled less privileged investigators to avail themselves of the knowledge of a select few and to use these methods in practice. "Thereby these operations are turned into a calculus (*Regning*), that can be learned and also carried out by those, who in terms of general insight lag behind the great masters, who until then had monopolized these investigations." These implications became clear when Zeuthen evaluated the dispute between Newton and Leibniz. He downplayed its importance because historical research had shown that both played an indispensable part in the discovery of the calculus. He added:

Whether one wants to thank the one or the other, depends on the primacy that one wants to give to the discovery of those deep mathematical truths which form the basis for the common method (*Methode*) of the two, or the discovery and easily accessible presentation of a technique (*Teknik*) that made the method immediately usable in the hands of other capable (*dygtige*) mathematicians and which has gradually made it even useful for such practical men, who lack deep mathematical insight or interest.[29]

Zeuthen's assessment of the Newton–Leibniz dispute was consonant with his world view. He felt that Newton was the deeper mathematician, yet he clearly recognized the value of Leibniz's awesome ability to synthesize. Whether in mathematics or in natural philosophy, Leibniz synthesized while simultaneously enriching the subject matter with his own fecund ideas. In Zeuthen's opinion, Leibniz's most important contribution to mathematics lay in the explicit rules for the calculus that generalized and encompassed those operations which hitherto had been seen as the solution of disparate mathematical problems. In this Leibniz set an example for later generations of mathematicians.[30]

The different historiographical positions of Cantor and Zeuthen – one chronicle-like and static, the other conceptual and practice-oriented – can be illustrated by their discussion of the absence of fluxions in Newton's *Principia*. If Newton had already invented and mastered the methods of fluxions when he wrote the treatise, why were there no traces of them in the *Principia*? "Why this silence" asked Cantor, if Newton already had the methods of fluxions at his disposal, when he composed the book? Fluxions had to be hidden somewhere underneath the surface and thus Cantor gave clear expression to the importance he attached to notation and formalism. Cantor asserted:

Only a single explanation seems possible which has also been given repeatedly. The laws of general attraction that are represented in the *Principia* were so new, so surprising, that Newton could fear a much stronger rejection of them, if he had deviated ever so slightly from the old-geometrical methods which all mathematicians accepted.[31]

In contrast to Cantor, Zeuthen saw Newton's development of the method of fluxions and the *Principia* as occurring along parallel tracks. The method of fluxions and the accompanying notation would not have sufficed as a tool for deriving the theorems in the *Principia* as could be done today with the aid of the calculus. The converse was rather the case; namely, the results in the *Principia* supplied material for the further development of the calculus, as it was extended to incorporate the new mechanics. Besides the concern for reticence and the questionable foundations of fluxions, Zeuthen argued that it would have been a detour for Newton to develop and present the main theorems of the *Principia* by using the calculus. It could be done much more directly by combining geometrical and limiting procedures. On the other hand, Zeuthen added that Newton would have done his contemporaries a service by presenting the investigations in the later parts of the *Principia* with the help of fluxions.[32]

For Zeuthen, mathematicians needed substantial results on which to hone their tools and skills. Their techniques were insignificant by themselves without important testing material. He recognized that Newton's reluctance to publicize less than polished results and to face stiff criticism and premature competition had also to be taken into account. In his extreme reticence Newton was simply peculiar (*Særling*).[33]

ZEUTHEN, ELITISM AND PRAGMATIC PLATONISM

Zeuthen's sociological remarks concerning the production of knowledge and the egalitarian nature of Leibniz's calculus, were not merely pedagogical; Zeuthen interpreted them more broadly. He saw the transition from individual investigations to studies based on general and comprehensive methods as analogous to the transition in industry from workshops to factories. He stressed that in individual investigations the individuality of the investigator was preserved. Thus he saw the mechanization of infinitesimal analysis made possible by Leibniz's version of the calculus at the end of the seventeenth century to be comparable to what occurred with respect to the rise of algebra earlier in the century.[34]

A nostalgic yearning for individual creation underlay Zeuthen's analysis, for mechanization entailed an aesthetic loss: the mass-made industrial pieces were uniform and practical, but often not very interesting. The same held for that part of mathematics, where the mechanized operations of algebra and Leibniz's calculus had moved mathematics to a new stage: it became important for the culture at large and it facilitated the generation of more mathematical knowledge. "A large part of the inner development of mathematics thus becomes nearly autonomous.... However meaningful the results of such investigations are, they could not interest us to the same degree as those, where new viewpoints would begin to be prevalent, and which themselves could not have been carried out with a finished method."[35]

Implicit in Zeuthen's emphasis on individual creativity was an elitist conception of mathematics. He made this explicit in his address as rector of the University of Copenhagen in 1896. There he discussed the historical development of mathematics from Greek antiquity until the end of the eighteenth century. He began by stating two criteria which scientific (*videnskabelig*) mathematics in general had to satisfy.

It can neither be derived mechanically by calculation that is learned from others, or by calculation that is formed by an unconscious combination of procedures, whose correctness has gradually been justified by experience. To be scientific it has to have a true foundation (*Begrundelse*) which makes the connection between the various facts, by which one thing follows from another, appear to consciousness.

Another sign of scientific mathematics is its ability to grow and develop. First through foundations it ties together facts that originally had pushed in from unknown directions; then it works further and discovers, that totally new facts are connected to old ones in a fashion similar to how the old ones were interconnected.

A scientific treatment of mathematics was therefore characterized by foundations and growth. Where one of these was totally absent, the other would also turn out to have been insignificant. These two aspects had not always been equally prominent in the history of mathematics, but the greatest mathematicians had combined both of them:

By penetrating into connections among things they already knew, they found procedures that took themselves much further. These procedures, frequently in an altered and for the less significant mathematicians more convenient form, have thereby also enabled the latter ones to contribute to the growth of mathematics.[36]

Zeuthen's views on the mechanization of mathematical investigations cast fresh light on the Second Scientific Revolution in the early nineteenth century, which signified such tremendous changes both cognitively and institutionally. It heralded the beginning of hierarchical social structures in science. The division of labor became increasingly pronounced, and the leaders of the emerging scientific disciplines stressed the divide that separated those who made discoveries from those who either had merely done the preliminary work or those who explored the implications of the elite's discoveries. One could learn to become a scientist, but not to make discoveries.[37]

The general outlines of this interpretation are convincing, yet the focus on social factors obscures those aspects which depended on internal technical developments, and were thus not directly related to the onset of professionalization and discipline building in the early nineteenth century. A better understanding of the role of mathematics in the manufacturing of new knowledge may provide a corrective.[38] The discovery of the calculus shows that at the technical level changes can occur which lead to a mechanized mode of knowledge production without being directly related to social strategies of control and centralization of power.

What happens is that an efficient algorithm or tool makes an increased division of labor possible. The irony is that although Zeuthen's two examples from the history of seventeenth-century mathematics – algebra and Leibniz's version of the calculus – seem to imply an increasingly democratic participation in science, the net effect is that such formalisms often will turn the scientific enterprise into a well-run factory. There the results found by individual workers will be trustworthy and useful, because they were made with the proper tools that were not produced by the workers themselves.

The impressive developments of the calculus, mathematical physics and mechanics in the eighteenth century demonstrate the importance of powerful and flexible mathematical methods and notation, which enabled a multitude of investigators to explore these closely related fields. Its mechanized aspects, together with the fact that the center of mathematical production had moved to France in the eighteenth century, ensured a rapid production of results and made it possible to utilize the efforts of many investigators during the ensuing period of normal scientific activity.[39]

The same occurred in the early twentieth century with axiomatization; namely the axiomatization of many fields was the end result of long and complicated negotiation processes, and once the leaders of the mathematical profession had agreed upon the axiomatic system, the daily laborers in the field of mathematics could spell out the implications of the axioms mechanically.[40] The employment of instruments in the physical sciences bears a certain resemblance to the use of formalisms and axioms in mathematics, namely the reliance on ready-made and standardized instruments.[41]

What mattered for Zeuthen in his historical practice was to recognize how mathematical truths and procedures had emerged and been utilized. Thus the readers of his history could discriminate between "the necessary kernel" in such procedures and the more contingent "outer form." From this pragmatic viewpoint, it was profitable to read old mathematical texts: "The mathematician encounters also here and there in the history of the subject viewpoints that were made superfluous by subsequent progress, but which mathematics could use again in its further development."[42]

Zeuthen's historiographical agenda was clearly recognized by his contemporaries. The mathematician Max Noether (1844–1921) characterized the second volume of Zeuthen's *Forelæsning* as follows:

> This work is primarily not an historical one in the customary sense: a determination and registration of individual facts; it is a *mathematical* work in an historical garb which is addressed to mathematicians, to show them how their methods have arisen and thus to bring those near from a new vantage point.[43]

Zeuthen was quite explicit about his pragmatic intentions, when he discussed the attitudes of a mathematician who, when reading old mathematics texts, would encounter results that could be derived by much more recent methods. Regarding the mathematician Zeuthen wrote:

> If he also becomes familiar with the investigation that has led to the result, he will understand it most easily in comparison with the viewpoints, with which he himself is intimately familiar, usually by a translation into the modern algebraic language, whereby he can turn the logical reasoning present into a calculation with letters (*Bogstavsregning*). Thus he will be able to penetrate into the basic thought, which usually will be exactly the same in the proof that has been handed down and in the one that he would have constructed with his own methods. This is a simple consequence of the fact, that the logical connections have a pure objective validity, independent of the various forms, by which they could come to our consciousness.[44]

Zeuthen's approach gave an opportunity to view old mathematical problems in fresh light and made it possible to explore their solutions anew in unknown directions. This directness can be an important ingredient in mathematical research. For example, it was integral to David Hilbert's style as a mathematician: "A characteristic feature of Hilbert's method is a peculiarly *direct attack* on problems, unfettered by algorithms; he always goes back to the questions in their original simplicity."[45] By studying old mathematical texts, mathematicians and teachers of mathematics can gain access to ideas of bygone days, as if they had stepped into the past and could peer over the shoulders of their predecessors as they sat at their desks. Present practitioners of mathematics may therefore have a chance to take different turns at important historical crossroads and explore untrodden paths into the future. Thus, historical excavation may free mathematicians from the yoke of algorithmic procedures.

Following Hardy I take Platonism to be synonymous with realism for mathematicians.[46] Erwin Hiebert has described Ludwig Boltzmann's philosophy of science as pragmatic realism.[47] Zeuthen's working philosophy can be called pragmatic Platonism. It has its dangers because it can lead to overly rational reconstructions in the history of mathematics. Over a period of years Cantor became increasingly sceptical of Zeuthen's way of exploring the history of mathematics and the disagreement led to a clash between them.[48] Furthermore it is to Zeuthen and Paul Tannery (1843–1904) that we owe the term "geometrical algebra" in studies of Euclid's *Elements*. This translation of geometry into algebra and subsequent misreadings in the history of Greek mathematics led to an explosive and eventually fruitful controversy in the 1970s.[49]

Zeuthen's historiography is an eloquent testament to the interest that mathematicians have in the history of their discipline. What may thus seem to be misreadings of old mathematical texts from the viewpoint of professional historians of science, is a characteristic feature of the behavior of the mathematical community.

This observation can be used to distinguish between mathematics and the neighboring sciences. That mathematicians regard old texts as a source of inspiration is visible in mathematical libraries, whose shelves contain many old texts, in marked contrast to physics libraries where old books and journals are quickly relegated to storage rooms. This difference between the practices of mathematicians and physicists is also visible in the longevity of unsolved problems. The mathematical ones may remain at the focus of discussion for centuries, whereas in physics such a state of affairs would be seen as highly anomalous. As Thomas S. Kuhn has noted: "For reasons and in ways that remain obscure to me, the sciences destroy their past far more thoroughly than do mathematics or the arts."[50]

The reason for this is the instrumental and ontologically aloof character of mathematics. It is exactly the instrumental character of mathematics which has made it so useful in the physical sciences and suggested fruitful analogies between different realms of experience, as demonstrated by the fruitful use that William Thomson (1824–1907) made of the analogies between the mathematical description of the transfer of heat and the behavior of electrical fluids.[51] It is the ontologically removed character of mathematics which accounts for the continued applicability of mathematical techniques. They are not tied to any one research program in the physical sciences and can maintain their utility and relevance, while the research momentum moves elsewhere because of new ontological commitments. It is this sharp contrast between endurance and transience which is manifested in mathematics and physics libraries. Old books and journals are of value to mathematicians whereas they become increasingly irrelevant to most physicists with the passage of time.

CONCLUSION

In this paper I have discussed what it means for theoretical formalisms to be considered equivalent. The example I used was the Newton–Leibniz discovery of the calculus in the second half of the seventeenth century. Other examples would be the alternative formulations of the law of gravitation and of the new quantum mechanics. In all cases the equivalence breaks down once it is realized that competing formalisms suggest separate directions for research and therefore generate different kinds of knowledge.

That Newton and Leibniz discovered the calculus independently is now generally accepted, but did they discover the same thing? Ivo Schneider has recently answered this question in the negative. He observes that although the combined efforts of Leibnizians, Newtonians and historians until the present have only confirmed that Newton and Leibniz created equivalent forms of the infinitesimal calculus, "the starting point, the main emphasis and the expectations of the two pioneers were not at all identical."[52]

Schneider sees the main characteristics of the infinitesimal calculus to be the inverse operations of differentiation and integration, or the formation of fluxions and fluents, and the encapsulation of these operations in a suitable algorithm. From this viewpoint Newton and Leibniz proceeded very differently. Newton expressed the reversibility either inconspicuously in geometrical form or he took it for granted; for Leibniz, on the other hand, the inverse nature of these operations was fundamental and formally built into the calculus. Thus, it comes as little surprise that the two versions turned to be non-equivalent in practice. The representational form of the *Principia* became the norm in England; the result was devastating for the development of English mathematics in the eighteenth century.[53]

What it means for formalisms to be non-equivalent in practice was made clear by the theoretical physicist Richard Feynman (1918–1988). Discussing the three different formulations of the law of gravitation, he wrote:

Mathematically each of the three different formulations, Newton's law, the local field method and the minimum principle, gives exactly the same consequences. What do we do then? You will read in all the books that we cannot decide scientifically on one or the other. That is true. They are equivalent scientifically. It is impossible to make a decision, because there is no experimental way to distinguish between them if all the consequences are the same. But psychologically they are very different in two ways. First, philosophically you like them or do not like them; and training is the only way to beat that disease. Second, psychologically they are very different because they are completely unequivalent when you are trying to guess new laws.[54]

Feynman pointed to the psychological cash value that potent formalisms can have. They can facilitate rapid dissemination of techniques and attract new practitioners to a field of inquiry by promising them "easy" discoveries. The combined effect can be the mechanization of knowledge production and simultaneous domination of a research field. The victory of the Leibnizian calculus in the eighteenth century is one example and the successes of the Schrödinger formulation of quantum mechanics vis-à-vis matrix mechanics from the late 1920s onwards is another.[55]

The still widespread belief that Newton must have used the calculus in order to develop the new mechanics indicates why equivalence continues to enjoy such popularity. Instead of struggling with the *Principia* in the original, it is easier to translate it into the language of the calculus which is the familiar idiom of classical mechanics. Equivalence enables the reader to appropriate the results of previous traditions in a convenient manner without doing the hard work that is necessary to understand and confront the strangeness of the past. Equivalence leads to a violation of the integrity of earlier scientific cultures. This is why Kuhn has described the historians' craft in anthropological terms. The historians must:

be prepared at the start to find that the natives speak a different language and map experience into different categories from those that they themselves bring to home. And they must take as their objective the discovery of those categories and the assimilation of the corresponding language. "Whig history" has been the term reserved for failure in that enterprise, but its nature is better evoked by the term "ethnocentric."[56]

Yet Whiggism has its virtues and may be exceedingly hard to avoid in practice. Besides, historians of science have a more important agenda than fighting the scourge of Whiggism.[57] I have called Zeuthen's mathematical world view pragmatic Platonism. Although he proceeded ahistorically by reading old texts in the light of modern knowledge, he understood that history could have happened differently. Thus in his awareness of the contingent nature of history, Zeuthen was less deterministic and therefore more historical then Cantor.

An integral part of Zeuthen's outlook was a distrust of mass-produced knowledge. He feared that by sheer reliance on calculations mathematics could lose its scientific character. He did not want to be rushed into the future; rather he wanted to travel slowly and decide at each step where he was going. This attitude was in accord with his elitism, but it furthermore indicated his realization that formalisms – here the Leibnizian calculus – could dominate scientific practice to such an extent that alternative ways of doing science might be throttled and forgotten.

Ian Hacking has argued that styles of reasoning become self-authenticating.[58] Zeuthen wanted to guard himself against this possibility. In order to break out of this confinement he acted Whiggishly. Hacking has furthermore discussed how difficult it is – if not impossible – to think outside the thought styles of each epoch. This is why the dream of a history of science free of Whiggism may be unrealizable.

What should matter instead is to approach the past in a sympathetic fashion, listen to the barely audible voices of its actors without forcing them to speak in modern tongues, pay attention to both elites and Kuhn's normal laborers in the lush vineyards of science, and recognize that what occurred in the past is often a matter of chance. The aim of history is not only to describe how the past really was, but also to make clear that it might all have happened differently.

Alexander von Humboldt Fellow
University of Göttingen

NOTES

* I thank Michael Becker, Deborah Coon, François De Gandt, Lenore Feigenbaum, Gerald Holton, Eberhard Knobloch, Joan L. Richards, Simon Schaffer, Silvan S. Schweber, and Sabetai Unguru for reading and criticizing earlier versions of this paper. All translations are mine, unless otherwise indicated.

[1] William James, 'Pragmatism's Conception of Truth (1907),' in James, *Essays in Pragmatism*, ed. Alburey Castell (New York: Hafner Press, 1948), pp. 159–176, quot. on p. 172.

[2] Robert K. Merton, 'Singletons and Multiples in Science (1961),' in Merton, *The Sociology of Science: Theoretical and Empirical Investigations*, ed. Norman W. Storer (Chicago: University of Chicago Press, 1973), pp. 343–370.

[3] Cf. A. Rupert Hall, *Philosophers at War: The Quarrel between Newton and Leibniz* (Cambridge: Cambridge University Press, 1980); Ivo Schneider, *Isaac Newton* (München: C. H. Beck, 1988); and Richard S. Westfall, *Never at Rest: A Biography of Isaac Newton* (Cambridge: Cambridge University Press, 1980).

[4] They were only published in the years 1711 and 1736, respectively.

[5] For further reading cf. H. J. M. Bos, 'Newton, Leibniz and the Leibnizian Tradition,' in I. Grattan-Guinness, ed., *From the Calculus to Set Theory, 1630–1910: An Introductory History* (London: Duckworth, 1980), pp. 49–93; I. B. Cohen, 'Isaac Newton (1642–1727),' *Dictionary of Scientific Biography* **10** (1974), 42–101, esp. pp. 45–53 and 84–85; and Joseph E. Hoffmann, 'Gottfried Wilhelm Leibniz (1646–1716): Mathematics,' *Dict. Sci. Biogr.* **8** (1973), 160–168.

[6] Steven Shapin, 'Licking Leibniz,' *History of Science* **19** (1981), 293–305, quot. on p. 296; and Shapin, 'Of Gods and Kings: Natural Philosophy and Politics in the Leibniz–Clarke Disputes,' *Isis* **72** (1981), 187–215.

[7] Cf. Timothy Lenoir, 'Practice, Reason, Context: The Dialogue between Theory and Experiment,' *Science in Context* **2** (1988), 3–22.

[8] Hall, *Philosophers at War* (ref. 4), quot. on pp. 257–258.

[9] G. H. Hardy, *A Mathematician's Apology* (Cambridge: Cambridge University Press, reprinted 1967 with a foreword by C. P. Snow (orig. publ. in 1940)), quot. on pp. 123–124; emph. in orig.

[10] Cf. also Jesper Lützen and Walter Purkert, 'Conflicting Tendencies in the Historiography of Mathematics: M. Cantor and H. G. Zeuthen,' København's Universitet Matematisk Institut, preprint series 1989, Nr. 22. I thank Jesper Lützen for making this preprint available to me. It has been submitted to *Historia Mathematica*.

[11] *Ibid.*, pp. 2–13; and Joseph E. Hoffmann, 'Moritz Benedikt Cantor (1829–1920),' *Dict. Sci. Biogr.* **3** (1971), 58–59.

[12] Dirk J. Struik, 'The Historiography of Mathematics from Proklos to Cantor,' *NTM-Schriftenr. Gesch. Naturwiss., Technik, Med.* **17** (1980), 1–22, quot. on p. 17.

[13] Moritz Cantor, *Vorlesungen über die Geschichte der Mathematik. III: Von 1668–1758* (Leipzig: B. G. Teubner, 2nd ed. 1901), esp. pp. 156–207 and 285–328. The third volume originally appeared in installments in 1894, 1896, and 1898.

[14] *Ibid.*, p. 166.

[15] *Ibid.*, p. 167.

[16] *Ibid.*, pp. 157 and 172.

[17] *Ibid.*, p. 328.

[18] Cf. Moritz Cantor, 'Wie soll man die Geschichte der Mathematik behandeln?', *Bibliotheka Mathematica: Zeitschrift für Geschichte der mathematischen Wissenschaften* **4**, 3rd series (1903), 113–117. This assessment is confirmed by Lenore Feigenbaum, 'Brook Taylor and the Method of Increments,' *Archive for History of Exact Sciences* **34** (1985), 1–140, on pp. 23–25.

[19] Ernst Mach, *The Science of Mechanics: A Critical and Historical Account of Its Development*, tr. Thomas J. McCormack (La Salle: Open Court, 1960), pp. 560–561.

[20] *Ibid.*, pp. 246–247.

[21] D. T. Whiteside, 'The Mathematical Principles Underlying Newton's *Principia Mathematica*,' *Journal of the History of Astronomy* **1** (1970), 116–138; and Schneider, *Isaac Newton* (ref. 4), pp. 60–61.

[22] François De Gandt, 'Le style mathématique des *Principia* de Newton,' *Revue d'histoire des sciences* **34** (1986), 195–222.

[23] D. T. Whiteside, 'The Prehistory of the *Principia* from 1664 to 1686,' *Notes and Records of the Royal Society of London* **45** (1991), 11–61, quot. on p. 11; emph. in orig.

[24] Lützen and Purkert, 'Conflicting Tendencies' (ref. 11), pp. 14–29; Karlheinz Haas, 'Hieronymus Georg Zeuthen (1839–1920),' *Dict. Sci. Biogr.* **14** (1976), 618–619; Johannes Hjelmslev, 'Hieronymus Georg Zeuthen,' *Matematisk Tidsskrift*, series A (1939), 1–10; and C. Juel, 'H. G. Zeuthen,' *Oversigt over det kongelige danske Videnskabernes Selskabs Forhandlinger*, June 1919–May 1920, pp. 66–77.

[25] H. G. Zeuthen, *Forelæsning over Mathematikens Historie. I: Oldtid og Middelalder. II: 16de og 17de Aarhundrede* (Copenhagen: Andr. Fred. Høst & Søns, 1893 and 1903), preface to vol. 1; cf. also Zeuthen, 'Ved Forelæggelsen af Mathematikens Historie i 16. og 17. Aarhundrede,' *Oversigt kong. danske Videnskabernes Selskabs Forh.: Videnskabelige Meddelelser* (1903), pp. 553–572.

[26] *Ibid.*, prefaces to both volumes.

[27] *Ibid.*, II p. 74.

[28] *Ibid.*, II pp. 93–96 and 563–564. The attention that Zeuthen paid to oppositions may have been inspired by his reading of Harald Høffding's history of philosophy. Cf. Gerald Holton, 'The Roots of Complementarity,' in Holton, *Thematic Origins of Scientific Thought: Kepler to Einstein* (Cambridge, Mass.: Harvard University Press, 1973), pp. 115–161; M. Norton Wise, 'How Do Sums Count? On the Cultural Origins of Statistical Causality,' in Lorenz Krüger, Lorraine J. Daston, and Michael Heidelberger, eds., *The Probabilistic Revolution. I: Ideas in History* (Cambridge, Mass.: MIT Press, 1987), pp. 395–425; and Jan Faye, 'The Bohr–Høffding Relationship Reconsidered,' *Studies in the History and Philosophy of Science* **19** (1988), 321–346. See furthermore Lützen and Purkert, 'Conflicting Tendencies' (ref. 11), pp. 16–17.

[29] Zeuthen, *Forelæsning* (ref. 26), II: pp. 96 and 101–102.

[30] *Ibid.*, II: pp. 103–106.

[31] Cantor, *Vorlesungen* (ref. 14), p. 199.

[32] Zeuthen, *Forelæsning* (ref. 26), II: pp. 537–541.

[33] *Ibid.*, II: pp. 80–82.

[34] Zeuthen, *Forelæsning* (ref. 26), II: pp. 303–308 and 583–585; cf. also Michael S. Mahoney, 'The Beginnings of Algebraic Thought in the Seventeenth Century,' in Stephen Gaukroger, ed., *Descartes: Philosophy, Mathematics and Physics* (Sussex: The Harvester Press, and New Jersey: Barnes and Nobles Books, 1980), pp. 141–155.

³⁵ *Ibid.*, II: pp. 303–308, quot. on p. 308.
³⁶ Zeuthen, 'Om den historiske Udvikling af Mathematiken som exakt Videnskab indtil Udgangen af det 18de Aarhundrede,' in *Indbydelsesskrift til Kjøbenhavns Universitets Aarsfest i Anledning af Hans Majestæt Kongens Fødselsdag den 8de April 1896* (Copenhagen: J. H. Schultz, 1896), pp. 1–90, quot. on pp. 1–2. I would like to thank Inger Jensen for having located this text in Copenhagen and made it accessible to me.
³⁷ Simon Schaffer, 'Scientific Discovery and the End of Natural Philosophy,' *Social Studies of Science* **16** (1986), 387–420; Timothy L. Alborn, ' "The End of Natural Philosophy" Revisited: Varieties of Scientific Discovery in Early Victorian England,' *Nuncius* **3** (1988), 227–250; and Richard Yeo, 'Genius, Method, and Morality: Images of Newton in Britain, 1760–1860,' *Science in Context* **2** (1988), 257–284.
³⁸ Cf. furthermore Eberhard Knobloch, 'Einfluß der Symbolik und des Formalismus auf die Entwicklung des mathematischen Denkens,' *Berichte zur Wissenschaftsgeschichte* **3** (1980), 77–94; and Knobloch, 'Symbolik und Formalismus im mathematischen Denken des 19. und beginnenden 20. Jahrhunderts,' in Joseph W. Dauben, ed., *Mathematical Perspectives: Essays on Mathematics and Its Historical Development* (New York: Academic Press, 1981), pp. 139–165.
³⁹ Cf. John L. Greenberg, 'Mathematical Physics in Eighteenth-Century France,' *Isis* **77** (1986), 59–78; and Michael S. Mahoney's essay review of the eighth volume of D. T. Whiteside's edition of the mathematical papers of Isaac Newton in *Isis* **75** (1984), 366–372.
⁴⁰ I am indebted for this observation to Res Jost.
⁴¹ Cf. Yakov M. Rabkin, 'Technological Innovation in Science: The Adoption of Infrared Spectroscopy by Chemists,' *Isis* **78** (1987), 31–54.
⁴² Zeuthen, *Forelæsning* (ref. 26), II: p. ii.
⁴³ Max Noether, 'Hieronymus Georg Zeuthen,' *Mathematische Annalen* **83** (1921), 1–23, quot. on p. 13. This article has an extensive bibliography of Zeuthen's writings.
⁴⁴ Zeuthen, *Forelæsning* (ref. 26), II: p. 305. He would reiterate this position later, see his 'Anvendelse af Regning og af Ræsonnement i Mathematiken,' *Oversigt kong. danske Videnskabernes Selskabs Forh.: Videnskabelige Meddelelser* (1914), pp. 271–286, esp. on pp. 278–280.
⁴⁵ Hermann Weyl, 'David Hilbert and His Mathematical Work (1944),' in Weyl, *Gesammelte Abhandlungen*, 4 vols. (Berlin: Springer, 1968), Vol. IV, pp. 130–172, quot. on p. 135; emph. in orig.
⁴⁶ As to the varieties of Platonism, cf. Paul Bernays, 'On Platonism in Mathematics (1935),' in Paul Benacerraf and Hilary Putnam, eds., *Philosophy of Mathematics: Selected Readings* (Cambridge: Cambridge University Press, 2nd ed. 1983 (1st ed. 1964)), pp. 258–271; and articles by John Fisher, John Charles Nelson and Ernst Moritz Manasse on Platonism in the *Dictionary of the History of Ideas* (1973).
⁴⁷ Erwin N. Hiebert, 'Boltzmann's Conception of Theory Construction: The Promotion of Pluralism, Provisionalism, and Pragmatic Realism,' in J. Hintikka, D. Gruender, and E. Agazzi, eds., *Probabilistic Thinking, Thermodynamics, and the Interaction of the History and Philosophy of Science* (Dordrecht: D. Reidel, 1981), Vol. II, pp. 175–198; cf. also Hiebert, 'The Scientist as Philosopher of Science,' *NTM-Schriftenr. Gesch. Naturwiss., Technik, Med.* **24** (1987), 7–17.
⁴⁸ Lützen and Purkert, 'Conflicting Tendencies' (ref. 11), pp. 30–41.

49 The article that sparked off the controversy was Sabetai Unguru, 'On the Need to Rewrite the History of Greek Mathematics,' *Arch. Hist. Exact Sci.* **15** (1975), 67–114. The ensuing debate is discussed in J. L. Berggren, 'History of Greek Mathematics: A Survey of Recent Research,' *Historia Mathematica* **11** (1984), 394–410, on pp. 397–398.

50 Thomas S. Kuhn, 'The Halt and the Blind: Philosophy and History of Science,' *The British Journal for the Philosophy of Science* **31** (1980), 181–192, quot. on p. 190.

51 Cf. Jed Z. Buchwald, 'Sir William Thomson (Baron Kelvin of Largs) (1824–1907),' *Dict. Sci. Biogr.* **13** (1976), 374–388, on pp. 374–376.

52 Schneider, *Isaac Newton* (ref. 4), pp. 142–144, quot. on p. 142.

53 *Ibid.*

54 Richard Feynman, *The Character of Physical Law* (Cambridge, Mass.: MIT Press, 1967), p. 53.

55 Cf. Norwood Russell Hanson, 'Are Wave Mechanics and Matrix Mechanics Equivalent Theories?' in Herbert Feigl and Grover Maxwell, eds., *Current Issues in the Philosophy of Science* (New York: Holt, Rinehart and Winston, 1961), pp. 401–425; cf. also E. L. Hill's commentary on pp. 425–428.

56 Thomas S. Kuhn, 'Afterword: Revisiting Planck,' in Kuhn, *Black-Body Theory and the Quantum Discontinuity, 1894–1912* (Chicago: The University of Chicago Press, 2nd ed. 1987 (orig. publ. 1978)), pp. 349–370, quot. on p. 364; cf. also Mario Biagioli, 'The Anthropology of Incommensurability,' *Stud. Hist. Phil. Sci.* **21** (1990), 183–209.

57 Paul Forman, 'Independence, Not Transcendence, for the Historian of Science,' *Isis* **82** (1991), 71–86.

58 Ian Hacking, 'Artificial Phenomena,' *The British Journal for the History of Science* **24** (1991), 235–241; cf. also Hacking, 'Language, Truth and Reason,' in Martin Hollis and Steven Lukes, eds., *Rationality and Relativism* (Cambridge, Mass.: MIT Press, 1982), pp. 48–66; and Hacking, 'Styles of Scientific Reasoning,' in John Rajchman and Cornel West, eds., *Post-Analytic Philosophy* (New York: Columbia University Press, 1985), pp. 145–165.

PART III

THE AIMS AND FOUNDATIONS OF PHYSICAL SCIENCE: THE CASES OF ELECTRICAL PHYSICS, PSYCHOPHYSICS, AND PHYSICAL CHEMISTRY

JED Z. BUCHWALD

THE TRAINING OF GERMAN RESEARCH PHYSICIST HEINRICH HERTZ

> Practical ideals – well – h'm – they don't appeal to me in the least. We have trade schools and technical schools and commercial schools springing up on every corner; the high schools and the classical education suddenly turn out to be all foolishness, and the whole world thinks of nothing but mines and factories and making money. [Old Johann Buddenbrook, 1835. From Thomas Mann, *Buddenbrooks*.]

1. HERTZ AND ELECTRIC WAVES

Just over a century ago a young German physicist of moderate though hardly overpowering reputation announced that he had successfully generated electric waves. In Germany and England replications of Heinrich Hertz's discovery rapidly followed; in France and Switzerland controversy over precisely what Hertz had found swirled for about five years. Scarcely a decade after the original finding, Hertz's laboratory devices were being rapidly transformed into technological apparatus as Oliver Lodge, Guglielmo Marconi and others concentrated on sending signals through, and extracting them from, the new world of the electromagnetic spectrum. By then Hertz was dead, having succumbed to septicemia in 1895 at the age of thirty-eight. He had however ceased experimenting nearly five years before, having turned his attention instead to abstract questions that had long bothered him concerning the foundations of mechanics, and indeed of all of physics.

Hertz's discovery was the first to have such a rapid and wide-ranging impact outside the confines of physics itself. Though followed in short order by Wilhelm Röntgen's discovery of X-rays, Hertz's generation and detection of electric waves permeated the popular press as well as the technical journals. In their fast transformation from laboratory artefact to technological appliance Hertz's oscillators and detectors had the kind of impact on social and economic relations that has today become a trite expectation of basic physical research but that was scarcely realized before the late nineteenth-century.[1] And, while undergoing technological mutation, Hertz's oscillators also became canonical devices for physical theory.

Despite the undoubted physical, technological, social and economic significance of Hertz's work it has received very little careful attention, excepting the usual (and certainly useful) memorial remarks by physicists and engineers. His apparatus has occasionally been discussed (and even reconstructed), but even here there has been little attempt to understand precisely how Hertz came to build these devices, how he learned to use them, or why it was *he* who succeeded where others (in particular British Maxwellians) thought failure was certain. And yet of all nineteenth-century German physicists Hertz is probably the most accessible because his diary and his letters home record, occasionally on a daily basis, how he became a competitive research physicist under the occasionally-irritating tutelage of Hermann (later von) Helmholtz. This essay sketches the early influences on Hertz to show how he became Helmholtz's reluctant golden boy, and it also points out some of the factors that impelled him by the mid-1880s to the frontiers of physical conception and experiment.[2]

2. THE MAGDALENENSTRASSE

A carefully-posed photograph taken in 1857 displays the infant Heins on his seated mother's lap. To Heins's left, standing but with left elbow casually leaning on the back of a heavy carved chair, Gustav Hertz, gaunt and broodingly handsome, stares intently, his right arm placed protectively behind wife and son. The mother, Anna Elisabeth, is plain and rather thin. Her long nose and broad cheeks are offset by lustrous dark hair. Heins's eyes stare quizzically from a round face. Portrait of a young bourgeois family in Hamburg during the 1850s in the midst of a sustained period of economic expansion and political retrenchment.

Gustav Hertz, an attorney, had married well and fittingly. Elisabeth, eight years his junior, was the daughter of a Frankfurt physician, Dr. Johann Pfefferkorn. Over time his career prospered, and the family moved from an apartment in the crowded Post-strasse to a solid three-story building on the wider Magdalenenstrasse, half-covered in ivy and with an impressive drive, where Heins grew up. Gustav's father was Jewish but his mother was not, and he was raised a Lutheran, though neither he nor Elisabeth were concerned to raise *their* children in a religious atmosphere. Formalities had nevertheless to be observed, and propriety respected, so Heins was confirmed, but the family engaged itself rather with the children's intellectual and moral character than with the well-being of their souls.³

Or, rather, Heins's mother concerned herself with such things. His father was for many years only an evening presence whose direct influence upon Heins and the younger children can perhaps be gauged from Frau Hertz's recollections:

> The greatest influence of all on the boys' development was their father. Not only could they feel most tender and affectionate care at all times, but because of his extensive knowledge and very good pedagogic disposition he became their teacher as well. At meals, *the only time during the week when my husband saw the children*, it was their interests that occupied us and our intimate chats made the hour dear to us.⁴

After he reached adolescence Heins, or Heinrich by then, became more interesting to his father, but until then he was in the hands of his mother, "the servants," and, at the age of six, the school-master. Anna Hertz did not however spend a great deal of time with the young Heins; an hour or so reading a book to him after dinner apparently sufficed.

Heins's early years were spent little differently from those of many other children of the North-German, Protestant bourgeoisie at mid-century. Home, play, servants, a concentrated moment of familial bustle at dinner, then a quiet hour with his mother over a picture book. Later there was music training, an essential for the properly cultured, and lessons in handicrafts (to elicit manual dexterity and to inculcate patience). Heins had no ear for music, but an excellent hand for building things, and he possessed what his mother termed "perfect memory." He was also enthusiastic and perhaps rather nervous, evidently requiring continual stimulation and absorption in activity.

At age six Heins was sent to Richard Lange's school where he remained for nine years. Lange had insisted that Heins not be prepared in any way, preferring no doubt to put his own impress on the unformed clay. That impress must have been rather forceful; at seventeen a worldly-wise Heinrich remarked:

> All of us, or at least the better scholars among us, were unusually fond of that school, despite the hard work and the great strictness. For we were ruled strictly; detentions, impositions, bad marks in neatness and behavior rained down upon us; but what particularly sweetened the strictness and profusion of work for us was, in my opinion, the lively spirit of competition that was kept alert in us and the conscientiousness of the teachers who never let merit go unrewarded nor error unpunished.⁵

Lange and his teachers were not alone in impressing discipline and a "spirit of competition" on Heins; his mother avidly participated. Anna Elisabeth, by then the mother of three boys, read and criticized everything, at least during the first year and a half. She was herself a strict and impatient taskmaster with a "quick temper," waiting avidly each Saturday for the weekly school reports. On the whole she gave the boys a conspicuous amount of freedom, but insisted on punctilious obedience.

Young Heins was at first rarely at the front of his class, according to Lange because of "slight fluctuations" in "diligence and neatness." But over time Lange and his mother damped Heins's fluctuations sufficiently that he became the acknowledged head of the class. This would not have endeared him to his fellows in a contemporary English public school, but the north-German bourgeois environment put a considerable premium on demonstrated and superior accomplishment. It was not enough to be adequate or even good; it was essential to be outstanding, to stand out from the crowd by virtue of dedicated application and natural ability. Still, the hot-house environment of a boy's school, even in mid-century Germany, tends not to reward classroom accomplishment with social esteem, something Heins no doubt realized:

"Well," [Heins] said, "the teacher also asked who was the brightest and most ingenious among us [students], and then they all pointed to me." "Oh, that must have made you happy?" "Yes," replied Heins, "but at the moment I like to have crawled *under* my desk."[6]

When he was about eleven Heins began also to attend the local *Gewehrbeschule*, or industrial high school, on Sundays where he learned drafting. From his earliest years Heins was absorbed to an unusual degree by modelling, drawing, constructing, in short by the manipulation, depiction and making of objects – "I often found a piece of his work," wrote Anna Elisabeth, "a mill that actually turned, a forge, country houses, and so on, and sometimes a pretty drawing." His parents encouraged Heins's infatuation with construction and depiction, to the extent that "on the advice of experts" they did not force him to follow the classical curriculum when the class at Lange's school divided. Little wonder that 'experts' had to be consulted in such a momentous decision. The class which he joined emphasized "arithmetic and the natural sciences;" the other emphasized Latin. Only the latter led to university.

Although the secondary-school system varied somewhat from principality to principality in pre-unification Germany, nevertheless since the early nineteenth century the division between practical and higher studies had been strongly maintained. During the 1810s Wilhelm von Humboldt, among others, had begun to forge a school system in Prussia that was based on the socio-cultural concept of *Bildung*. This powerful, yet slippery, notion had originally a strong democratic component, in that it was to guarantee individual freedom by infusing in students a concern for (and selecting among them on the basis of) an intellectual and aesthetic embrace of learning in its widest possible sense – but learning that determinedly excluded practical training.[7] Over time this neo-humanistic *Bildung* became equated with training in ancient languages, and in short order a social divide of nearly unbridgeable proportions developed between those trained in the classical *Gymnasia* and those trained in the practical *Realschule* and the technical *Gewehrbeschule*.[8] By the late nineteenth century, particularly after a period during which the boundaries between the schools were redrawn, the antagonism had become sufficiently intense that contemporary memoirs refer to insults and fist-fights between students from the different kinds of schools.[9] "The antipathy," notes J. C. Albisetti, "of the neo-humanists for anything even remotely tainted with practical training meant that the *Gymnasium* neglected the contribution that manual skills might make to *Bildung*; even drawing played at best a tertiary role in a *Gymnasium* education" – and the *Gymnasium* had a monopoly, through its *Abitur*, or leaving, examination on training for university education, which meant in particular on the training of all higher civil servants.

Hamburg was a commercial city, and despite his training as a lawyer, there can be little doubt that Heinrich's father Gustav must have had considerable ongoing contact with that practically-oriented world, and he and Elisabeth may have decided that Heinrich need not follow the usual classical path. This was not by that time quite so limiting a decision as it had been a decade earlier (or in Gustav's own youth), because after 1870 graduates of first-class *Realschulen*, which had greatly expanded following new regulations in 1859, could matriculate in university philosophical faculties and could teach mathematics, the sciences and modern languages at *Realschulen* (but not at *Gymnasia*).[10] There was a great uproar over this, but it perhaps eased the way for the Hertzs to permit Heins's not immediately following a classical curriculum.

Heins however came soon to feel that the classical students were learning more than he was, perhaps because they were learning things that he had never been introduced to formally, whereas they at least had learned something of "arithmetic and the natural sciences" before the class had been divided. There is moreover little doubt that he encountered some of the disdain which incipient *Gebildete* directed at the "practical hack."[11] It was in any case hardly likely that Hertz's parents did not intend him eventually to obtain the *Abitur*, even if Heins did become an engineer (as he intended), since with it his military service would be reduced to one year in the officer's corps. Gustav was by this time beginning to take a deeper interest in him and arranged for private classes with a Dr. Köstlin, with the result that Heins would have the advantage of both kinds of training from his twelfth through his fifteenth year.

The next two years were critical ones for Heins's development. He did not attend school but continued his studies with Köstlin and began to take mathematics with a Herr F. Schlottke, in addition to persevering at the *Gewehrbeschule* and attending a daily exercise class. Most of the day he studied at home. His mother paints a vivid portrait of this period:

When he sat with his books nothing could disturb him nor draw him away from them. His desk stood in a room through which I often had to pass, but I always saw him bent over his books in the same way, deep in work. We never exchanged a word. At 12:30 we both had lunch with our little Otto [his youngest brother]. Half an hour's play with the little boy was his only relaxation. Then he studied afresh, mostly until dinner time at 5 o'clock.[12]

During these years Heinrich's father displaced his mother as the major influence on him:

... he found full support and inspiration in his father, who followed Heinrich's studies with great interest. Their lively conversations gave me great pleasure. Unfortunately I understood very little myself. I remember one evening he wanted to tell me about his mathematical studies and I exclaimed, "Oh Heins, I am too stupid for that." He put his arms around me tenderly and said with deepest feeling. "Poor Mama, that you have to miss this pleasure!"[13]

In the evenings after dinner he worked at a turner's lathe, forgoing all study. As usual he "received instruction", this time from a "master turner" named Schultz. And he soon began to produce "various types of physical apparatus, cutting each brass screw, pouring the little weights, and making all the essential parts by himself, with incredible patience."

The separation we see here between mathematical study and physical manipulation endured for over a decade. It strongly distinguished Hertz from other, later German physicists, who seated physics in mathematical abstraction, and throughout Hertz's life powerfully molded his approach to physical theory. Very much unlike his contemporaries in the classical *Gymnasia*, Hertz passionately embraced a universe of devices. His mother recalled a significant incident:

> He tried to make a spectroscope. His father had promised him the prisms for it and had written about it to Herr Schroeder, who was famous for his optical glass. The reply was that Heinrich should come to see him on Sunday, he would be in his office until 12 o'clock and would fix the prisms for him. Heinrich's joys and hopes were great. He worked on his apparatus with fiery zeal, got up at 5 o'clock on Sunday, hardly permitted himself time for breakfast in order to be ready, and then set out with his father. But because they had apparently miscalculated the time required, they did not get there until five minutes past 12 o'clock and found the place closed and a sign saying that Herr Schroeder would be away for several weeks. Heins came home inconsolable, and it cut me to the quick to see him crying in his quiet way, one big tear after another rolling down his cheeks.[14]

For many years Hertz continued to separate his passion from his interest, mathematics.[15] This is quite apparent in his earliest publications, where the mathematical structure remains subordinate to the demands of experiment and is indeed chosen precisely to fit the necessities of the laboratory. Had Hertz been able – as he was not – to remain enmeshed in the laboratory after his training under Helmholtz then he might never have merged passion with interest, in which case the interpenetration of theory with laboratory practice that more than any other single thing characterizes his mature work would never have evolved. Conversely, the early separation between the universes of mathematics and of devices was itself essential for Hertz to achieve his rooted understanding of laboratory electrodynamics untroubled by the deeper problems of contemporary theory.

When he was seventeen Heinrich returned to the classroom for one year and excelled in the classics curriculum which he had striven privately for two years to master. The school was difficult because a new director (Hoche), seemingly determined to make his mark, failed nearly half the class. Heinrich did well, but despite his mastery and even enthusiasm for languages[16] he did not go on to university. Instead he decided at first to pursue his original goal – a career in engineering – by apprenticing to a Prussian architect in Frankfurt am Main. For, having satisfied his need to prove himself the master of classics as well as mathematics and science, Heinrich remained enthralled by the concrete.

The tension we see here between, as it were, the practical (engineering) and the abstract (mathematics) was particularly sharp among the mid-century German bourgeois. On the one hand there was the continuing pull of the traditional classical curriculum, of the lofty old cultural concept of *Bildung*, which emphasized the superiority of unsullied contemplation.[17] But, as old Johann Buddenbrook's remark quoted above suggests, by mid-century this was rather heavily counterbalanced by a practical, though still highly formal, alternative. The resulting stress between *Bildung* and the concrete betrayed itself in the *Gymnasium* and in the university. Hertz's father, Gustav, the successful attorney (and later senator) was no doubt educated in a thoroughly traditional manner. However he evidently regretted the rigorous discipline of his youth,[18] which he perhaps associated with the old educational system, and indeed Gustav encouraged Heinrich's practical bent. Accordingly Heinrich, having demonstrated that he could master the classics (and, more importantly, having obtained the requisite credentials) remained committed to a practical career and decided to apprentice himself for a time to architecture. "Only," he wrote at the time, "if I were to prove unsuited for the profession or if my interest in the natural sciences were to increase further, would I devote myself to pure science."

A boy who, like Hertz, had a strong passion for building and using devices could not possibly have found a comfortable home in the contemporary *Gymnasium* for any length of time. The classical curriculum, which had evolved over the course of the century to embrace a disdain not only for natural science but also for contemporary history, had no place for someone like Hertz, who adored monkeying with apparatus. However, until 1870 the *Gymnasium* was also the only route to the philosophical faculty at the universities, and hence to a career in natural science. Consequently the division between the *Gebildete* and the *Ungebildete* also embraced, until that date, the division between the scientist and the engineer. Once, however, the first class, or semi-classical, *Realschulen* could also send students to philosophy then the barriers, both formal and social, were no longer quite so firm. It was consequently possible for Heinrich to conceive that starting out in engineering did not forbid his turning to physics, provided only that he obtain an appropriate *Abitur*, which he did. Moreover, the career in engineering that Heinrich initially followed in any case required similar credentials. A career in science would certainly have been more prestigious socially than one in engineering at this time, because the natural scientist was still a member of the philosophical faculty at the university and admirably *Gebildete*, but it was also more risky because the chances of success (i.e. of becoming an Ordinarius at a prestigious university) were not large.[19]

3. CIVIL ARCHITECTURE, THE MILITARY AND MUNICH

Hamburg
home and hearth
1856–1875

Frankfurt am Main
civil architecture and boredom
spring, 1875–spring, 1876

Dresden
the Polytechnic: engineering, mathematics, Darwin and Kant
spring, 1876–fall, 1876

Imperial Berlin
the military, discipline and more boredom
fall, 1876–October, 1877

Munich
October, 1877–November, 1878: *Polytechnic*
November, 1878–August, 1879: *the University and physics*

Helmholtz's Berlin
September, 1879

The young Hertz's Wanderjahren

Heinrich arrived in the spring of 1875 and went immediately to work under the Baurat Behnke on redrafting designs for public works buildings. Since his mother came from the town, and her brother-in-law (Emil von Oven) was a local prominence, Heinrich had scarcely left the nest altogether. From the outset, however, he felt restless and occasionally irritable. The office work neither challenged nor interested him, and in any case filled at most six hours a day: from nine in the morning until noon, and then from three until six. He tried to occupy his luncheon time with reading but found it hard to find an appropriate place to do so. "How I could make some good use of my day", he wrote home, "even with only 6 working hours, I do not know." His parents perhaps sensed the homesick boy's ill ease and visited him in May, which revived his spirits. He plunged into a campaign to master whole new areas of learning and craftsmanship, and to pursue old ones further. Sculpture, Greek literature and philosophy, and architectural history were gobbled in turn. But nothing satisfied him. "If only I were educationally or usefully occupied here I would just as soon stay" he complained in September.

But then he rediscovered the allure of physics. In his increasingly desperate search for something interesting and satisfying to do Heinrich tried to join the local Physics Club, which was run by Professor Rudolph Boettger, who was an able producer of demonstration experiments.[20] At first he had trouble contacting Boettger but nevertheless plunged ahead with his usual enthusiasm, reading Adolf Wüllner's *Physics* (a widely-used elementary text of the day filled with diagrams of elaborate instruments) and the German translation of John Tyndall's *Heat as a Mode of Motion*.[21] "I have regained a strong inclination for natural science through reading Wüllner," he wrote in late October, "but I cannot convince myself to give up what I have set as the most desirable goal for myself." Boettger's public lectures, with elaborate demonstrations, enthused him: "My outlook, my thoughts of the future, change with every day. Since I read a great deal in Wüllner's *Physics* I am again turning very much towards natural science." The signs must have been unmistakable to Gustav and Anna over the Christmas holidays, but Heinrich did not in any case have long to spend in Frankfurt. In the early spring of 1876 he left for Dresden to study engineering at the local Polytechnic. Here at least Heinrich could grapple directly with science, but he remained lonely and homesick. For the most part, and apparently for the first time, mathematics captivated him – "sometimes marvellous things come in view that make one's head swim" – and his expertise in it grew rapidly. He also read Immanuel Kant's *Critique of Pure Reason* and studied 'Darwinism.' Then, in September, Heinrich began his compulsory year in Berlin serving in the military of Chancellor Bismarck's five year-old German empire.

Hertz found military discipline a good antidote to laziness, but he was soon bored – "each day and each hour is like every other, and every day that has passed is regarded as a day conquered." By the end of February he was brooding over his future, gloomily writing home that "day by day I grow more aware of how useless I remain in this world." His mood fluctuated, but he was in good form by summertime. In October he left Berlin for Munich, where he intended to study at the Polytechnic to further the formal engineering background that he had begun in Dresden. His determination did not last long. By the end of the month he had decided that engineering was not for him, that he was gripped by the natural sciences, and therefore that he had to enrol instead at the University. Such a major decision naturally required his parents' consent, and on November 1 he wrote them a long and touching letter.

I cannot understand why I did not realize it before now, for even in coming here it was with the best intention of studying mathematics and the natural sciences and with no thought at all about surveying, building construction, builders' materials, etc., which were supposed to be my main subjects. I would rather be an important scientist than an important engineer, but rather an unimportant engineer than an unimportant scientist; yet now, as I stand on the brink, I think that what Schiller said is also true: "And if you don't dare to stake your life, you can never hope to win the strife," and that too much caution would be folly.22

His father's permission arrived on November 7, and Hertz decided to "stake his life" on physics.

He at once contacted the Munich professor of physics, Philip von Jolly, a well-known experimentalist who urged mathematics, mechanics and home study on him. Hertz took this advice to heart, including Jolly's recommendation to concentrate on the "old sources," and he spent the next four months intensely engaged with Joseph-Louis Lagrange ("or sometimes another mathematics text, for Lagrange is dreadfully abstract") and with Jean Étienne Montucla's ancient and lengthy history of mathematics. He found little useful in the "new mathematics" (including non-Euclidean geometry) for the physicist "for I find it so abstract, at least in parts, that it no longer has anything in common with reality." Already he itched for the laboratory though he was not even studying experimental physics: "ideas for a hundred experiments I should like to undertake now occur to me, and I already look forward with pleasure to the distant homecoming in the long vacation, when the turner's lather will really turn again." His "main wish" was for a copy of Wüllner's *Physics*, the text he had first examined in Dresden. Wüllner's work was filled with diagrams of intricate experimental devices and careful, though not deeply theoretical, explanations of how they work. But Hertz immersed himself for the present in Lagrange and Montucla, occasionally turning for relief to "a significantly easier mechanics text by Poisson." He read slowly and carefully, trying to reason out for himself the implicit meanings of physical and mathematical concepts and of their relationships with one another – "I need much time to ponder over matters myself, and particularly the principles of mechanics (as the very words: Force, time, motion indicate) can occupy one sufficiently; likewise, in mathematics, the meaning of imaginary quantities, of the infinitesimally small and infinitely large and similar matters."

In the next semester, at the beginning of May, laboratory courses began. Hertz joined the group that "had never done it before" and had to learn at an elementary level how to work a balance and so on – he had to learn what sort of a place the laboratory was, and how people generated meaningful knowledge in it. He progressed fast and buried himself in it, working in Jolly's laboratory at the University for six hours a week and in Wilhelm Beetz's at the Polytechnic for eight hours. "I start at 7 o'clock in the morning," he wrote, "and when I return from work at 6, I am quite tired, especially my eyes, so that I cannot do very much more, and almost the whole afternoon is spent." Professor Wilhelm von Bezold (director of the University's physics institute) tried to restrain Hertz's obvious propensity to submerge himself in the laboratory, urging him "not to turn to physics too early."

By the end of July Hertz had decided not to stay at Munich. He does not tell us why, but he probably felt that the courses in experimental physics available there were not sufficiently powerful.[23] He had a long conference with Beetz at the Polytechnic about where to go, and he must have emphasized his overwhelming desire to find laboratory work since Beetz (exaggerating the reality[24]) told him that he "would find a laboratory and a chance to work in it anywhere." Beetz offered introductions, and Hertz eventually decided on Berlin, or, more precisely, on Helmholtz's renowned laboratory, which was not at the time a typical place for doing physics. Its unique characteristics had a powerful, lasting impact on the young, eager and competitive Hertz, and not only on him. German physics as a discipline underwent important transformations at the hands of Helmholtz and those he trained or influenced during the 1880s, and so before continuing our discussion of Hertz's career we shall pause briefly to consider the laboratory-based science that Helmholtz had begun to construct in 1880s Berlin.[25]

4. THE YOUNG PROFESSIONAL IN HELMHOLTZ'S BERLIN

> Of the many students now scattered over the earth there is not one who will not to-day think of his master with love as well as admiration. [Hertz on Helmholtz in 1891]

a. *Two Kinds of German Laboratories*

In the summer of 1874 Arthur Schuster, having just obtained his Ph.D. in Gustav Kirchhoff's laboratory at Heidelberg, visited Göttingen for a few months, where the 70 year-old Wilhelm Weber had just handed over his laboratory to Eduard Riecke.[26] "There was," Schuster recalled, "only one student at work beside myself; for the purposes of his doctor dissertation he was magnetising ellipsoids and testing magnetic formulae in the orthodox fashion." The laboratory that Riecke had inherited from Weber was for measuring constants and for testing properties. At the end of the summer Schuster travelled on to Berlin, where he found "a laboratory of very different character and ambitions," a place where "promising students from all parts of Germany," crammed into a few small rooms, "were preparing their doctor dissertation on some subject arising directly out of Helmholtz's work." Here, Schuster felt, exciting new work was being done, work that went far beyond mere measurement and testing. In Helmholtz's laboratory "no efforts were made to push numerical measurement beyond its legitimate limits, and though most of the work done was quantitative in character, qualitative experiments were not discouraged."

By the early 1870s laboratories concerned with physical research were quite common throughout Germany, but very few among them were either well-housed or well-funded, and physics as a discipline still had to make claims on resources by emphasizing its usefulness to students in such areas as medicine, pharmacy and the natural sciences.[27] Despite the poverty of material foundations, and the ever-present demands of often elementary teaching, the "research ethos" was widespread by this time, particularly in laboratories whose directors had the energy and forcefulness to fight for resources, and the craftiness to play one university off against another.[28] Schuster's vivid contrast of Göttingen with Berlin reveals that considerably different kinds of work went on within the overall ambit of German laboratory research.

In Weber's and Friedrich Kohlrausch's laboratory at Göttingen, as well as in Kohlrausch's influential later establishments, *measurement* was the goal, the elemental purpose of research.[29] Even when Kohlrausch did eventually formulate a new hypothesis (concerning electrolytic conductivity), it was precisely tailored to the demands of measurement.[30] He scarcely used the laboratory as an engine for discovery; he used it as a generator for constants. Helmholtz's "two or three rooms" in early '70s Berlin had a different ethos. There measurement *per se* engaged little interest.[31] One doubts, for example, that Helmholtz would have looked with great interest on a project to pursue deviations in the accuracy of Ohm's law, though Schuster had done just that at Göttingen, where Weber had "entered with great spirit into the question."[32]

The division between Berlin-trained or associated physicists and those who took their guide from the Weberean ethos became acute by the early 1880s. Max Planck, for example, who took over from Hertz at Kiel in 1885, and who had been much impressed by Helmholtz and Kirchhoff during the year he spent as a student in Berlin,[33] was bruised by it in 1887, though he himself had nothing directly to do with the laboratory. Planck, like Hertz, strongly felt the competitive pressures of the discipline, but because he had turned exclusively to theoretical physics the difficulty of establishing a reputation, and of getting a position, was particularly sharp. "All the more compellingly," he later wrote, "grew in me the desire to win, somehow, a reputation in the field of science."[34] This led him to compete for an 1887 prize offered by the Philosophical Faculty of Göttingen on "The Nature of Energy." Though he won second prize, no first was given, and the reason he had failed of complete success, Planck noted, was to be found in a single sentence of reservation in the judges' decision, to wit that ". . . the Faculty must withhold its approval from the remarks in which the author tries to appraise Weber's Law." Planck continues:

Now, the story behind these remarks was: W. Weber was the Professor of Physics in Göttingen, between whom and Helmholtz there existed at the time a vigorous scientific controversy, in which I had expressly sided with the latter. I think that I make no mistake in considering this circumstance to have been the main reason for the decision of the Faculty of Göttingen to withhold the first prize from me. *But while with my attitude I had incurred the displeasure of the scholars at Göttingen, it gained me the benevolent attention of those of Berlin, the result of which I was soon to feel.*[35]

The 'results' Planck referred to were quite concrete indeed: in 1889 he was invited to replace the recently-deceased Kirchhoff at Berlin as an

Extraordinarius, or associate professor, and in only three years he became an Ordinarius, or full professor. According to his own testimony Planck's early work on the foundations of thermodynamics before the late 1880s had aroused little interest,[36] so that his call to Berlin almost certainly reflected Helmholtz's conviction that Planck was a powerful ally in an important endeavor.

Hertz began attending courses at Berlin in October, 1878, the fall after Helmholtz's institute moved to spacious new quarters which included lecture halls, laboratories designed for specialist work, and a library.[37] Under construction since 1873, the new physics institute was and remained the costliest in Germany until well after the turn of the century,[38] and its facilities were very nicely adapted for specialized work.[39] At first Hertz did not intend to sign up for the laboratory, but then he discovered that "one of this year's problems more or less falls into my field," to wit electrodynamic measurement.[40] From the very beginning, then, Hertz was captured by electromagnetism, and almost immediately Helmholtz took an interest in him. "I have already talked about it [the prize problem] with Prof. Helmholtz." Hertz wrote his parents, and "he has kindly given me some information about the literature." Helmholtz of course included his own articles among "the literature," and Hertz began by "spending most of [his] time in the reading rooms."

The time was well spent, though Hertz was a quick study, for by November he was busy arranging things in the laboratory. The prize question concerned a subject of great interest to Helmholtz, namely whether or not the electric current possesses inertia: for on his understanding it should not, whereas on Weber's electrodynamics it must.[41] Hertz, we shall see, was already well aware of the highly competitive environment he had joined, and he was at first reluctant fully to commit himself to the project "since I may fail." Helmholtz encouraged him: "I reported to Prof. Helmholtz yesterday that I had thought over the matter up to a point and would like to begin. He went with me to see the demonstrator and was kind enough to spend another 20 minutes in discussion as to how best to begin and what instruments I should need." On that day, November 5, Helmholtz began to mold Hertz into a *Helmholtzian*, a process that developed extremely rapidly as Hertz tackled the prize problem, and that was, if not complete, then thoroughly formed by early February, when Hertz decided that he had finished.

The close link that early formed between Helmholtz, the renowned and powerful *doyen* of German physical science, and Hertz, the young, eager and competitive aspirant to status, was indubitably furthered by the culture of research that uniquely characterized German universities. But, perhaps more importantly even than the emphasis upon research, the peculiar *guild-like* training that characterized the German seminar conduced to developing driven scholars of the kind that Hertz soon became. The research seminar was presided over by a knowledgeable Ordinarius, but it was not based on learned authority; it was rather based on skilled authority. A recent history of German universities puts the point particularly well. "The successful seminar participant," C. E. McClelland remarks, "would end by surpassing the authority of the professor at least in a limited sector of his expertise, rather than merely absorbing the authoritative information provided by the professor. Yet the student retained deference to the residual, traditional authority of the teacher, who remained a valued critic and, in a particular sense, a professional guide and patron."[42] Though in a strict sense Hertz did not attend a seminar, nevertheless the cultural environment that Helmholtz built at his Berlin laboratory was quite similar to one. His relationship with Hertz perfectly reflects the inevitable tensions between the skilled master who imparts technique to the young apprentice, hoping that he will in the end create something outside the master's own ken, and the apprentice, who admires and fears the master, wishing to acquire his skill and patronage, and wishing eventually to set himself apart. Such an intense cultural and social structure strongly and enduringly impressed the neophyte Hertz, who made its values and goals his own.

The sort of physics that Helmholtz had striven for the past decade to realize in his Berlin laboratory, and the cultural dispositions that went along with it and that deeply influenced Hertz, was thoroughly wedded to instrumentalities, to the controlled manipulation of objects. Although *Helmholtzian physics* was a complicated structure that can be thoroughly grasped only through a myriad of examples, its guiding ethos – never explicitly stated – was extremely simple.[43] From the late 1860s (and perhaps with roots dating to ideological issues that first became pressing in the 1840s) Helmholtz had sought to purge physics of the *a priori* and to embed it in the laboratory. To do so he envisioned a world in which objects relate directly to other objects without anything intervening between them. This meant, to take a typical example, that charged conductors interact with other charged conductors *simply as charged conductors* – they do not interact with one another because, e.g., they contain hidden things (such as electric particles) that do the interacting and with which they in turn interact. For Helmholtz and the group of apprentices and students that he trained in the '70s and early '80s, physics had to be built directly on unmediated links between irreducible objects – and these objects had to be the very things that are manipulated in the laboratory: they were such things as metal wires, iron cores, sulphur plates, whatever fills a Geissler tube, and so on. There is no room in this vision of physics for creating abstract structures out of which instrumentalities are themselves built, nor is there room for considering the world to be a pale reflection of an ideal mathematical realm. There is the laboratory; there are the objects that furnish it. All else is fantasy.

Helmholtz's vision carried with it a galaxy of implications for how to do physics, for how to think about what being a physicist meant (and, naturally, for how to judge other physicists). That vision contrasts strikingly with another one that became increasingly important in the early 1890s. Lewis Pyenson[44] reveals how apostles of mathematics gained control of teaching, particularly in the *Realanstalten*, where most German physicists *circa* 1914 were educated, thereby providing a fertile ground for the flowering in Germany of an intensely mathematical physics of a kind that Hertz, and many people who passed through Helmholtz's molding hands in Berlin, could never have countenanced. For them mathematics remained always and entirely a useful tool for understanding laboratory-based physical relations. Hertz's and Helmholtz's mathematics had to fit physics; for pure mathematicians like Felix Klein, whose views were immensely influential in the early 1890s, it was very nearly the other way around.

b. *Learning How to Work in the Laboratory*

During the fall Hertz settled into a routine: "an interesting lecture every morning: then I go to the laboratory, where I stay until 4, with a short break; afterwards I work at home or in the reading room; till now I have been kept busy gathering material on extra currents."[45] As well as ingesting how to do laboratory work Hertz also began to understand that he had entered an extremely competitive profession, which both stimulated and frightened him. He wrote:

> ... when I went to sign up with Prof. Borchardt [for his lectures on analytical dynamics], I took the opportunity of looking into the several registration books of the other students and I was really shocked by their zeal and stamina. Most of them had signed up for 2 hours Weierstrass, elliptical functions, 2 hours Kummer, number theory, etc., one thing after the other, as if it were nothing, so that I was properly ashamed of myself with my two courses [the other being Kirchhoff's on electricity and magnetism]. Then when I took my seat, I could see my neighbor's hand and he too had signed up for the most difficult branches of mathematics every day from 8–12 and in the afternoon as well. When Prof. Borchardt opened his lectures with the words, "Mathematics has been called, more in jest than in truth, the science of what is self-evident," a pen started writing rapidly near me, and as I turn around I see that he takes down every word in shorthand and is already on the second sentence. To my horror, he went on like this the whole two hours. He looked quite healthy. Even though this zeal is, perhaps, just a little exaggerated, it is true that people work tremendously hard here; and that it is more or else contagious and that I already have some qualms about whether I should not have taken more courses. Anyway this spirit pleases me very much, when it is not exaggerated; it is a comfort that there is no need to apologize when one has some work to do.[46]

During these early months Hertz, who already had a good measure of competitiveness – not, as we see from his remarks, unusual for German students of the time – learned that he would have to work very hard indeed to make a career in physics. The rapid series of publications that he produced during the next five years show how thoroughly he assimilated this intense research ethos.

Hertz was awarded the prize in early August, and this event marked his proper entry into the profession:

> ... The evaluations will be printed and you will be able to read them. Another paper was entered on the same subject, but it had no luck and received such a devastating judgment that it was greeted with general merriment. The author must have been possessed of great innocence when he sent it in.[47]

He had lost his own 'innocence' sometime the previous fall, and the prize award stamped him in his and his colleagues' eyes as a promising new force in the profession. Already somewhat imbued with the sentiments of the German university student, Hertz's early success powerfully embedded in him the profession's elite, competitive ethos. That, we shall see, undoubtedly stimulated him to pursue laboratory discovery, but it also exacerbated his tendency to self-doubt and melancholia, a rather common one among contemporary German professionals.

With success came pressure. Helmholtz's eye was focused very closely on Hertz. The first prize competition tested Hertz's competence in the laboratory, as well as his ability to deploy Helmholtzian methods and concepts. Having brilliantly demonstrated his competence, Hertz could now be trusted to do much more, to probe issues that had a direct bearing on Helmholtz's own electrodynamics. Early in July the Berlin Academy proposed a new prize of exactly this kind. Helmholtz pressed Hertz to compete, only this time the project would take quite some time, and Helmholtz evidently wanted it undertaken in secrecy:

... when I had waited for Prof. Helmholtz in his laboratory last week to thank him [for his help in the first competition], he was exceptionally friendly and congratulated me repeatedly and suggested a new project to me, one that would however occupy me completely for two or three years, and that I would not undertake if it were not for the particular honor of Helmholtz's invitation and the way he offered it, and by which he promised me his continued support and interest.[48]

Hertz had in the meantime written up his prize paper and given it to Helmholtz to read. But by this time the master was much more interested in convincing Hertz to begin the investigation for which the first one had merely been a testing ground. The apprentice became distressed and a bit testy, for he did not wish to devote such a long time at such an early stage of his career to a project with uncertain outcome. If the results, like his first ones, were again negative then this would hardly advance his reputation much further. It is worthwhile quoting in full the letter he wrote home expressing his anxiety, for it dates the beginning of Hertz's attempt to distance himself a small amount from the master who was seeking to bind him even closer:

I am working in the laboratory again now [November 4], but not on the [new] prize problem of the academy. When I first went to see Helmholtz, I showed him my paper and asked him to look it over, which he promised to do; but when I came back a few days later, he said that he had not got around to it yet. Since he has not returned it to me, nor otherwise referred to it, he must simply have put it aside. So I could have spared myself the labor of copying it. Otherwise he was very kind and pointed out which experiments he thought practicable [for the new prize questions]. However, since these experiments [the ones Helmholtz himself proposed] required too much equipment, and since I had no chance to talk with him about simpler experiments that I had in mind, I said that I should at least like to think the matter over first and would rather undertake something else. I could have wished that he had taken a closer look at my paper, if only because of the calculations that I had carried out in addition and that take up most of the space.[49] As far as the immediate purpose of this work is concerned, as preparation for experiments which I have postponed, I am very glad of the way it all turned out; for I was properly afraid that Helmholtz would approve the experiments and would thus force me to go on with them. The thought of working in *secret* for three years was a nightmare. If these experiments can be performed easily, then they will not require three years, and there will still be time to turn to them.[50]

Hertz succeeded in deflecting Helmholtz's pressure and turned, perhaps in reaction, away from the laboratory, to work on a "rather theoretical subject" for his doctoral dissertation – the currents induced in conducting spheres that rotate under magnetic influence. Although the subject was an intricate and difficult one, requiring much clever approximation (at which Hertz was becoming expert, having honed his talents on the first prize competition), he took care to connect his results very firmly to laboratory measures. The techniques he learned here were turned to good measure years later in his computation of the field of a radiating dipole. Of more immediate significance, the investigation suggested questions with important Helmholtzian meaning.

Hertz progressed with extraordinary rapidity – so fast that he had to request special permission to be examined, which he was on the evening of February 5. This experience, like the prize judging the previous August, again brought out the intensely competitive nature of the profession: "doctorates in my class [*magna cum laude*]," he wrote home the next day, "from this university are very few in number, especially Helmholtz and Kirchhoff are said not to have awarded many." He lamented that "the fear of a bad outcome takes away the pleasure of a good one." His intense immersion in "theoretical work" sated his appetite for it, and by the end of February he was back in the laboratory "from 9 in the morning until 9 in the evening," but not working on Helmholtz's problems.

During the spring Hertz rummaged around for a project, and though he began one he was unhappy by mid-summer. "Staying on here is becoming very disagreeable; I regret to say I am suffering from weariness and distaste for work in general and my specialized work in particular, which would disconcert me if I had not the experience of last year and two years ago, when I felt the same at this season but got everything going again in the autumn." What truly bothered Hertz was that he did not have a position of any kind. This changed when, on August 8, Helmholtz offered him a two-year, renewable assistantship at the Institute vacated by Heinrich Kayser. When he accepted Hertz became a fledgling physicist.

By early October Hertz had not yet had time to begin his own laboratory work, but he was being pulled into Helmholtz's social and professional circle, dining with the family and attending Physical Society meetings. "I grow increasingly aware," he wrote home, "and in more ways then expected, that I am at the center of my own field; and whether it be folly or wisdom, it is a very pleasant feeling." But he still kept a wary eye on Helmholtz, whose slow speech now irritated him, and whose opinion Hertz "did not care to put forth" his own against. Hertz was rather oppressed by his mundane duties as a laboratory teacher, but he wanted very much to please Helmholtz, and to have "confidence in [his] accomplishments vis-à-vis" the master.

During the early months of his tenure in Helmholtz's laboratory Hertz became increasingly familiar with apparatus, but he also became taken with theory for a time. "All these papers," he wrote home about his research, "are theoretical ones; I find it almost impossible to do experimental research when I have something else in mind, and I have resigned myself to regarding the beautiful laboratory as a luxury for the moment." During this period Hertz deepened his theoretical insight into two areas.[51] First, he now began to grasp the subtleties of Helmholtzian electrodynamics itself.[52] Second, he struck off in a new direction by applying elasticity theory (in the form he learned from Kirchhoff) to the compression by contact of two isotropic bodies.[53] The first work contains many important clues to Hertz's Helmholtzianism. Hertz's novel application of elasticity theory was immediately influential and yielded as well a new way to define the hardness of a body.[54] Here, in pursuing elasticity, Hertz for the first time had to deal professionally with someone other than Helmholtz at the lofty level of professor, namely Kirchhoff. And Kirchhoff was apparently not undividedly impressed.

Hertz had not actually dealt directly with Kirchhoff because he had sent the paper to Leopold Kronecker for inclusion in *Borchardt's Journal*, where it eventually did appear. This had been in late January or early February, and when he did not receive an answer by late April he went to see Kronecker, which must have seemed rather forward to the distinguished mathematician. Hertz described what had happened:

. . . he said the delay arose because the paper had been sent to Prof. Kirchhoff for review; the latter was very interested in it, and they would be glad to take it, but Prof. Kirchhoff had some criticisms of the form of the paper, and since it would be several months before it was set up in print, he would like to let me have the paper back for now, so that I could improve it in the meantime. He then showed me how Kirchhoff had thoroughly annotated the paper and had rewritten three or four pages in another form in the margins. At first I was surprised and even flattered that Kirchhoff had gone over it so thoroughly, but apart from a wrong sign that I had indeed overlooked, his comments seemed only to say the same thing (and by no means better) that was in the paper. In part the points were expressed in a manner peculiar to Kirchhoff which I do not like at all, and which I should be very unwilling to have imposed on me.[55]

So, having bothered Kronecker, Hertz decided to confront Kirchhoff directly. Kirchhoff, he wrote home, was also very friendly, but (according to Hertz) persisted in thinking that the paper contained critical errors "and seemed to believe that [Hertz] had reached the right results so to speak by accident". Hertz dug in his feet and within a few weeks he had forced Kirchhoff to back down:

> ... in looking over my paper, I found that Prof. Kirchhoff himself had made the main error with which he had reproached me (what I had written was merely not quite clear) *and I have demonstrated it to him.*[56]

Having already deflected Helmholtz's pressure to perform secretive experiments, Hertz had now also forced Kirchhoff, the master of mechanics, to confess error. Little wonder that he flourished in the highly competitive atmosphere of the physics community, rejoicing in "testing his strength" against that of his colleagues. For the next two years Hertz learned how to make his way in the profession as his understanding of contemporary physics – particularly in its Helmholtzian form – gradually deepened.

Institute for the History and Philosophy of Science and Technology
University of Toronto

NOTES

[1] One might object that electro-technology, particularly power transmission and telephony-telegraphy, were well advanced by the 1880s, and that they also constitute the sort of applied physics that Hertz's work mutated into in others' hands. However none of these three areas emerged rapidly after, *and directly out of*, research that was widely regarded as fundamental. Moreover the scale of the popular reaction to Hertz's discovery is simply out of all proportion to that of anything previous.

[2] I will, however, not here consider theory or experiment at all, nor will I discuss in great detail the *Helmholtzian* system of belief that settled like a mist over most German investigations in the 1880s and 1890s. The full story will appear in my next book, *The Creation of Scientific Effects, Heinrich Hertz and the Discovery of Electric Waves.*

[3] This was certainly not unusual for a bourgeois family in mid-century Hamburg: see, e.g., H. Holborn, *A History of Modern Germany 1648–1840* (New York: Alfred A. Knopf, 1964), pp. 493–4, who remarks that often "these *Kulturprotestanten* did not retain anything of the historic faith of the Reformation except an anti-Catholic attitude."

[4] J. Hertz, *Heinrich Hertz. Memoirs. Letters. Diaries.* Second enlarged edition, prepared by M. Hertz and C. Susskind. English translation by L. Brinner, M. Hertz, and C. Susskind (San Francisco: San Francisco Press, 1977).

[5] J. Hertz, *Heinrich Hertz* (ref. 4), p. 15.

6 *Ibid.*, p. 7.
7 J. C. Albisetti, *Secondary School Reform in Imperial Germany* (Princeton: Princeton University Press, 1983) provides a detailed discussion of German secondary schools during the nineteenth century. He notes (p. 19) that according to Wilhelm von Humboldt, mixing practical with classical training produces "neither wholly developed human beings nor fully integrated members of the separate classes."
8 L. Pyenson, *Neohumanism and the Persistence of Pure Mathematics in Wilhelmian Germany* (Philadelphia: American Philosophy Society, 1983), p. 4 quotes George Steiner on the social purpose of schooling in dead languages: "Power relations, first courtly and aristocratic, then bourgeois and bureaucratic, underwrote the syllabus of classic culture and made of its transmission a deliberate process."
9 Albisetti, *Secondary School* (ref. 7), p. 31.
10 Albisetti, *Secondary School* (ref. 7), pp. 62–6. Pyenson, *Neohumanism* (ref. 8), p. 35 discusses the creation in 1859 of two "orders of Realschulen, the first order teaching Latin but not Greek and the second order substituting sciences and modern languages for classics. Graduates of a nine-year course in the Realschulen of the first order – known later as *Realgymnasien* – enjoyed substantially the same privileges in the governmental hierarchy as Abiturienten from the Gymnasien, but Gymnasien in Prussia and in most other German states retained a monopoly on sending students to the university."
11 Cf. Albisetti, *Secondary School* (ref. 7), p. 31.
12 J. Hertz, *Heinrich Hertz* (ref. 4), pp. 15–17.
13 *Ibid.*, p. 17.
14 *Ibid.*, p. 17.
15 The director of *Gewehrbeschule*, Jessen, felt that Heins had a superior aptitude for mathematics and should study it intensely. But Heins told his mother that "I should not like that, mathematics is such an abstract science in which one must immerse himself completely, and I should like so much to be involved with people." Hertz was rather gregarious, but his dislike of pure abstraction arose in equal measure from his powerful attachment to physical manipulation and construction. Even his most abstract work – his *Mechanics* – betrays the impress of the laboratory and so the distant effect of his early passion for the concrete.
16 He had even begun to study Arabic under the tutelage of a Professor Redskob, whose enthusiasm for Heinrich was unlimited, stimulated no doubt by the great joy of finding anyone, much less one as intelligent as Heinrich, who showed an interest.
17 C. Jungnickel and R. McCormmach, *Intellectual Mastery of Nature*, 2 vols. (Chicago: The University of Chicago Press, 1986).
18 Anna Elisabeth wrote that "Their father gladly gave [the children] the freedom of which he had been deprived as a child, and I certainly did not begrudge it, but sometimes I should have liked to have been far away."
19 Indeed, not a few professors in the natural sciences were opposed to accepting *Ungebildete* into their precincts. See Pyenson, *Neohumanism* (ref. 8), chap. 4.
20 Jungnickel and McCormmach, *Intellectual Mastery* (ref. 17), vol. 1, p. 126.
21 Which was also read early in his career by Oliver Heaviside.
22 J. Hertz, *Heinrich Hertz* (ref. 4), p. 63.
23 On June 4 he had written his parents that "I have not yet said goodbye to mathematics, yet I should first like to get a general impression of experimental physics, and a better one than is given in the course on experimental physics."

24 D. Cahan, 'The Institutional Revolution in German Physics, 1865–1914,' *Historical Studies in the Physical Sciences* **15** (1985), p. 4.

25 I will not here discuss in detail the conceptual and ideological foundations of Hemholtzian physics, though I will explain them in broad terms. See my 'The Background to Heinrich Hertz's Experiments in Electrodynamics' in T. H. Levere and W. Shea (eds.), *Nature, Experiment, and the Sciences* (Dordrecht: Kluwer Academic Publishers, 1990), pp. 275–306 and (forthcoming) 'Helmholtzianism in Context: Object States, Laboratory Practice, and Anti-Idealism' for an introduction to the subject, which forms a major part of my forthcoming book on Hertz.

26 A. Schuster, *The Progress of Physics During 33 Years (1875–1908)* (Cambridge: Cambridge University Press, 1911), pp. 15–16.

27 See Cahan, 'The institutional revolution' (ref. 24) and Jungnickel and McCormmach (ref. 17) for details.

28 Friedrich Kohlrausch – the son of Weber's collaborator, Rudolf – to take but one from a myriad of examples, first used an offer from Zurich in 1870 to obtain more from Göttingen, and then used Göttingen's very nice response to get a better deal from Zurich, where he went, but only for a year, preferring to return to Germany where he felt so much more at home than among the alien Swiss, who rather objected to the establishment of the Imperial German Reich. Although Kohlrausch was certainly no Zöllner, nevertheless he always felt that "a German is preferable to a Jew" [D. Cahan, 'Kohlrausch and Electrolytic Conductivity. Instruments, Institutes, and Scientific Innovation,' *Osiris* **5** (1989), 167–85], and he was not overly enchanted with Catholics. After his father's death in 1858 Kohlrausch's education was guided by Weber and by Wilhelm Beetz (who told Hertz two decades later that he could find a laboratory almost anywhere).

29 Cahan in 'Instruments' (ref. 28) translates the following remark that Kohlrausch made in 1890: "Measuring nature is one of the characteristic activities of our age. Without the [measuring of nature] the progress made during the last century in the natural sciences and technology would not have been possible."

30 *Ibid.*, p. 2, for details.

31 Nor was Helmholtz deeply enamored of sensitive instruments designed for practical ends. Visiting Britain in 1884, he wrote his wife that he had "an impression that Sir William [Thomson] might do better than apply his eminent sagacity to industrial undertakings; his instruments appear to me too subtle to be put into the hands of uninstructed workmen and officials, and those invented by Siemens and Hefner v. Alteneck [less sensitive but more robust] seem much better adapted for the purpose" [L. Koenigsberger, *Hermann von Helmholtz*, translated by F. A. Welby (New York: Dover Publications, 1965. Reprint of the original English translation published in 1906 by the Clarendon Press, Oxford), p. 349].

It is hardly that Helmholtz was uninterested in, or uninvolved with, accurate instrumentation. Far from it – he developed new or more accurate devices in several areas, including galvanometry. But he was not personally gripped, as William Thomson was, by *measurement* and industrial economics. By 1887, when the immense Physikalisch-Technische Reichsanstalt was taking shape under his direction, Helmholtz had however forged a precision industrial research and testing laboratory (though he was replaced after his death in 1895 by the preeminent apostle of measurement, Rudolf Kohlrausch).

However Helmholtz's involvement with the Reichsanstalt does not reflect a direct concern with exact measurement for whatever purpose, but a complex of motives including a hoped-for release from teaching, his close friendship with Werner Siemens and his intense German patriotism, as well as a sense that he was unlikely to produce much more fundamental research.

Moreover, Helmholtz treated the Reichsanstalt rather like an inflated Physics Institute, but with indifferent success since, Cahan writes, despite the research spirit "under [Helmholtz's] administration the [Scientific Section] spent most of its time creating physical standards, instruments, and measuring methods" [*An Institute for an Empire. The Physikalisch-Technische Reichsanstalt 1871–1918* (Cambridge: Cambridge University Press, 1989), p. 109]. The Reichsanstalt was in part at least an attempt to synthesize the measuring goals of the Weberean tradition with the very different aim of physical discovery epitomized by Helmholtz's Physics Institute, a synthesis made pressing and desirable by the rapid development of science-based industry in Germany.

32 The question did in fact bear on Weber's comprehensive theory.

33 He remarked in his autobiography that "... it was in Berlin that my scientific horizon widened considerably under the guidance of Hermann von Helmholtz and Gustav Kirchhoff, whose pupils had every opportunity to follow their pioneering activities, known and watched all over the world" [M. Planck, *Scientific Autobiography and Other Papers*, translated by F. Gaynor (London: Williams & Norgate, 1948), p. 15].

34 Planck, *Scientific Autobiography* (ref. 33), p. 20.

35 *Ibid.*, pp. 22–3. Emphasis added.

36 "Helmholtz," he wrote concerning his doctoral dissertation on applying entropy to irreversible processes, "probably did not even read my paper at all. Kirchhoff expressly disapproved of its contents" [*Ibid.*, p. 19]. Nevertheless Berlin – which certainly means Helmholtz – would never have appointed him had his work not achieved a considerable measure of respect by the late 1880s. But respect does not always carry with it enthusiastic acceptance, and Helmholtz's early indifference would not likely have turned itself into the effusive welcome that brought Planck to Berlin.

37 Jungnickel and McCormmach, *Intellectual Mastery* (ref. 17), vol. 2, pp. 28–9.

38 See Cahan, *An Institute for an Empire* (ref. 31).

39 Schuster did not approve of the change, feeling that it destroyed the intimacy of the old place: "In place of a small number of badly furnished and crowded rooms, there was now a noble building, impressive on the outside and perhaps also inside to the casual visitor, but all the soul and scientific spirit of the old place had gone. The laboratory was constructed on the principle so dear to beginners, that every one should have a private room in which he could set up his apparatus without fear of outside interference. In consequence, Helmholtz no longer found the time personally to see and to advise the students, who worked without a common bond and generally without scientific impulse, and mainly for the purpose of completing a dissertation of sufficient merit for their degree" [Schuster, *The Progress of Physics* (ref. 26), pp. 17–18].

Schuster's gloomy description of the changed atmosphere at Berlin is rather hard to reconcile, as we shall see, with Hertz's experience at least, since he received a great deal of personal attention from Helmholtz. Schuster may perhaps have been reacting to the inevitable consequences of having many more students present than was previously possible, and to the emerging social patterns of the professional German research physicist, immersed in a highly specialized, esoteric topic and acutely aware of the discipline's competitive structure.

[40] All quotations from Heinrich Hertz are in J. Hertz, *Heinrich Hertz* (ref. 4), pp. 93–129. I will give precise references only for the longest ones.

[41] Buchwald, 'Helmholtzianism in Context' (ref. 25) examines Hertz's revealing efforts to attack the problem; here we will concentrate on the impress that the Berlin experience had upon Hertz as a fledgling physicist.

[42] C. E. McClelland, *State, Society, and University in Germany 1700–1914* (Cambridge: Cambridge University Press, 1980), p. 179.

[43] Perhaps the best example of Helmholtzianism at work is to be found in Hertz's unpublished manuscripts, now held at the Science Museum in London.

[44] In *Neohumanism* (ref. 8) and see especially his *The Young Einstein. The Advent of Relativity* (Bristol and Boston: Adam Hiller Ltd., 1985), chaps. 6 and 7.

[45] For his experiment.

[46] J. Hertz, *Heinrich Hertz* (ref. 4), p. 97.

[47] *Ibid.*, p. 133.

[48] *Ibid.*, pp. 114–5.

[49] These involve the inductance of a doubly-wound spiral.

[50] J. Hertz, *Heinrich Hertz* (ref. 4), pp. 116–7.

[51] Though, as was usual with Hertz, both applications of *theory* are closely linked to experiments that he either performed or that could be done.

[52] H. Hertz, 'Über die Verteilung der Elektricität auf der Oberfläche bewegter Leiter,' *Ann. Phys. Chem.* **13**: 266–75. Translated as 'On the Distribution of Electricity over the Surface of Moving Conductors' in H. Hertz, *Miscellaneous Papers*, translated by D. E. Jones and G. A. Schott (Leipzig: J. A. Barth, 1896), pp. 127–45.

[53] H. Hertz, 'Über de Berührung fester elasticher Körper,' *Jl. f. d. rein und angwndt. Math.* **92** (1881), 156–71. Translated as 'On the Contact of Elastic Solids' in H. Hertz, *Papers* (ref. 52), pp. 146–62; see also 'On the Contact of Rigid Elastic Solids and on Hardness' in *Papers*, pp. 163–83.

[54] In his words: "The hardness of a body is to be measured by the normal pressure per unit area which must act at the centre of a circular surface of pressure in order that in some point of the body the stress may just reach the limit consistent with perfect elasticity" [Hertz, 'On the Contact' (ref. 53), p. 180].

Only the first of Hertz's two papers on elasticity (on collisions), which contains results that are used in the second, is discussed (uninformatively) in I. Todhunter, *A History of the Theory of Elasticity*, 3 vols (New York: Dover Publications, Inc., 1960). Originally published by Cambridge University Press in 1886 (vol. 1) and 1893 (vol. 2, parts 1 and 2)), vol. 3, secs. 1515–17. S. P. Timoshenko, *History of Strength of Materials* (New York: Dover Publications, Inc., 1983), pp. 347–9 remarks that Hertz's first paper (on compression due to impact) "attracted the interest not only of physicists but also of engineers, and, at their request, he prepared another version of his paper [this is Hertz, 'On the Contact' (ref. 53)], in which a description of his experiments with compression of glass bodies and with circular cylinders is added" and that "In recent times [1930s], Hertz's theory of compression of elastic bodies has found wide application in railway engineering and in machine design." However, he continues, "Hertz's method [for measuring hardness] did not find acceptance, since in ductile materials it is very difficult to find at what load permanent set begins." Charles Susskind [in O'Hara and Pricha, *Hertz and the Maxwellians* (London: Peter Peregrinus in association with the Science Museum, 1987), p. xiv] however remarks that Hertz's discussion of *hardness* "remains the theoretical basis for all subsequent developments in the understanding of contact and wear in engineering structures and measurements."

[55] J. Hertz, *Heinrich Hertz* (ref. 4), p. 147.
[56] *Ibid.*, p. 149. Emphasis added.

RICHARD L. KREMER*

FROM PSYCHOPHYSICS TO PHENOMENALISM: MACH AND HERING ON COLOR VISION

At the end of his famous 1672 paper in which he presented his new theory of light and colors, Isaac Newton admitted: "But to determine more absolutely, what light is, ... and by what modes or actions it produceth in our minds the phantasms of colours, is not so easie. And I shall not mingle conjectures with certainties."[1] Although Newton later in the *Optics* would occasionally "mingle conjectures" about how the human eye "sees" color, his theory was primarily a theory of light, a physical theory with light rays, prisms, angles of refraction, lenses and barycentric diagrams as its chief working objects. Not until the nineteenth century did color become a fully subjective phenomenon, an aspect of nature impossible to consider apart from human verbal reports about visual experience. This shift in the conceptual and disciplinary location of color, from the physical to the physiological and psychological, began perhaps with Johann von Goethe's *Zur Farbenlehre* (1810) but it became canonical only with the appearance of the second section of Hermann von Helmholtz's *Handbuch der physiologischen Optik* (1860). Yet even after 1860, sorting out the relative roles for physical, physiological and psychological language in describing color experiences and in explaining such experiences remained exceedingly controversial. Despite its subjective foundations, color and its theory refused to fit easily into the increasingly distinct disciplinary boundaries being erected across the German scientific landscape of the late nineteenth century.

Disagreement over such disciplinary boundaries best explains, I think, the controversy between Helmholtz and Ewald Hering, and their respective disciples, over color theory. Launched in 1872–74 when Hering published a stinging attack on Helmholtz's treatment of color and proposed an alternative approach,[2] this battle continued even after 1905, when one of Helmholtz's students tried to combine the two theories by allocating each to a specific level or "zone" in the visual process.[3] As is well known, Helmholtz and Hering disagreed about nearly every aspect of physiological optics, from binocular vision to spatial perception, the horopter, optical illusions and contrast effects. Since 1867, when

Helmholtz in the third section of his *Handbuch* first sought to organize these disagreements with Hering, and especially since Edwin Boring's account in his *Sensation and Perception in the History of Experimental Psychology* (1942), these battles usually have been described in terms of "empiricism versus nativism." Although some version of such a polarity may characterize their disputes over spatial perception, for color theory their clashes are better understood as deriving from different disciplinary orientations in the explanation of sensory phenomena.

Elsewhere I have examined Helmholtz's efforts through 1860 to develop a theory of color.[4] Here I wish to summarize Hering's theory, and to suggest that those points where it most sharply contrasted with Helmholtz's may have derived, in part, from the influence of Ernst Mach. Although Mach himself in 1886 claimed priority for Hering's theory of colors and several historians have noted the connection, no one to the best of my knowledge has sought to explore the genesis of Mach's ideas on color, or their relation to Hering's later program.[5]

Mach spent the most productive period of his life, from 1867 to 1895, in Prague holding the chair for experimental physics. Hering was professor of physiology at the same university from 1870 to 1895. In the 1880s, both men actively participated in the struggle to preserve a "German" university in Bohemia. After the university split into German-speaking and Czech-speaking components, Hering served as the former's first rector, Mach as its second.[6] More importantly, together they made Prague a leading center for research on sensory physiology, and more than once publicly announced their mutual agreement on matters of method.[7] In 1874, Hering described his program as follows:

The great tasks, which face physiology and especially nerve physiology, are most effectively taken up from two sides at the same time, like the construction of a tunnel, namely not only from the physical-chemical but also from the psychical side .... The goal is knowledge of the causal relation of all physical events on the one side, of all psychical events on the other; the presupposition is the lawlike dependence of both types of events on each other.[8]

In *Knowledge and Error* (1905), his last and most comprehensive book on scientific epistemology, Mach praised Hering's tunnel analogy as identical to his own view of the relation between the physical and the mental, a view which Mach had begun to work out in the 1860s during his early psychophysical researches.[9]

CONSTRUCTING THE TUNNEL: HERING'S THEORY OF COLOR VISION

By 1872 when he began publishing his *Lehre vom Lichtsinne*, Hering had emerged as the most vociferous critic of Helmholtz's empiricist account of spatial perception.[10] This existing controversy prompted Hering to turn his *Lichtsinne* essays, from the opening paragraph onward, systematically and vituperatively against Helmholtz. He opened his first communication with a stinging denunciation of Helmholtz as a "spiritualist" who answered questions psychologically which belonged in the realm of physiology.[11] Just as earlier physiologists, asserted Hering, had deployed the word *"Lebenskraft"* to explain what they could not understand physically, now there "appears on every third page of a physiological optics [i.e., Helmholtz's *Handbuch*] 'soul,' '*Geist*,' 'judgment' and 'conclusion' as deus ex machina to overcome all difficulties." Against such spiritualism Hering set his own physiological version of Gustav Fechner's psychophysical law: "phenomena of consciousness are seen as conditioned and carried by organic processes ... as far as this is possible." Nativism and empiricism, the terms with which Helmholtz in 1867 had characterized his disagreements with Hering, represent no fundamental opposition, Hering argued, as long as both restrict themselves to explanations of physiological function. A deeper divide, however, separates spiritualist and physiological approaches. The latter seeks to derive the laws of consciousness from motions of organic materials; the former avoids this difficulty by attributing these laws to peculiarities of the soul. The latter requires two steps. "Philosophical psychology" secures, orders and describes the phenomena of consciousness. "Physiological psychology" then relates these phenomena to physical processes in the organism. At issue for Hering was not merely the doctrine of the *Lichtsinne* but the deeper issue of which disciplinary strategies – linguistic, experimental and theoretical – should guide the study of sensation.[12]

Hering explored phenomena of the *Lichtsinne* primarily by self-observation of the effects of simultaneous contrast. For example, he fixed his eyes on a small round, bright field laying on a large dark ground (see Figure 1). After a minute, he tightly closed and covered his eyes, and observed the negative afterimage of a darker field against a dark ground, with the field surrounded by a halo of brighter light that decreased in

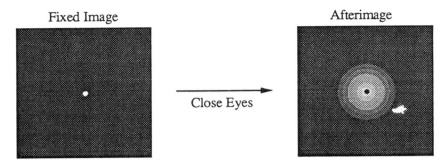

Fig. 1. Simple light "halo" around a negative afterimage. Adapted from Hering, *Lichtsinne* (ref. 2), XXXVII, 9–10.

intensity as distance from the field increased. He warned against confusing "accidental" features with the "essential" phenomena reported above. If for some reason one would fail to see what Hering and his unspecified number of witnesses had seen, Hering explained that failure to hold the retinas steady during the initial stare, or too intense objective illumination could ruin the observation.[13] Such observational events required no measurements, no complex apparatus, and no specialized language to describe the experiences; the observer merely had to compare patches of light and dark scattered across the visual field of an afterimage, avoid attending to the "accidental," and keep the eyes within their range of "normal" activity.

Deciding what was "accidental", of course, depended on what was required to refute spiritualism. Helmholtz too had observed light halos around negative afterimages, had attributed the negative afterimage to the physiological processes of retinal fatigue and the "inner light" of the eye, but argued that the halo resulted from contrast generated by "false judgments." Hering thus designed a second observation intended to reduce Helmholtz's psychological explanation to "absurdity." If instead of fixing on a small round field, one placed two small rectangles only four mm apart on the ground, and fixed on the narrow strip of ground visible between them, the resulting negative afterimage showed the light halo around both rectangles, but twice as bright on the strip between the rectangles (see Figure 2). After some time, the darker afterimages of the rectangles disappeared to the closed eyes; yet the

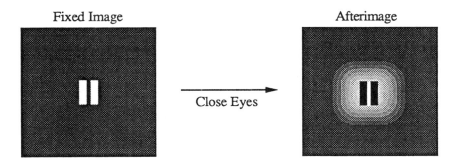

Fig. 2. Compound light "halo" around a negative afterimage. Adapted from Hering *Lichtsinne* (ref. 2), XXXVII, 13–15.

bright strip between them still remained on the afterimage. Helmholtz had not reported such an observation, but Hering ridiculed a "spiritualist" explanation of such phenomena as false judgments. This would be as if one would believe false news all the more if received from two persons, or as if one would keep believing lies long after the liar had departed. Why not, argued Hering, attribute contrast effects to physiological interactions among neighboring retinal particles? A consistently physiological explanation would be at hand for the "essential facts" of contrast and afterimages if one combined Helmholtz's mechanisms of retinal fatigue and "inner light" of the eye with Mach's idea of lateral retinal interactions.[14]

Not until Hering's third *Lichtsinne* essay, presented late in 1873, did color appear in the monograph, albeit very briefly. After exploring more phenomena of contrast and afterimages, all designed to refute "spiritualist" explanations in terms of judgments, Hering mentioned, almost as an afterthought, that all such observations could be repeated using green and red or blue and yellow instead of black and white fields and grounds.[15] Admitting, however, that achieving universal agreement on color sensations is more difficult than it is for sensations of brightness and darkness, Hering in 1874 founded his theory for the *Lichtsinne* beginning with the sensations of whiteness and blackness.

Hering began this formulation with "philosophical psychology," seeking "suitable and strict concepts" to use in talking about one's sensations. It was, he argued, "self-evident" that such concepts must be "drawn merely from the sensations themselves," and not from their physical or physiological causes (aether vibrations or retinal structures). Artists thus usually provide better descriptions of visual sensations than do physicists or physiologists. Hering's reasoning then became an extended thought experiment, probing how one can most accurately talk about visual experiences of brightness and darkness. Since every sensation "reminds us" simultaneously of white and black, he argued, one could talk of two "intensity scales," for whiteness and blackness, and could assume that all sensations are mixtures of both except for the "simple sensations" of pure white or pure black. The scales are not the mathematical negatives of each other, for the sensation of neutral grey does not result from a cancellation of white and black, but rather from equal intensities of both. Red and yellow, continued Hering, form two opposing intensity scales similar to black and white, for in orange and all intermediate colors one clearly sees both red and yellow.[16] Hering applied this same principle to white, Newton's compound light and by the Young–Helmholtz theory the result of an equal stimulus of the three color receptors in the visual system.[17] One cannot call the white a mixed sensation composed of complementary colors, argued Hering, because one *sees* neither red or green, nor yellow or blue in white. For the same reason, one cannot call white a mixture of red, green and violet (Young's primaries). Such talk confuses physical causes for descriptions of the sensations. Instead, for those "completely without physical and physiological assumptions," white is a "sensation of its own type," as is black, red, green, yellow and blue. The core of Hering's color theory, where it differed most fundamentally from Helmholtz's, is thus founded on white-black sensations.[18] Most non-colored and colored sensations are compounded, and for both the components can be "seen out" or at least "remembered."

Hering's final step in devising a theory for the *Lichtsinne* required "physiological psychology," or Fechner's psychophysical principle of correlating mental events with hypothesized physiological events in the visual organ. And those latter events must be conceived as chemical, asserted Hering. Contemporary nerve physiology had "sufficiently proved" that any activity in nerves alters their substances chemically; common terms used to describe nerve activity – such as excitability, fatigue, or recovery – all imply underlying chemical processes. Other physical explanations of nervous action, such as Emil Du Bois–

Reymond's electricity or Fechner's oscillations of ponderable or imponderable particles, can easily be reduced to chemical changes. And *Stoffwechsel* is the most general property of living bodies. Thus, any physiological theory of sensory processes is justified in reducing these to metabolic chemical changes, even though, Hering admitted, one cannot say whether such changes are only intermediate steps in a more fundamental process, or whether they "really" are the psychophysical events registered in the consciousness. The simplest assumption, concluded Hering, would be to correlate the opposing "intensity scales" of whiteness and blackness with opposing metabolic processes of assimilation and dissimilation (anabolism and catabolism). Since white sensations fatigue the eye more than black ones, dissimilation must correspond to white and assimilation to black sensations.[19]

From this hypothesis Hering derived a series of laws. If the type of a colorless sensation is determined by the ratio of dissimilation to assimilation, then neutral grey arises when $D = A$, brighter sensations when $D > A$, and darker sensations when $D < A$. Furthermore, the amount of D provoked by a D-stimulus depends not only on the size of the stimulus but also on the quantity of intact materials still available, a law of mass action by which effects of fatigue could be explained. The "purity, clarity and distinctness" of a sensation depends on the ratio of its "weight" to the "total weight" of all simultaneous sensations $[D/(D + A)]$. And in another hypothesis, Hering suggested that the "direct" stimulus of a retinal element (say by D) "indirectly" stimulates its neighboring elements to the opposing action (hence A). He could then explain all phenomena of contrast and afterimages in the language of assimilation and dissimilation, thereby providing a strictly physiological alternative to Helmholtz's semi-psychological explanations.[20]

In his final *Lichtsinne* communication of 1874, Hering extended the above theory to the complete range of color sensations. Beginning again with the principle of "philosophical psychology," that light and color sensations can be ordered only by their "inner relatedness," he asserted that four series of color sensations exist (red to blue, blue to green, green to yellow, and yellow to red), in which both end members can be seen simultaneously in any intermediate member, so that the latter always "appear to contain" the former. The four end members, which never appear to have any other colors mixed in, Hering called "fundamental colors." Furthermore, no sensations appear to mix red and green simultaneously, or yellow and blue. These pairs are thus unlike the white-black pair, and comprise what Hering called "opposing colors." Central to this classification is Hering's claim that in all but four color sensations, one can "see out" more than one color, just as the ear can "hear out" the individual pitches of a chord. Helmholtz, Fechner and most other observers had flatly denied that the eye could "see out" individual responses of the three different receptors or the colors of two monochromatic rays when mixed additively in the eye.[21]

For his physiological psychology of color, Hering proposed that the visual organ consists of three separate substances, each of which can undergo processes of dissimilation and assimilation independently. In the red-green substance and the blue-yellow substance, a given stimulus can produce either dissimilation or assimilation to yield the opposing colors. In the black-white substance, assimilation and dissimilation occur simultaneously. Ordering the sensations along a spectral sequence, Hering then described (but did not graph) the response curves for his three substances as a function of wavelength of incoming light. Any colored sensation is comprised of white and two other colored sensations (except for the fundamental sensations of yellow, green and blue; red however has some yellow mixed in). The eye "sees" the separate pure colors in different intensities; yet whichever of the fundamental colors has the greatest "weight" gives a sensation its "character" and its name. If several rays of light are sent into the retina so that they create equal amounts of A and D in one of the substances, that substance is neutralized and produces no sensation. From these mechanisms, Hering derived several different explanations of lateral interactions across the visual field. Only further investigations, he allowed, would permit a decision as to which explanation is correct.[22]

Hering's program of color research thus clashed at many points with Helmholtz's. First, Hering began with the phenomena of color sensations, all but four of which he claimed are mixed. Helmholtz began with the mixing of monochromatic rays in the eye, and claimed that all color sensations are unitary and simple, and that components are never recognizable. Second, Hering envisioned an opponent system, in which a given stimulus could excite a given basic sensation but never its opponent. For Helmholtz–Young, any stimulus simultaneously excites all three basic sensations. Third, Hering hypothesized a non-functionally differentiated visual organ, in which non-localized metabolic changes of assimilation and dissimilation occur everywhere from the retina to the brain. Helmholtz, on the other hand, placed functional differentiation in his model, with transducers in the retina, passive transmission in the optic nerve, and psychological processors in "higher" parts of the system. Finally, Hering restricted his language strictly to physical and physiological terms; Helmholtz, when faced with phenomena (like simultaneous contrast) not amenable to such explanations, turned to the language of psychology. At each of these clashes, Hering's approaches approximate those taken by the young Mach, nearly a decade earlier.

BECOMING A PHENOMENALIST: MACH AND
THE ORIGINS OF HERING'S PROGRAM

Mach's potential influence on Hering derived from three roots: Mach's early embrace of Johann Friedrich Herbart's mechanical psychology, Mach's experimental psychophysical researches which culminated in his 1867 discovery of "Mach bands," and Mach's rather scattered remarks on color vision. After laying bare these roots, I shall consider to what extent each may have shaped Hering's program as summarized above.

Mach's early academic career and interests during the 1860s placed him squarely within the overlapping and slowly coalescing disciplines of physics, physiology, psychophysics and psychology. After habilitating in 1861 at the University of Vienna for "physics in an unrestricted sense" (the philosophy faculty had rejected his first attempt to qualify for "physics as an introduction [*Vorschule*] to physiology"), Mach began trying to scratch out a living by lecturing at Ernst Brücke's physiological institute on physics and sensory physiology. He later remembered being attracted to the latter field, "which led to [my] critical epistemological investigations," because it required no costly apparatus.[23] These lectures, two of which were published in 1863 as *Compendium der Physik für Mediciner* and *Vorträge über Psychophysik*, brought the young *Privatdozent* into contact with Herbart's psychology, Fechner's psychophysics and Helmholtz's studies of physiological optics and sensations of tone.[24]

Until at least 1865, Mach's contemporaries considered him an Herbartianer.[25] In this semi-Kantian philosophy of logic, metaphysics and aesthetics developed a half century earlier, Herbart had established the soul as a real simple being, existing in neither time nor space and possessing no categories, forms, faculties or any other qualities.[26] This soul exists solely through the interplay of what Herbart called "presentations" [*Vorstellungen*]. "Just as physiology constructs the body from fibers, so the psychologist constructs the soul from presentations."[27] Explicitly copying the procedures of other sciences, especially "empirical physics," Herbart argued that one must begin with fictional entities like "forces" and derive fictional laws which can then be used to organize the phenomena of psychology. Hence he divided his psychology into two parts, empirical and rational, i.e., the natural history of psychological phenomena and their explanation in terms of presentations. Following the method of physics, Herbart developed an elaborate statics and mechanics of presentations, treated as colliding forces. Metaphysically, he argued that "between several different simple beings there exists a relation, that one with help of an analogy from the world of bodies can call *pressure* and *counter-pressure*." Psychologically, if presentations did not resist, they would all merge in the mind and consciousness would be unitary rather than varied as we experience it. Hence, the most basic characteristic of presentations is their mutual inhibition [*Hemmung*] and resistance. Some presentations are not opposed to each other, like pitch and color, and may combine without reducing each other. Opposing presentations, like red and blue, seek to displace each other. If two presentations interact, neither will be able to drive the other completely below what Herbart called the "threshold of consciousness." But in these interacting presentations, one may become completely inhibited and forced below that threshold.[28] Manifolds of presentations in the mind thus come into equilibrium, partially inhibit each other, force one another below the threshold of consciousness, or draw each other up again above that threshold. To describe such interactions, Herbart proposed an elaborate set of differential equations to characterize changes over time in what he called the "total amount of mutual inhibition" [*Hemmungssumme*].[29]

Much to the dismay of his critics, Herbart presented his mechanics of mind at a very abstract level. Despite their quantitative relationships, Herbart's presentations could not be measured empirically. And nowhere did he discuss in any detail the process of sensation or the behavior of "sensory presentations." Only from scattered remarks is it apparent that he accepted three primary colors. Tones form a one-dimensional continuum of presentations, Herbart noted; colors, however, form continua of two dimensions, as red, yellow and blue "appear" to form an equilateral triangle, with three sets of "nuances" between the vertices. If one adds further the opposition of dark and bright, then a third dimension is required to depict presentations of color. Such figures, however, have nothing to do with "physiological, physical, [or] chemical theories of color," Herbart asserted, but concern only the presentation of the soul.[30]

Precisely the quantitative nature of Herbart's mechanical psychology first attracted Mach's attention. One of the major themes in Mach's 1863 lectures on psychophysics was that psychology was on the way to becoming an "exact" science. As examples, he cited Adolphe Quetelet's social physics, Fechner's psychophysics and Herbart's psychology. The latter, he argued, could be considered as a "respectable scientific theory" quite apart from all "metaphysical questions." Even though he praised Fechner's psychophysics for providing "facts of experience" without hypotheses, Mach also defended Herbart (against recent attacks by Wilhelm Wundt) for deriving psychical phenomena from a few "basic laws" [*Grundsätze*]:

The reproach that one can make of Herbart, which however is no reproach for a young science, is that he ignored the organization of the body and began with too simple assumptions. The entire work still has incalculable value, even if the assumptions are proved partly false. For the value of an hypothesis consists primarily in the fact that as a type of *regula falsi* it always leads closer to the truth.[31]

Yet since the mathematical consequences of Herbart's theories "so closely correspond to experience," his hypothetical assumptions cannot be far from the truth.[32] Mach thus surveyed for his auditors Herbart's views of presentations, including the concept of mutual inhibition, and noted that blue and yellow (two of Herbart's primaries) are "fully" opposed whereas blue and green are only "partly" so.

Herbart's psychology became even more important for Mach as he began his own experiments in sensory physiology. In 1863, Mach published a study of the acoustics of the middle ear, inspired by the appearance in that year of Helmholtz's *Die Lehre von den Tonempfindungen*. Quoting Helmholtz's suggestion that we hear partial tones by "attending" to them,[33] the young *Privatdocent* showed no reticence in challenging the master. If we want to understand attention not as some "mysterious thing" (i.e., Helmholtz's approach) but rather according to the principles of psychophysics, by which all mental activity is assumed to be based on physical changes, then one might consider attention as variable tension in the tensor tympani. This muscle, Mach speculated, might accommodate the ear, just as the eye possesses mechanisms for accommodation. Indeed, perhaps attentiveness is a more general feature of the organism. "Some bodily processes exclude others. In the amount one is present, anothers retreat. Bodily processes, sensations mutually repress each other. On a physiological path we have found a law similar to that governing sensations, just as Herbart long ago expressed it for presentations."[34] Mutual inhibition was becoming for Mach, the Herbartianer, a major tool not only for psychological but also for physiological explanation.

A second source of potential importance for Hering's color theory derived from Mach's early psychophysical experiments. Already in 1860 Mach had begun seeking to confirm Weber's law of "just noticeable differences," which Fechner had made the centerpiece of his *Elemente der Psychophysik* (1860), for a sensory phenomenon hitherto ignored – the "sense of time" [*Zeitsinn*]. In repeated trials, Mach found that just noticeable differences of audible time intervals, created by metronomes or other pendula, deviated considerably from Weber's logarithmic law. He tried to explain these results with Herbart's theory of the *Zeitsinn* as the internalization of a chronological series of presentations. This prompted a further series of speculations which led Mach finally to color sensations. He asked first whether the "sensation of rhythm," which is independent of sensation of pitches in a musical melody, might be another mechanism of accommodation or attention, deriving from the alternation of fatigue and recovery. Wundt had shown, for example, that the astronomers' "personal equation," closely related to the *Zeitsinn*, can be varied with the help of attention. Extending the analogy even further, Mach wondered whether an eye might have different personal equations for different colors, presented in short intervals. To test this idea, he began experiments with rotating colored disks, which unfortunately had yielded no "positive results."[35] Nonetheless, Mach again had deployed the Herbartian mechanism of inhibition in seeking to explain temporal sensation, and within the context of Fechner's psychophysics had for the first time considered color experimentally.

After failing to confirm Weber's law for the *Zeitsinn* measured with the ear, Mach turned his psychophysical experiments to vision, hoping to explore *Zeitsinn* not by means of just noticeable differences but rather with the phenomena of "*Abklingen*" and "*Anklingen*" of sensations (i.e., the less than instantaneous fading in or fading out of sensations after an objective stimulus has instantaneously started or ceased).[36] Again Mach selected a sensory phenomenon accessible to exact quantitative measurement. Instead of Weber's law, he now sought to test the Talbot–Plateau law, i.e., that the intensity of a sensation of a regularly interrupted light is identical to that of a sensation of non-interrupted light, if the same amount of light enters the eye in a given time and if the frequency of the interruption is high enough so that fusion occurs.[37] Finding slight deviations from this law, Adolf Fick in 1863 had speculated on the form of the response curves of the retina, as a function of time, when objective light stimuli are instantaneously turned on and off.[38] To explore similar phenomena, Mach employed rapidly rotating belts on which he could vary (i) the percentage of time the stimulus (alternating white and black or differently colored fields) was off and on in a cycle and (ii) the overall frequency of the cycle (see Figure 3). Once again, however, he was forced to break off his experiments without achieving his goal of testing the Talbot–Plateau relation. As the ratio of "on" to "off" stimuli in a cycle neared unity (i.e., equally spaced white and black fields), the flickering sensations refused to fuse into a uniform gray, and Mach observed rather a shimmering of white as if seen through black, a phenomenon similar to stereoscopic luster. Upon photographing this flickering stimulus, however, Mach found that photographic paper followed the Talbot–Plateau law exactly; i.e., photochemical processes displayed no phenomena of fade-in or fade-out. Again, Mach did not miss the opportunity to speculate on the meaning of this observation. If the effect of light on the outermost retinal layer is primarily chemical, then the visual receptors must act as the photographic paper. "Fade-in

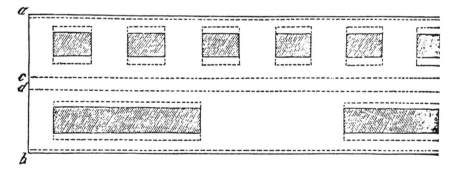

Fig. 3. A segment of Mach's rotating belt, used to study visual *Zeitsinn*. Reprinted from Mach, "Lichtreize" (ref 36), 630.

and fade-out then arise first from the interaction of further [beyond the retinal surface] nerve elements. The matter is clear mechanically. Only greater masses could show a noticeable fade-in and fade-out . . . . The outermost nerve elements are excited; it depends on the disposition of further elements how and when they conduct this excitation further."[39] Once again, a failed psychophysical experiment provoked Mach to hypothesize about chemical and mechanical processes within the visual system. Following Fechner's psychophysical principle, Mach tried to correlate a mental experience (Fick's fade-in and fade-out) with a hypothetical physical event (interacting nerve elements).

Also in 1865, Mach announced his "discovery" of Mach bands, the subjective excitation or inhibition of brightness sensation across a field of evenly changing objective intensity.[40] Noticed by chance as he was experimenting with rotating disks to explore *Zeitsinn*, the two bands of more intense and less intense illumination across the uniformly varying field also appeared on photographs. Mach bands thus represented a rather peculiar type of simultaneous contrast, subjective phenomena which had puzzled students of vision for a century and on which, as noted above, Helmholtz and Hering had disagreed. Comparing the intensity curves of the objective stimulus with that of the sensation (see Figure 4), Mach

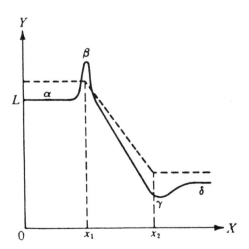

Fig. 4. Mach bands. The dotted line represents the brightness of an objective stimulus across a visual field, the solid line the intensity of the subjective sensations. Reprinted from Ratliff (ref. 40), 57.

realized he could easily express the latter curve mathematically by adding another term to Fechner's formula for Weber's law.[41] And again he proposed a psychophysical explanation of the mathematics in terms of lateral interactions among retinal points. Citing recent anatomical studies which indicated that up to one-hundred rods are connected to a single retinal ganglion cell, Mach hypothesized the existence of networks of retinal elements, able to "determine among themselves which sensations they will further conduct" to the fibers of the optic nerve.[42] That is, the sensation produced by a given retinal element depends not only on its own illumination but also on that of its neighbors. Mach's retina is not a passive transducer, but rather "schematizes and caricatures."[43]

Such an explanation, Mach argued, is far more plausible than the "unconscious conclusions and judgments" to which "many physiologists" (i.e., Helmholtz) retreat. Why should not the ganglion cells in the retina possess the same power of "logic" that is allocated to cells of the brain or the higher nervous system?[44] In a later essay, Mach defended the equality of retinal and brain cells with a flourish of political analogies:

The state is an organism, formed of elementary organisms – of men. It cannot be denied that the whole state behaves on a grand scale, in a manner similar to individual man behaving on a small scale. The individual soldiers, craftsmen, civil workers (the muscle and nerve fibers) think many things which do not enter the consciousness of the government (the brain). Every class has its skills, its traditions, of which only a small part comes to the consciousness of the government.

Or in a slightly different analogy, Mach asked who could believe that a monarch would concern himself with checking passports of foreigners at the border. "Is it not equally absurd to assume that the brain itself is concerned with every single light ray incident on the retina? Rather than giving the brain/state alone the power of making unconscious judgments, Mach urged the "more scientific view that the light stimuli are already sorted out and organized in the retina and that, thus already sorted out and organized, they enter consciousness."[45] The young psychophysicist thus freely hypothesized hidden physiological processes – retinal interactions, mutual inhibitions – as he sought to correlate physical events in the nervous system with various subjective sensory effects generated in his experiments.

A third area where the young Mach would influence Hering, and where Mach began to assume a more phenomenalist outlook, was color and color vision. As with his responses to Herbart and Fechner, Mach first assimilated the theories and programs of others, especially Thomas Young's theory as elaborated by Helmholtz, and then began to criticize and to reformulate those approaches. Unfortunately, Mach's published papers, extremely terse in this period and often replete with unrelated yet extremely significant asides, do not always illuminate the steps in his reasoning as he moved from assimilation to critique.

In his first published experimental work, Mach attempted to mediate an earlier controversy between Christian Doppler and Mach's Viennese mathematics professor, Joseph Petzval, over the reality of the Doppler effect and its applicability to the color of light emitted by stars. Since both sound and light are wave phenomena, argued the young Mach in 1860, one can extend by analogy effects proved experimentally in acoustics to optics.[46] By 1863, however, after becoming familiar with Fechner's psychophysics, Mach backed away from this analogy and began to examine hypotheses concerning the nature of the visual system which could account for color vision as a process unlike auditory sensation.

Fechner devoted a lengthy section in his book to a comparison of the sensations of sound and light, identifying more differences than similarities.[47] For example, sound sensations lack spatial components, unlike those of light. In sounds, a continual change of frequency produces a continual change in pitch; continual changes of the frequency of light do not always produce proportional changes in the sensations of color. And for mixed frequencies of sound, the ear can hear the component pitches, unlike the eye which cannot distinguish between a sensation produced by a monochromatic ray and by a properly selected mixture of such rays. Fechner then offered five mechanical hypotheses concerning the sensory apparatus, by which these differences might be explained. He began with the well-worn idea, discussed since at least Descartes and Newton, that the "activity" of the nervous system, as well as light and sound stimuli, might be considered as forms of oscillatory motion.[48] A given sound wave excites sympathetic vibrations in a given auditory fiber, probably the Corti fibers, tuned to a specific frequency.[49] Every optical fiber, however, can be excited to vibration by any ray of the visible spectrum; if several monochromatic rays fall on the same fiber, their sympathetic vibrations add by the principle of interference; and any monochromatic ray excites multiple vibrations in a given optical fiber.[50] Fechner suggested that optical nerve fibers might be able to receive various frequencies if different atoms in their molecules vibrate with various natural frequencies. Given such promiscuous nerve elements, it is not clear how Fechner thought the eye could distinguish colors at all. After reading Helmholtz's 1860 elaboration of Young's theory, Fechner admitted that the restriction to three fundamental sensations of color deviated "essentially" from his own view. But in this area, "assumptions must supplement the conclusion" and any argument over this point would be superfluous.[51]

Mach in his 1863 lectures on physics presented positively Fechner's hypotheses for the visual process, including the idea of optical fibers as "system of atoms."[52] Indeed, if Fechner's hypotheses were correct, remarked Mach, then one might learn about the physics of atoms from subjective experiences of color. Giving no sign that he had yet read Helmholtz's *Handbuch*, Mach confusingly attributed the idea of three basic color sensations, each of which are stimulated in different degrees by incoming monochromatic rays, not to Young but to David Brewster. This Scottish natural philosopher's anti-Newtonian theory that sunlight consists of only three objective rays (red, yellow, blue) had been refuted by Helmholtz in 1852.[53] The idea of three basic sensations, thought Mach, is highly probable since any hue can be produced by mixing three colors. In his earliest statement on color, then, Mach accepted Fechner's mechanical approach and Young's theory (misassigned to Brewster).

By the time he published his lectures on psychophysics later in 1863, Mach had read Helmholtz, from whom he extracted an accurate rendering of Young's theory.[54] Reporting to his auditors that Helmholtz had hypothesized three different nerve fibers for Young's basic sensations, and that Fechner has suggested the same fiber undergoes three different processes,[55] Mach opted for the latter's hypothesis, suggesting that nerve ends are comprised of three atoms, which can oscillate with different periods. In these lectures, Mach also refused to accept Helmholtz's analogy between the telegraph wire and nervous action, by which both passively conduct signals. Despite Helmholtz's "great authority," wrote the young Mach, electrical studies of nerves are still very "coarse" and shed little light on underlying molecular processes. Mach clearly favored the mechanical hypotheses of Fechner's *Psychophysik*.

By the 1865 psychophysical experiments on the "sense of time," Mach abandoned Young's theory, apparently because of an increasingly phenomenalist approach to the study of sensation. In these unsuccessful attempts to explore the "sense of time" by visual experiments, Mach at one point replaced the black and white rectangles on his rotating belts with different colors. In each case, colors gave results identical to those yielded by the black and white stimuli, with a mixed color rather than grey produced in those cases where fusion occurred, or with one color seen through the other where luster appeared. Hence, concluded Mach, black acts on the retina not as a lack of excitation but "precisely as a color, precisely as a positive light sensation."[56]

In a move not made clear by his published text, Mach then offered a major criticism of Young's theory, proposing the foundations of what would become Hering's theory of color vision several years later.[57] The theory of three fundamental color sensations, Mach asserted, appears "not completely satisfactory." Just because all color mixtures can be produced with three *physical* frequencies of light does not mean that an hypothesis of three fundamental sensations explains the process *psychologically*. The theory does not exclude, for example, the positing of more than three fundamental sensations. Fick, a year earlier, also had emphasized the "dubious advantage" of a theory which would allow various assumptions about what color sensations are to be considered fundamental, and Mach cited Fick for support as he asserted that violet (one of Young's and Helmholtz's three basic sensations) is not fundamental.[58]

Mach then offered an extraordinary and nearly unprecedented claim. "I believe one must assume as many basic sensations and basic processes as there are simple colors, in which no others are recognized. These colors would be red, yellow, green and blue." Against Fechner, against Helmholtz, and indeed against most of those who had written about colors since Newton, Mach claimed that the vision produces two types of color sensations, simple and compound. Sensations of orange or violet, for example, are not simple because the eye can recognize in them more than one color, regardless of whether they are produced by monochromatic or mixed rays. Only in the four specified colors can one recognize no other colors.

Mach defended his claim that the eye can "dissolve" [*auflösen*] out the basic components of non-complementary mixed colors by several very laconic arguments. Common linguistic usages, such as "blue-green" for turquoise or "red-yellow" for orange, accurately describe colors seen by the eye. The psychophysicist had begun to become a phenomenalist. Rather than *measuring* sensations in the laboratory, Mach now proposed to base a new theory of color vision on how ordinary people *talk* about their experience of color. Simultaneous contrast, Mach continued, also proves that the four basic color sensations comprise four basic and independent visual processes. Indeed, in his essay announcing the discovery of Mach bands, also published in 1865, Mach mentioned his hypothesis of four basic colors, as a result of the "heuristic principle of psychophysical research . . . , [that] every psychical event corresponds to a physical event and vice versa."[59] Yet as far as I can tell from Mach's published articles, the experiments by which he discovered Mach bands would not have suggested the conclusion that all color sensations consist of four basic sensations. Finally, Mach suggested that phenomena of luster, which had derailed his experiments on visual *Zeitsinn*, also "perhaps prove" the four-color hypothesis. Again, Mach did not make explicit his reasoning, and I cannot see any obvious connections between luster and the four-color hypothesis.

Yet four basic sensations and processes were not enough. For sensations of white, Mach admitted that the eye normally cannot dissolve the complementary or otherwise multiple rays which produce that sensation. Hence the sensation of white requires a special process.[60] And for black another process is required. "Excitation by a lack of stimulus [i.e., black] is not so strange if one considers it with phenomena of inhibitory nerves. The stimulus must not necessarily come from without, but can be exerted by the organism itself."[61] Mach apparently referred here to the fact that stimulating the vagus nerve can inhibit the heartbeat, a discovery made in the 1840s, by Eduard Weber and others.[62] In such cases, stimulus (p) produces inhibition (q) of an autonomic response. For the sensation of black, Mach in an odd twist of logic suggested that "non-stimulus" ($\sim p$) of the visual system could be understood as producing "non-inhibition" ($\sim q$) or sensation.

Clearly, Mach's thought was driven by the logic of Fechner's psychophysical heuristic and Young's theory. Yet unlike his earlier embrace of nerve fibers and atomic motions to explain physical processes, Mach now referred only vaguely to "basic processes" as the physical events. And he provided few clues for how he arrived at the six simple color sensations. In his *Analysis of Sensations* (1886) Mach briefly reviewed the history of the six-sensation theory, noting that Hering had "adopted" the idea which, it is "frequently asserted" had been "first proposed by Leonardo da Vinci, and later by Mach and Aubert." Mach rejected Leonardo's priority, for the painter had listed the six "simple" colors while discussing pigment mixing and not color sensations. "My own scattered remarks concerning the theory of color-sensation, were perfectly clear," concluded Mach, implicitly awarding himself the laurel since Hermann Aubert received no further mention.[63]

Indeed since the seventeenth century, most authors writing about color mixing had assumed the existence of three "primary" colors, by which they simply meant those colors from which all other colors could be produced by (usually pigment) mixing. They did not elevate these primaries to any special status as fundamental sensations, or postulate special processes in the visual sensory system.[64] As is well known, Newton although his theory had postulated an infinitude of solar rays, each with a unique refrangibility and color, nonetheless had named only seven colors within that multitude, a move he defended by analogy to the seven tones in the musical scale. Arthur Schopenhauer, in a largely ignored work on color first published in 1816 with an expanded edition in 1854, tried to refute Newton on linguistic grounds. Despite the fact that all spectral colors merge smoothly together with none more privileged than the other, Schopenhauer argued:

> all peoples, at all times find special names for red, green, blue, orange, yellow, violet which are everywhere understood as designating quite definite colors, even though these seldom appear purely and perfectly in nature; these must so to speak be recognized apriori (similar to geometrical figures, which are never given in reality): and if those names are mostly attributed to the real colors aposteriori, i.e., every present color is named according to which of the six it mostly closely resembles, yet everyone knows how to differentiate this color from that bearing the name and how the former deviates from the latter, e.g., if an empirically given yellow is pure, or whether it moves into green or orange: there must therefore be a norm, an ideal, an Epicurian anticipation of yellow and every color independent of experience, with which every actual color can be compared.[65]

Aubert, writing several years before Mach, named only four fundamental sensations. Praising Young's theory as the only means to avoid the requirement for an infinitude of nerve fibers if one accepted Johannes Müller's doctrine of specific sense energies, Aubert nonetheless asserted that we classify color sensations into four "principal" colors – red, yellow, green and blue – simply because from the earliest times these have been the most repeated sensations in human experience.[66] What for Schopenhauer had been an apriori norm became for Aubert simply an historically contingent fact. Mach with his psychophysical heuristic made the basic sensations a result of physiological processes, and his version would become elevated in Hering's theory.

Mach's brief discussion of color in 1865 represented several critical transitions in his thought. First, by basing his analysis on the pure sensations of color rather than the causes of these sensations, he for the first time in his published work had privileged sensation, that fundamental concept which by 1886 would dissolve into Mach's "elements" and would become the basis of Mach's phenomenalist epistemology.[67] Second, by identifying the simple sensations via linguistic criteria, Mach in 1865 had foreshadowed what would become an important feature of his subsequent historical analyses of science. In his *Science of Mechanics* (1883) Mach suggested that all sciences begin with the "instinctive gathering of experiential facts." Names then arise when regularities in these experienced facts are described and communicated to others. Hence for Mach, names became powerful tools to explore not only the content of experience but also the historical origins of the natural sciences.[68]

CONCLUSION

The above analysis has indicated a number of features in Mach's early work which appear in one form or another in the theory of color which Hering developed several years later. Mach's interests in Herbart's mathematical if fictitious psychology, in Herbart's emphasis on mutually inhibiting psychical processes, in Fechner's psychophysical heuristic principle of hypothesizing physical correlates for all psychical events, and in grounding color theory on sensations rather than on light rays all find a resonance in Hering's later work. To what extent, however, did Mach actually influence the development of Hering's program?

Of course, such a question can never be answered conclusively. In his monograph on *Lichtsinne*, Hering frequently cited Mach's essays on Mach bands (but none of his other early essays) and acknowledged Mach's influence, even as he criticized his predecessor for not reaching further.[69] According to Hering, Mach rightly recognized the importance of Fechner's psychophysical principle, yet described the latter merely as an "heuristic" principle rather than the "conditio sine qua non" of psychophysical research. Or Mach's explanation of simultaneous contrast in terms of interacting retinal particles did not go far enough, for Hering, in providing a physiological basis for such interactions. And Mach's idea of four fundamental color sensations, clearly a "progressive" step declared Hering, nonetheless did not include opponent actions among these sensations.[70]

Neither actual citation nor similarity in ideas, however, exhaust the possible avenues of influence, especially since Mach and Hering saw each other nearly daily in Prague during the time Hering was creating his color theory and Mach his phenomenalist epistemology. Undoubtedly, Hering had read Fechner and Herbart on his own, without any urging from Mach. Nonetheless, it seems impossible to deny any mutual influence between Mach and Hering, the physicist and physiologist each of whom refused to remain within the boundaries of his respective discipline and both of whom increasingly privileged sensation alone as the primary source of knowledge.

Department of History
Dartmouth College

NOTES

[*] To Erwin N. Hiebert, in grateful appreciation for introducing me to Ernst Mach in a Harvard seminar on "Scientists as Philosophers."

[1] Isaac Newton: (1672), 'New Theory about Light and Colours,' reprinted in I. Bernard Cohen, ed., *Isaac Newton's Papers & Letters on Natural Philosophy*, 2nd ed. (Cambridge, Mass.: Harvard University Press, 1978), p. 57.

[2] Ewald Hering, *Zur Lehre vom Lichtsinne* (Vienna: Gerold's Sohn, 1878), first published in six installments between 1872 and 1874, essays which were reprinted in the second volume of Ewald Hering, *Wissenschaftliche Abhandlungen*, 2 vols. (Leipzig: Georg Thieme, 1931). Since these volumes number Hering's essays sequentially and paginate each separately, I shall hereafter cite the essay number in Roman, the page number in Arabic numerals.

[3] Johann von Kries: (1905), 'Die Gesichtsempfindungen,' in Wilibald Nagel, ed., *Handbuch der Physiologie des Menschen*, 5 vols. (Braunschweig: Vieweg, 1904-10), **3/2**: 109–282.

[4] Richard L. Kremer, 'Innovation through Synthesis: Helmholtz and Color Research,' in David Cahan, ed., *Herman von Helmholtz: Scientist and Philosopher*, forthcoming.

[5] Ernst Mach: (1886), *Contributions to the Analysis of Sensations*, transl. C. M. Williams (Chicago: Open Court, 1897), p. 34. Cf. Richard Jung, 'Ernst Mach als Sinnesphysiologe,' in Wolfgang Merzkirch and Frank Kerkhof, eds., *Symposium aus Anlass des 50. Todestages von Ernst Mach* (Freiburg: Ernst-Mach-Institut, n.d. [1967]), p. 134, John T. Blackmore, *Ernst Mach: His Work, Life and Influence* (Berkeley: University of California Press, 1972), pp. 58–60; Mitchell G. Ash, 'The Emergence of Gestalt Theory: Experimental Psychology in Germany, 1890–1920,' unpublished Ph.D. dissertation, Harvard University, 1982, pp. 96–97.

[6] Paul Molisch, *Politische Geschichte der deutschen Hochschulen in Österreich von 1848 bis 1918*, 2d enl. ed. (Vienna: Braumüller, 1939), pp. 49–53; *Die deutsche Karl-Ferdinands-Universität in Prag* (Prague: Calve, 1899), p. 16; Blackmore, *Mach* (ref. 5), pp. 73–83.

[7] Erna Lesky: (1964), *The Vienna Medical School of the 19th Century*, transl. L. Williams and I. S. Levij (Baltimore: The Johns Hopkins University Press, 1976), pp. 484–85.

[8] Hering, *Lichtsinne* (ref. 2), XLI, 216–17.

[9] Mach: (1926), *Knowledge and Error*, Introduction by Erwin N. Hiebert, transl. from the 5th German ed. by Thomas J. McCormack (Dordrecht: Reidel, 1976), pp. 12–14.

[10] See Leo M. Hurvich and Dorothea Jameson, 'An Opponent-Process Theory of Color Vision,' *Psychological Review* **64** (1957), 384–404; Leo M. Hurvich and Dorothea Jameson, 'Introduction' in Ewald Hering: (1920), *Outlines of a Theory of the Light Sense*, transl. Hurvich and Jameson (Cambridge, Mass.: Harvard University Press, 1964), pp. vii–xxvii; Leo Hurvich, 'Hering and the Scientific Establishment,' *American Psychologist* **24** (1969), 497–514; R. Steven Turner, 'Consensus and Controversy: Helmholtz on the Visual Perception of Space,' in Cahan, ed., *Helmholtz* (ref. 4). By *Lichtsinne*, Hering referred to those phenomena of vision deriving solely from sensations of light, darkness and color. In his *Handbuch der physiologischen Optik* (Leipzig: Voss, 1856–67), Helmholtz had distinguished such phenomena, which he called "visual sensations," from "visual perceptions" by which presentations of the existence, form and position of external bodies are created; Hering implicitly accepted these distinctions.

[11] Hering was criticizing Helmholtz's explanation of the phenomena of simultaneous contrast in psychological terms of "unconscious judgments" and "habits." See Kremer, 'Innovation' (ref. 4), pp. 50–55. This would not be the last time Helmholtz would be linked to spiritism. See Wayne H. Stromberg, 'Helmholtz and Zoellner: Nineteenth-century Empiricism, Spiritism, and the Theory of Space Perception,' *Journal of the History of the Behavioral Sciences* **25** (1989), 371–83.

[12] Hering, *Lichtsinne* (ref. 2), XXXVII, 5–9.

[13] *Ibid.*, XXXVII, 9–16; XXXVIII, 193–4.

[14] *Ibid.*, XXXVIII, 200.

[15] *Ibid.*, XXXIX, 244.

[16] *Ibid.*, XL, 85–97.

[17] Kremer, 'Innovation' (ref. 4), pp. 32–34.

[18] Hering, *Lichtsinne* (ref. 2), XLI, 179–81.

[19] *Ibid.*, XLI, 184–89.

[20] *Ibid.*, XLI, 190–213.

[21] Helmholtz, *Handbuch* (ref. 10), p. 272; Gustav Theodor Fechner: (1860), *Elemente der Psychophysik*, 2 vols. 3d unchanged ed. (Leipzig: Breitkopf & Hartel, 1907), **2**: 274.

[22] Hering, *Lichtsinne* (ref. 2), XLII, 169–97.

[23] K. D. Heller, *Ernst Mach* (Vienna: Springer-Verlag, 1964), pp. 12–14; Blackmore, *Mach* (ref. 5), pp. 14–15; John T. Blackmore, 'Three Autobiographical Manuscripts by Ernst Mach,' *Annals of Science* **35** (1978), 401–18, at 410, 415. Before assuming in 1864 his first academic chair in Graz, Mach lectured in Vienna on physics for medical students, methods of physical research, higher physiological physics, principles of mechanics and mechanistic physics in its historical development, Fechner's psychophysics and Helmholtz's sensations of tones.

24 For recent analyses of these lectures, see Erwin N. Hiebert, 'The Genesis of Mach's Early Views on Atomism,' in Robert S. Cohen and Raymond J. Seeger, eds., *Ernst Mach: Physicist and Philosopher* (Dordrecht: Reidel, 1970), pp. 79–106, and Wolfram W. Swoboda, 'Physics, Physiology and Psychophysics: The Origins of Ernst Mach's Empiriocriticism,' *Rivista di filosofia* **73** (1982), 234–74.

25 See Fechner to Mach, 18 April 1864 and 11 December 1865, in Joachim Thiele, *Wissenschaftliche Kommunikation: Die Korrespondenz Ernst Machs* (Kastelbaum: Herm, 1978), pp. 41–42, 44, 250. Swoboda, 'Physics, Physiology and Psychophysics' (ref. 24), p. 261, suggested that Mach became acquainted with Herbart in 1858 through a course taught by Franz Lott, Vienna's leading Herbartianer of the 1850–60s.

26 See Johann Friedrich Herbart, *Lehrbuch der Psychologie* [1816, 2d ed. 1834], reprinted in idem, *Sämtliche Werke*, ed. Karl Kehrbach and Otto Flugel, 19 vols. (Langensalza: Beyer, 1887–1912), **4**: 295–436, and idem, *Psychologie als Wissenschaft neu gegründet auf Erfahrung, Metaphysik und Mathematik*, 2 vols. [1824–25], reprinted in *Werke* **5**: 177–434, **6**: 1–338. For the intellectual and pedagogical heritage of Herbart's psychology, which I here ignore, see Harold B. Dunkel, *Herbart and Herbartianism* (Chicago: University of Chicago Press, 1970); David E. Leary, 'The Philosophical Development of the Conceptions of Psychology in Germany, 1750–1850,' *Journal of the History of Behavioral Sciences* **14** (1978), 113–21; idem, 'The Historical Foundation of Herbart's Mathematization of Psychology,' *Journal of the History of the Behavioral Sciences* **16** (1980), 150–63; Gary Hatfield, *The Natural and the Normative: Theories of Spatial Perception from Kant to Helmholtz* (Cambridge, Mass.: The MIT Press, 1990), pp. 117–28.

27 Herbart, *Werke* (ref. 26), **5**: 180.

28 Herbart, *Werke* (ref. 26), **4**: 364, 371–2.

29 For example, for the static case of two fully opposed presentations a and b, the total amount of mutual inhibition equals the weaker presentation b, and $(a + b)/b = b/(b^2(a + b))$, and the remaining non-inhibited amount of $a = a - b^2/(a + b)$. For a case of movement, Herbart's simplest law is $s = S(1 - e^{-t})$, where S is the total amount of mutual inhibition, t the elapsed time since the interaction began, and s the supressed portion of the total sum over t. If a completely inhibited presentation Π is being raised above the threshold of consciousness with the help a new presentation P, and if ρ is the remainder of Π not inhibited, and r the remainder of P not inhibited, then $(r\rho/\Pi) [(\rho - \omega/\rho] dt = d\omega$, where ω is that portion of ρ already brought into consciousness. Herbart, *Werke* (ref. 26), **5**: 288, 339, 368–69.

30 Herbart, *Werke* (ref. 20), **5**: 299–300, 415–16; **6**: 192. Such a three-dimensional graphic representation of all possible color hues and saturations had been long discussed by artists and natural philosophers. See Paul D. Sherman, *Colour Vision in the Nineteenth Century* (Bristol: Adam Hilger Ltd, 1981), pp. 60–80.

31 Ernst Mach, 'Vortäge über Psychophysik,' *Oesterreichische Zeitschrift für praktische Heilkunde* **9** (1863), cols. 146–8, 167–70, 202–4, 225–8, 242–5, 260–1, 277–9, 294–8, 316–8, 335–8, 352–4, 362–6, at cols. 169–70, 363, 365–6; cf. Hiebert, 'Atomism' (ref. 24), p. 99.

32 Mach, 'Vorträge' (ref. 31), col. 204.

33 Hermann L. F. Helmholtz: (1887), *On the Sensations of Tone as a Physiological Basis for the Theory of Music*, transl. from the 4th German ed. by Alexander J. Ellis (New York: Dover, 1954), pp. 49–65.

³⁴ Ernst Mach, 'Zur Theorie des Gehörorgans,' *Sitzungsberichte der mathematisch-naturwissenschaftlichen Classe der Kaiserlichen Akademie der Wissenschaften*, **48/2** (1863), 283–300, at 297 (hereafter *SW*). See Swoboda, 'Physics, Physiology and Psychophysics' (ref. 24), pp. 245–47.

³⁵ Ernst Mach, 'Untersuchungen über den Zeitsinn des Ohres,' *SW* **51/2** (1865), 133–50.

³⁶ Ernst Mach, 'Bemerkungen über intermittirende Lichtreize,' *Archiv für Anatomie, Physiologie und wissenschaftliche Medicin* (1865), 629–35. For an overview of previous research on the persistence of afterimages, see Helmholtz, *Handbuch* (ref. 10), pp. 336–56.

³⁷ Joseph Plateau, 'Betrachtungen über ein von Herrn Talbot vorgeschlagenes photometrisches Princip,' *Annalen der Physik und Chemie* **35** (1835), 457–68; Helmholtz, *Handbuch* (ref. 10), pp. 339–40.

³⁸ Adolf Fick, 'Ueber den zeitlichen Verlauf der Erregung in der Netzhaut,' *Archiv für Anatomie, Physiologie und wissenschaftliche Medicin* (1863), 739–64.

³⁹ Mach, 'Lichtreize' (ref. 36), p. 633.

⁴⁰ Ernst Mach: (1865), 'On the Effect of the Spatial Distribution of the Light Stimulus on the Retina,' transl. in Floyd Ratliff, *Mach Bands: Quantitative Studies on Neural Networks in the Retina* (San Francisco: Holden-Day, Inc., 1965), pp. 253–71. Similar phenomena had been noticed at least once before, without attracting any further discussion. See James Jurin, 'An Essay upon Distinct and Indistinct Vision,' in Robert Smith, *A Compleat System of Opticks*, 2 vols. (Cambridge: Printed for the Author, 1738), **2**: 115–71, at 168 and Plate 20, Figure 68.

⁴¹ By adding the second derivative of the intensity of the stimulus as a function of lateral position across the retina, Mach rendered Fechner's law as $e = a \log [i/b \pm k \, (d^2i/dx^2)^2/i]$, thereby making the intensity of sensation (e) a function not only of the intensity of stimulus (i) but also of the rate of curvature of that intensity. See Mach, 'Spatial Distribution' (ref. 40), pp. 262–64.

⁴² Mach, 'Spatial Distribution' (ref. 40), p. 269.

⁴³ Ernst Mach: (1868), 'On the Physiological Effect of Spatially Distributed Light Stimuli', transl. in Ratliff, *Mach Bands* (ref. 40), pp. 299–306, at 306.

⁴⁴ Mach, 'Spatial Distribution' (ref. 40), pp. 267–69.

⁴⁵ Ernst Mach: (1868), 'On the Dependence of Retinal Points on One Another,' transl. in Ratliff, *Mach Bands* (ref. 40), pp. 307–20, at 316–17.

⁴⁶ Ernst Mach: (1860), 'Über die Änderung des Tones und der Farbe durch Bewegung,' reprinted in *Annalen der Physik und Chemie* **112** (1861), 58–76; **116** (1862), 333–38. In the second article, however, Mach agreed with a critic that stellar speeds are too slow for any Doppler shifts in color to be detected. See Swoboda, 'Physics, Physiology and Psychophysics' (ref. 24), pp. 236–39.

⁴⁷ Fechner, *Psychophysik* (ref. 21), **2**: 265–308.

⁴⁸ See Kremer, 'Innovation' (ref. 4), pp. 20–2.

⁴⁹ Fechner, *Psychophysik* (ref. 21), **2**: 285–8, supported this hypothesis by referring to recent anatomical studies of the cholea and the Corti fibers, and noting that Helmholtz in 1859 had extolled its virtues.

⁵⁰ Fechner, *Psychophysik* (ref. 21), **2**: 284, 289–90, 299.

⁵¹ *Ibid.*, **2**: 555–60.

⁵² Ernst Mach, *Compendium der Physik für Mediciner* (Vienna: Braumüller, 1863), pp. 233–4; see Hiebert, 'Atomism' (ref. 24), pp. 86–94.

53 Fechner, *Psychophysik* (ref. 21), **2**: 252, 276, discussed Helmholtz's refutation of Brewster, which makes it all the more interesting that the young Mach mistook Brewster for Young.

54 Mach, 'Vorträge' (ref. 31), cols. 295–8, 316–8, 335–6.

55 Fechner nowhere had restricted the basic color sensations to three, and indeed as noted above had explicitly rejected this feature of Young's theory.

56 Mach, 'Lichtreize' (ref. 36), p. 631. Helmholtz, *Handbuch* (ref. 10), p. 281, also had described black as a "real sensation" and not as the "lack of all sensation."

57 Mach, 'Lichtreize' (ref. 36), pp. 633–5. "Auch zur Young'schen Farbentheorie stehen einige der angeführten Bemerkungen in Beziehung," began Mach (p. 633). The "remarks" in question might refer to the outermost nerve elements, cited above at n. 39. Yet Mach's comment about interacting, mutually inhibiting nerve elements would not provide the basis for his rejection of Young's theory.

58 Adolf Fick, *Lehrbuch der Anatomie und Physiologie der Sinnesorgane* (Lahr: Schauenburg, 1864), pp. 291–4. Fick named red, green and blue as the fundamental sensations, and unlike Helmholtz gave the red curve a double peak in his graph of the response curves for the three receptors. Despite this modification, Fick's textbook strongly supported the remaining elements of Young's theory as elaborated by Helmholtz.

59 Mach, 'Spatial Distribution' (ref. 40), pp. 269–70.

60 Mach, 'Lichtreize' (ref. 36), p. 634, did note that contrast effects can make visible complementary components of a sensation of white; but that fact did not allow the white to be folded into the four simple color sensations.

61 Mach, 'Lichtreize' (ref. 36), p. 635.

62 Hebbel E. Hoff, 'The History of Vagal Inhibition,' *Bulletin of the History of Medicine* **8** (1940), 461–96.

63 Mach, *Analysis of Sensations* (ref. 5), pp. 34–5.

64 For several scattered counter-examples, see Kremer, 'Innovation' (ref. 4), pp. 20–2, 43–4.

65 Arthur Schopenhauer, *Ueber das Sehn und die Farben* [1816, 2d ed., 1854], reprinted in Paul Deussen, ed., *Arthur Schopenhauers sämtliche Werke*, 13 vols. (Munich: Piper, 1911–42), **6**: 28; for the 1854 nearly identical version, see *idem*, **6**: 156–57. Schopenhauer, **6**: 145–60, also related these six apriori colors to a vaguely Goethe-like hypothesis of retinal action. Color sensations, he suggested, arise from the "qualitatively divided activity" or the "polarity" of the retina, and the fundamental six colors correspond to the simplest numerical ratios specifying the degree of retinal activity. Black (0) results from total retinal inactivity; white (1) from complete retinal activity. The complements red and green are equally "distant" from black and white, and thus represent an equal division of retinal activity and inactivity (or 1/2). Yellow represents 3/4 of the full activity of the retina; its complement violet only 1/4 of that activity. Orange represents 2/3 of the full activity, with its complement of blue representing 1/3. For a similar treatment relating fractions of retinal action to the sensations of five "primitive" colors, a number agreed on by "all the world," see Victor Szokalski, *Essai sur les sensations des couleurs dans l'état physiologique et pathologique* (Paris: Cousin, 1841), pp. 29–33, who acknowledged a debt to Schopenhauer.

66 Hermann Aubert, *Physiologie der Netzhaut* (Breslau: Morgenstern, 1861), p. 186.

[67] See Robert S. Cohen, 'Ernst Mach: Physics, Perception and the Philosophy of Science,' in Cohen and Seeger, eds., *Mach* (ref. 24), pp. 126–64.
[68] Ernst Mach: (1933), *The Science of Mechanics*, transl. from the 9th German ed. by Thomas J. McCormack (LaSalle, Ill.: Open Court, 1960), pp. 1–9.
[69] See Hering, *Lichtsinne* (ref. 2), XXXVIII, 196, 200; XLI, 186, 201, 203; XLII, 198. Hering rarely cited other sources in his monograph; among authors he did cite, Mach appeared about as frequently as anyone, along with Helmholtz, Fechner and Aubert.
[70] Hering, *Lichtsinne* (ref. 2), XLI, 186, 201; XLII, 198.

DIANA KORMOS BARKAN

A USABLE PAST: CREATING DISCIPLINARY SPACE FOR PHYSICAL CHEMISTRY

> This is physical chemistry, formerly a colony, now a great, free land.
> J. H. van 't Hoff 1891

The traditional narrative of the founding years of physical chemistry begins during the late 1880s, when the profession became established as part of modern physical science primarily through the determined efforts, and achievements, of a troika of scientists: Wilhelm Ostwald, Jacobus Hendricus van 't Hoff and Svante Arrhenius. By the turn of the century, physical chemistry grew into a lively, well-populated and well-organized scientific discipline. Recently, the historiography of physical chemistry has been criticized because it has purportedly relied too heavily on the self-image of turn-of-the-century physical chemists, who saw themselves as distinctive specialists. It has been said that in general the history of physical chemistry has been written "by the public relations writers of the winning side."[1]

Rather than settling the debate over the role of the old thermochemistry versus the newer chemical thermodynamics, I would like to propose that the birth of modern physical chemistry was constructed in a mold which bears the characteristics of what a group of historians have termed the "invention of tradition." In the early 1980s, Eric Hobsbawm, Hugh Trevor-Roper and others published a volume of essays under the title *The Invention of Tradition*.[2] This volume examines what Hobsbawm termed the mass production of traditions, primarily in Europe, that was most "enthusiastically practiced" in the years 1870 to 1914. The essays were devoted to political movements and social groups, and addressed intriguing issues such as the invention of public ceremonies, the annexation of the revolutionary tradition in France, the mass production of public monuments, or the provision of historical legitimacy for the Bismarckian era by a "merger of Prussian and German history." The forging of a national identity, by the invention of the Scottish tartan or the rediscovery of Druids and Celts as the heroic ancestors of the Welsh people, or the establishment of a "usable past" to serve the British monarchy in India are more closely related to traditional political history. Other less obvious topics were the proliferation of fraternities, old boys dinners, and the invention of the 'school tie' in 1890; the invention of national holidays in the USA, the invention of imperial jubilees and the issuing of historical commemorative stamps; and the invention of international rituals for political movements such as the workers' May Day rallies and demonstrations.[3]

While all the above could be easily categorized as prominent representations of modernity and change, their inherent feature and purpose were the construction of a usable past, of stressing a historical dimension and connection to an uninterrupted tradition. The invention of traditions came in great part as a response to dramatic social change, which was accompanied by a need to reshuffle and reorder hierarchies, ideologies, and customs. This change encompassed to a significant degree academic and scientific life. It is therefore not incidental that the emergence of specialized departments and chairs in scientific disciplines in Germany came not only at a time of rapid industrialization, as the traditional historical explanation would have it: they also came at a time when student enrollment rose dramatically as the result of the embourgoisement of the student population, the growing appeal of higher education for the rising propertied middle classes, a change motivated in part by precisely the same drive towards tradition, towards the expansion of an "upper-middle-class elite socialized in some suitably acceptable manner."[4]

The birth of new scientific disciplines, or subdisciplines, has been dated as of these same generations. The 1880s witnessed the expansion of physical chemistry, biochemistry, experimental psychology with the attendant landmarks: the proliferation of chairs, institutes and journals, the influx of foreign students into the European, primarily German, academic graduate seminar and laboratory, the establishment of professional organizations and more generally, the beginnings of international meetings, congresses and networking.

I would like to suggest that the events linked to the self-definition and professionalization of physical chemists were a characteristic episode at the turn of the century, characteristic precisely of the changing social fabric of the scientific community and of the invention of tradition. It was reflected in contemporary literature pertaining to this self definition, with its need for justification. This literature certainly influenced later historians of chemistry, whose work however is not necessarily a deliberate "whiggish" misinterpretation.

In the mid-nineteenth century, early positivists had postulated the desirability for the social sciences to emulate the methods and practices of physical sciences. By the last third of the century, uneasiness with such reductionism led to a call for a unique methodology of the social sciences from several directions. Among others, neo-Kantians such as Wilhelm Windelband and Heinrich Rickert, framed the distinction between idiographic and nomothetic sciences, namely between historical and social specificity as opposed to scientific universality of natural laws. While these antitheses have been the subject of extensive scholarship, the use of historical models and insights for the development of a physical scientific discipline has been overlooked.[5]

It is therefore of considerable import that several physical chemists at the turn of the century sought to justify the construction of a space for their new discipline along historical lines, emulating recent precedents set by Ernst Mach in his positivist program and by Ludwig Boltzmann in his effort to bolster atomism. The traditional physical sciences – physics and chemistry – had supposedly reached by the 1870s and 1880s a satisfactory stage at which no further fundamental theoretical advances seemed possible or even necessary, while their remaining task called for precision measurements and detailed descriptionism.[6] Thus it was essential for physical chemists to articulate an adequate justification for the elaboration of a new science which would address foundational issues. Historians have often claimed scientific rationality as a guide to historical methodology. In our case, it was the insights gained from historical studies that helped physical chemists justify a scientific enterprise.

The new "science" was created by its founders on three levels: the historical, institutional, and theoretical. The practitioners of physical chemistry sought to find professional recognition by highlighting historical circumstances and ties to the past of established scientific disciplines; by forming schools of physical chemistry, specialized journals and organizations, and by promoting a number of novel theoretical conceptions which claimed fundamental ontological progress in the understanding of physical and chemical processes. By examining several programmatic texts written by influential physical chemists between 1890 and 1915, I will show how each of them used history to construct a scientific tradition and thereby justify and promote the new field. Their historical accounts vary: according to their timing, to the author's background and agenda, each writer claimed different scientific ancestors. Yet, their texts reflect a progressively increasing reliance on physics as the legitimate ontological precursor to physical chemistry.

Perhaps the first comprehensive effort at such an enterprise was made by Ostwald in his impressive history of *Electrochemistry*, published in 1896, a study which was to be emulated during the first half of the twentieth century in numerous programmatic declarations by van 't Hoff, by Nernst to a certain extent, and by many of their students.[7]

Wilhelm Ostwald belongs to the group of leading representatives of German positivism. In his numerous publications, Ostwald was mapping out a grand strategy: his foremost goal was to formulate, if possible, natural laws for the development of scientific thought. These universal systems were popular endeavors among nineteenth-century thinkers, the more famous among them being those of Auguste Comte, Herbert Spencer and Karl Marx. For Ostwald the history of science parallels its logical development, and although he conceded that the historical method may not provide the shortest route to scientific study, Ostwald still considered it to be "the most successful and attractive" mode of scientific education. Ostwald's purpose in the *Electrochemistry* was not only to provide a coherent history of a specific discipline but also to answer the question whether "it be at all possible to establish general laws for history." The book represented an "experiment" in this direction, and a successful one in Ostwald's final analysis.[8] The core of Ostwald's historical work aimed at showing that the historical development of scientific ideas, theories and concepts occurs according to certain laws, similar to scientific laws. He compared the process of scientific development to the process of crystallization of substances from a liquid solution, *Auskristallisation*: a new scientific theory will crystallize because it is isomorphous with other similar theories. But only rarely will this product be pure. It will contain numerous impurities and will continue to remain such for many decades, due to the persistence of old, sometimes even mistaken conceptions, and only conscious, active interventions will ultimately succeed to bring about a fresh reformulation and a definite theoretical breakthrough.[9]

Secondly, Ostwald aimed to develop in the spirit of positivism a completely hypothesis-free chemistry as laid down in his *Prinzipien der*

Chemie (1907), and as postulated earlier in his *Überwindung des Wissenschaftlichen Materialismus* (1895). For Ostwald, the subject of scientific investigation at its most fundamental was the exploration of human sense perceptions, namely our knowledge of phenomena. Such knowledge Ostwald judged to be inherently relative, while the search for essences – for the true causes of phenomena – is futile and misplaced. In line with Mach, Henri Poincaré and other contemporary positivists, Ostwald deemed metaphysical preoccupation fruitless, since no avenue of correlating knowledge thus derived to observation could be found. His anti-atomistic stance and his later preoccupations with the development of energetics as an overarching conceptual scheme for all sciences, physical and otherwise, made Ostwald into an admired, revered and much contested scientific and cultural figure of the German-speaking world, an apostle and a heretic in Europe, England and America alike. In contradistinction to early Comtian positivism, Ostwald and Mach postulated a science "completely free of any arbitrary hypothesis," related to empiricism, which would allow for the existence of real objects of which we gain knowledge through perception.[10]

However, what is rarely spelled out by historians of the period is Ostwald's tolerance towards a multiplicity of explanatory schemes: he emphatically denied the conclusion that energetics is the only and a sufficient vehicle for the understanding of nature:

Even though the energetic world view has immense advantages over the mechanistic or materialistic one, a few points which are not covered by the laws of energetics indicate that other principles exist which transcend energetics. *Energetics will remain side by side with these new laws . . . and will constitute a special case of more general conditions*, of whose future shape we hardly have any premonition at the present time. [emphasis added][11]

And thirdly, Ostwald's history of electrochemistry constituted the first major undertaking to create space for the new discipline of physical chemistry through historical exposition. Entitled *Electrochemistry. Its History and Theory*, this massive tome is a serious and probably still unsurpassed attempt at a total disciplinary history. Ostwald's main line of argument was that:

An ever recurring experience as researcher and teacher has convinced me that there are no more effective means for the stimulation and deepening of study than the penetration into the historical creation of [scientific] problems. We often see that conceptions which at the present seem to us imperturbable pillars of the scientific edifice have been subject to violent attacks at the time of their inception – the greater their importance, the more they were attacked – while other notions that were treated at their time as unquestionable truisms seem nowadays so preposterous that we cannot understand how they could have ever been conceived.

This passage points to the core of Ostwald's disputes with the scientific community. Ostwald used history as a vehicle to legitimate scientific knowledge and his own professional position, which was contested at the time mostly on account of his energeticist program. One of the main regularities which Ostwald detected in the history of scientific ideas was the fact that novel theories or hypotheses are unfailingly met with hostility, a process which occurs in two stages: at first comes total negation, primarily justified by the crass contradiction with well established views; to be followed by the invariable claim that the new ideas are in fact old, and that forerunners exist to most any innovation. Ostwald had begun work on his *Electrochemistry* in 1893, but by the time of its publication in 1896 one could not avoid but associate its ideology with that most shattering experience which had been the Lübeck meeting of the German Association of Scientists and Physicians at the end of 1895, at which his new science and principle of energetics had been vehemently opposed by the foremost physicists of the time.[12]

The self-definition of physical chemistry at the time rested heavily upon the acceptance of the Ionist program, of persuading the community of physicists and chemists that free ions exist in solution and that electrical and chemical phenomena can be accounted for only if such ions are assumed and proved to exist. Therefore – for programmatic reasons, I believe – a major part of Ostwald's book is devoted to the work of Alessandro Volta around the year 1800, and the ensuing rivalry between those supporting a physical as against a chemical interpretation of Volta's theory of the galvanic cell – the so called voltaic pile. Volta had held that the electric current between two or more metal plates immersed in a solution is solely the result of the differences in potential between the two metals, ignoring the conductive role of the solvent. His views were further explored by a number of physicists, such as G. S. Ohm, whereas another group, including Humphry Davy and Michael Faraday, explored the source of the electric current and found it to lie in the chemical processes taking place in solution.[13]

In his history of electrochemistry, in which he decries the pernicious effect which Volta's contact theory had had on the subsequent development of the field, Ostwald explicitly denied that Volta's 'inability' to perceive the correct role of the solvent in producing an electric current between two metal plates in a conducting solution should be criticized. He considered Volta to have interpreted "the phenomena such as they presented themselves to him in the context of his time [and] . . . corresponding completely to the regular pace of scientific development." He reminded his readers that there had once been no good reasons to doubt that the sun circles the earth solely on the basis of appearances. It was only the scientific examination of the problem, that is "the need to understand this phenomenon in the context of other related phenomena" which led to a reversal in our conceptions.

Similarly, Ostwald contended, only the more recent necessity to *explain the processes* taking place between metals and solutions resulted in a reconsideration of Volta's seemingly persuasive experiments. Volta was therefore not to be imputed scientific error, since he had persisted in his contact theory primarily due to prior commitments to a physicalist interpretation of electrical phenomena, in his opposition to the vitalistic views. However, Ostwald emphasized, *posterity had to be faulted* for persisting in supporting Volta's theory in the face of "sufficient and decisive" evidence to the contrary. A great number of scientists had been willing to vehemently defend rather than thoroughly examine an inadequate theoretical interpretation.[14] The longevity of the dispute, which lasted until the latter part of the nineteenth century, was in Ostwald's opinion due to the lack of a rigorous theoretical basis of chemistry. This gap, he considered, had only lately been filled by "an all-round satisfactory chemical theory of the Voltaic pile" provided precisely by the new physico-chemical theories of the 1880s.[15]

Ostwald thus placed the debates over the validity of the ionic dissociation theory, in which he was embroiled, within a sweeping historical perspective: its origins reached into a significant past, although they could only with difficulty be associated with present conceptions. But, according to Ostwald, the history of electrochemistry contains two parallel chains of events which have pitted against each other two opposing scientific world-views – the physicalist against the chemical. And although the great themes, or world-lines of scientific thought are continuous, and are only in rare instances reconcilable, it was only with the new physical chemistry that old enmities would now be resolved.

Ostwald's 1896 history explicitly claimed that physical chemistry was a natural outgrowth and culmination of the development of electrochemistry. The nineteenth and final chapter, entitled "The Theory of

Electrolytic Dissociation," makes interchangeable references to electrochemistry and the "new physical-chemical theory" and concludes with the remarkable statement that, by virtue of the Ionic theory of Arrhenius and van 't Hoff, the future holds in store *a "revolution" (Umwälzung) comparable to that brought about by the steam engine.*[16] The seeds of self-conscious historical writing, later picked up by students of the Ostwald school, were sown when Ostwald compared the work that "emerged from the rooms of the Leipzig physical-chemical institute" by a handful of scientists to the "events connected with the elaboration of the antiphlogistic theory by Lavoisier and his collaborators." "This research found solutions for old puzzles," he wrote, "and annexed region after region to the empire of scientific electrochemistry."[17]

Thus, almost a decade after the foundation of the *Zeitschrift für physikalische Chemie*, Ostwald was able to close the circle: In both cases "A few men, at first opposed with scorn and anger, eventually made their way through the world and gave science a new face."[18] In 1887 he had written that "physical chemistry is not only a branch but it is the flower of the tree." Physical chemistry was to many readers of Ostwald's rhetoric many things at the same time: the rational theoretical foundation of chemistry; a branch of chemistry; a direct descendant of electrochemistry; the crowning achievement in the history of chemistry.

And yet, this historical view was only partially shared, that same year, by Ostwald's most successful pupil in Germany at the time, Walther Nernst. In 1896 Nernst became the director of the first research institute devoted explicitly to electrochemistry and physical chemistry. At the opening ceremonies of the Göttingen institute, Nernst gave a lecture entitled: *The Goals of Physical Chemistry*. The greatest part of his lecture was devoted to a historical exposition of the relationship between chemistry and physics during the nineteenth century. Nernst's main thesis was an elaboration upon Helmholtz's dictum that physics forms the theoretical basis of all other branches of scientific research, which should be expanded and modified in the sense that the collaborative activity of physics and chemistry were at the time the foundation of the scientific world view. Nernst supported this thesis by adducing the historical argument that it was essentially between the years 1835 and 1885 that physics and chemistry developed as separate sciences. Before 1835, he claimed, physics and chemistry were still enmeshed in a struggle for their existence against Naturphilosophie, while after van 't Hoff's famous paper on dissociation in 1885 a new, rich, and rewarding field opened up, that of physical chemistry.[19]

While praising the extraordinary progress achieved precisely through this "division of labor" by both sciences, Nernst argued that after 1885, physical chemistry assumed essentially the role of a "diplomatic mediator" between physics and chemistry. Physical chemistry was to fill the unexplored no-man's-land which lies between the two, by applying simultaneously the methods of both sciences. What had contributed to these transformations? The recent conceptions, Nernst argued, arose because during the preceding fifty years "a series of general and, in their simplicity, exceptionally important and useful natural laws had been established which made it possible to encompass a very large number of facts in few words or formulae."[20] These were the principles of thermodynamics, the Maxwell–Hertz equations of electrodynamics, the theory of electrolytic phenomena, and the principles of the kinetic theory of gases; at the same time chemistry brought forth the tables of atomic weights, the periodic system, the constitution of organic substances etc.

Nernst's rhetoric was addressed at the new generation of scientists: attention was to be paid to advances in both physics and chemistry, in order to achieve a general, new conception of natural phenomena.[21]

Only two years later another well-known chemist, Pierre Duhem published an extensive, two-part article entitled "Une science nouvelle. La chimie physique."[22] Duhem had been a major contributor, together with Boltzmann, Max Planck, and J. W. Gibbs, to the formulation of theoretical, physical-mathematical thermodynamics, in his case with particular regard to chemical phenomena. His position within the French scientific community was controversial, since his views were largely opposed by a large segment of Parisian chemists, who generally belonged to the Berthelot chemical school. Duhem was resentful towards those establishment physicists and chemists who had excluded him from the city of lights and condemned him to a life in the provinces and no access to the prominent forum of French science. His historical exposition of a "new science," the development of physical chemistry, was thus largely an exhortation not only against each science in particular, but also a sarcastic pastiche of the destructive enmities and intolerance among physicists and chemists.[23] Duhem's aims in writing "Une science nouvelle" are obvious, if not explicit: to bolster his own position and that of the science and the ideas which he represented. Like Ostwald, he sought to "create space" for a neglected scientific domain by tracing the history of physical chemistry, but his search focused on French origins. By describing the demise of an earlier physical chemistry in France and its concurrent acceptance and flourishing in other countries, especially Germany, Duhem stressed the great economic and national advantages brought to foreign nations as a result of their support for the new science.

Duhem devoted the first installment of his essay to a history of physical chemical concepts, of thermochemistry, specifically in France, of "[T]he plant which had grown on our soil, withered and dried, but [whose] seed, dispersed to the four winds, did not take long to germinate and bring an abundant harvest" in Holland, Germany and later the United States. Duhem explicitly tied national prosperity to investment in science and, quoting Alexander von Humboldt, Duhem accentuated the French shortcomings: "The countries who neglect to rely on scientific luminaries will see their prosperity endangered, while the neighboring nations develop and strengthen themselves under the influence of the arts and sciences."[24] Humboldt, the great promoter of German science and admirer of France, had warned the Germans against scientific backwardness. Duhem returned the compliment, warning the educated French that the prowess of the German people was a result of concerted scientific policies. The threat was explicit, made less than 30 years after the debacle that Bismarck had brought upon the French, and in the midst of imperialist competition among European nations.

In fact, Duhem's entire piece is designed according to the prescription "divide and conquer," utilizing a traditional scientific rivalry between physics and chemistry as a justification for the new physical chemistry. Presenting the methods of physics and those of chemistry as being "profoundly divergent," and the two disciplines as having been rivals at least since the seventeenth century, Duhem, similarly to Ostwald, argued that since the time that Newton had asked his last query: "Could gravitation also be an explanation of chemical reactions?," two lines of thought had developed during the 18th century: the elaboration of mechanical physics by Pierre-Simon Laplace (and its expansion into electrical mechanics by Simon-Denis Poisson, Augustin Cauchy, Charles Coulomb, André-Marie Ampère) and of a new mechanical chemistry. The new school, initiated by Pierre Macquer and Louis Guyton de Morveau, "professed that chemistry must be treated by physical methods; that chemical reactions can be reduced, in the ultimate instance, to mechanical effects; that they [the reactions] are explainable in the forces of *affinity* which act, according to Newton, between the very close particles of bodies."[25] But because the experimental data of chemistry were still vague, these mechanical applications were at first misplaced: affinity tables did not succeed to match the facts. Therefore, Duhem continued, the chemists of the empirical school, G. F. Rouelle's students, triumphed amidst these contradictions, ridiculing the 'theoretical chemists, the systematicians, the rationalists, the makers of tables.' What Macquer had termed "our modern physical chemistry" (notre chimie-physique moderne) was eventually victorious because of Antoine Lavoisier, whom Duhem completely appropriates for the new science:

Lavoisier was a physical chemist because he was constantly concerned with clarity and logical rigor; because with the help of the physicist's instruments – the balance, the thermometer, the calorimeter – foreign until then to the chemist, he introduced into chemical experiments an unthought-of precision; because the last adepts of the phlogiston theory, the most irreconcilable adversaries of the new theory of combustion, are precisely the empiricists, the bitter opponents of the doctrine of affinities; he belongs, finally, because his first partisans, his collaborators are precisely those who developed the system of molecular attraction, Guyton de Morveau, Fourcroy, Berthollet, Monge, and especially Laplace.[26]

In Duhem's view it was Claude Louis Berthollet, working in a laboratory adjacent to Laplace's garden in Arcueil, who achieved the crowning synthesis of chemical and physical mechanics, but was deposed from the heights of scientific achievement by Proust's law of definite proportions. This victory established an "impassable frontier between chemical reactions and physical changes of state, and became the dominant law of all chemistry." Affinity was forgotten and physical chemistry in France withered away. From then on, the new science developed elsewhere and France fell behind.

Duhem's heroes are the great synthesizers of ideas. Newton, Laplace, and Berthollet are followed by van 't Hoff, who explored the consequences of energetics (i.e., the thermodynamics of Rudolf Clausius, J. W. Gibbs, and Helmholtz) for chemical reactions, drawing analogies between gas laws and dilute solutions. Again the term "new science" is used, and continuity with other "new sciences" is implied: mechanical physics, chemical thermodynamics, and physical chemistry compose a continuum, a historical tradition.[27] Arrhenius and Ostwald were considered by Duhem to be "van 't Hoff's respected emulators," who tied together the laws of electrochemistry and the ideas of the Dutch scholar, followed by a whole "legion" of workers who tap each vein of the opened mine. From here on, the account turns into a barrage of institutional achievements, primarily in Germany, Holland and the United States.

It is again by historical and national example that Duhem reached out to impress upon his public the nefarious consequences of an indifferent, centralized, atrophied, and obstinate scientific establishment. The opposition to the "reform of organic chemistry" in the mid 1800s left the field wide open to the rise of German chemical industry: French scientists refused to cooperate with industry; they "opposed the adoption of novel conceptions introduced into chemistry by Gerhardt and his successors." That part of chemical industry dependent on the new theories "has failed in France while it came to full fruition in Germany."[28] And then Duhem issued a final warning:

> Beware! If the same routine, the same hatred of all that contains a novel idea, the same horror of all that demands a new intellectual effort, will hamper any longer the rise of physical chemistry and electrochemistry in France, truth, outraged by those whose profession it is to serve her, will revenge herself once more by inflicting another industrial disaster ... [29]

Despite the establishment of electrochemical industries in the Savoie, for example, its management and technical staff were trained abroad, and were followed by foreign workers and foreign machines, because "the state has created in France only one course of physical chemistry" by appointing the young physicist Jean Perrin to a Paris position.[30] Duhem suggests that, besides the electrochemical institute in Nancy, the other university centers such as Lyon, Marseille and Grenoble should receive equal attention. Physical chemistry should become one of the required examinations for those preparing themselves for a career in chemical industry, insists Duhem; it is such a policy which made Göttingen and Leipzig successful and which would assure French industrial advancement. Rather than developing "another few kilometers of local railways," it is investment in higher eduction which will bring with it national influence in the world.[31]

The essay stressed nationalist elements, and also highlighted Duhem's frustration as a provincial, isolated scientist, his disdain and mistrust of the centralized French academic system, its neglect of industrial necessities, its staid and conservative attitudes towards new scientific ideas.

Both arguments – in fact the entire canvas composed with characteristic French love of metaphors combined with rational precepts and reverence for learning – were meant to impress the public, the scientists and the Paris legislature. What seemingly began as a historical and mundane journalistic piece of general interest concluded as a powerful appeal to the strengthening of France's position in the outside world. Duhem, who from the beginning had contributed and joined the editorial board of the *Zeitschrift für physikalische Chemie*, and had been in close touch with most European and American practitioners of the new "discipline," was a central member of this group at least in his historical missionary style, if not in its conceptual tenets.

In 1888, Ostwald and van 't Hoff had solicited Duhem's collaboration to the journal. Duhem wrote to van 't Hoff that he had wished to do so since the inception of the journal, a journal which in the future "will give such real services by calling the attention of chemists to the study of mechanical chemistry." Ostwald and van 't Hoff were very appreciative of Duhem's willingness to collaborate, since he had also just published a mathematical treatment of the dissociation of solutions.[32] Through this collaboration van 't Hoff saw the emergence of a real "dynamicists" party, and jokingly exclaimed "May God protect our souls."[33] While praising its achievements and virtues, Duhem explicitly used the ascendancy of physical chemistry abroad for his own purposes within the French context, as a warning to French society that former enemies were threatening France not only in terms of pure science, but with attendant international economic and industrial repercussions. The historical method, used to extol the virtues of past French science and nationalistic themes, served to advance the new physical chemistry.

In contrast to Duhem's nationalistic motivation for developing physical chemistry, his Dutch counterpart van 't Hoff underscored the universal scientific benefits of the new science and its applicability to numerous other disciplines and domains.

In June of 1901, van 't Hoff delivered to the University of Chicago a series of lectures on "Physical Chemistry in the Service of the Sciences."[34] The lectures were divided into four parts, attempting to present the achievements and contributions of physical chemistry from four perspectives: the application to pure chemistry; its relation to applied or technical chemistry; its relation to physiology; and to geology. It is evident from the design and intent of the presentation that van 't

Hoff was bringing new knowledge to the United States, performing what at the time had become a traditional role: that of a European ambassador of science to the young and vibrant American scientific world in which physical chemistry was beginning to take roots.

A number of American scientists had by then received European training, in particular in Ostwald's laboratory, including A. Noyes, H. C. Jones, and J. Loeb, who had worked with Ostwald and with Nernst. By then, physical chemistry had been introduced in a number of courses, in particular by Wilder D. Bancroft and J.-E. Trevor at Cornell, and the *Journal of Physical Chemistry* had begun publication in 1896. Already by 1899 G. N. Lewis – in an important paper on chemical equilibria read to the American Academy of Sciences by Harvard's T. W. Richards – had exhibited the universality and strength of the new ideas bolstered by self-confidence, using scientific examples coupled with historical arguments:

The advance of modern physical chemistry has been largely due to the application to physico-chemical problems of the first and second laws of thermodynamics and the gas laws ... Upon this basis the whole theoretical treatment of chemical equilibrium rests at present.[35]

In a characteristic enlightened and self-conscious manner, van 't Hoff depicted the new science of physical chemistry to his American audience by providing a rich historical perspective. He took his cue from the chemist-historian Alfred Ladenburg, whom he considered one of the foremost historians of chemistry, and who had shown that the "most characteristic feature of chemistry" in the years 1885 to 1900 was the "continued increase of this physical, or general, chemistry."[36] Van 't Hoff recalled that, when he had been introduced to chemistry in 1870 in Bonn as a student of Auguste Kekulé, the esteemed professor had held chemistry to be "without visible prospect of new advance."[37] This attitude, alluded to earlier, is once again the source of historical justification. For the young, aspiring, and extremely idealistic van 't Hoff, an admirer of Byron, the stiff atmosphere in Kekulé's laboratory and the perception of drudgery and routine without prospects for intellectual satisfaction apparently did not lead to disillusionment with chemistry in general, but with chemistry such as it was presented to him at the time. Van 't Hoff defined his own work as innovative and utilized historical narrative to highlight novelty and rejuvenation as concrete motives for the birth of a new science. For chemists in the 1870s the belief in the existence of atoms "seemed to be well-founded," claimed van 't Hoff, a belief enhanced by Kekulé's development of stereochemistry to such an extent that "the real existence of the atoms being assumed, not only was their mode of union described, but also their relative positions in the molecule was determined."[38] And even though van 't Hoff's brief training in Kekulé's laboratory seemed to have left little imprint on both teacher and pupil, van 't Hoff was soon to make major contributions to the further elaboration of spatial representations and formulae of chemical substances.

Van 't Hoff had gone to Bonn in 1872 to study in Kekulé's laboratory, checking out the possibility of an academic career in chemistry. But he had been discouraged by his parents, to whom he was extremely devoted, to pursue a metier which was of little prospects. Consequently he resolved to link chemical experience with a technical engineering education, to be later acquired in Utrecht. Van 't Hoff did little to distinguish himself among the twelve other students engaged in organic chemistry research in Kekulé's laboratory. His letters home and his anguished relationship to his teacher reflect the slow progress and the lack of creativity which both van 't Hoff and Kekulé seemed to perceive during those brief ten months of apprenticeship. We know little about the precise research goals of the group, and it seems that what the student was supposed to achieve was "to find something new."[39] When in October 1874 van 't Hoff completed his doctoral research at the University of Utrecht, after several months' work in the laboratory of Adolphe Wurtz in Paris, he had already published an eleven-page essay in which he showed that structural formulae of organic molecules as employed at the time did not satisfactorily account for isomerism, a subject which was to introduce the later famed tetrahedral spatial representations in three dimensions – as opposed to the customary plane constructs.[40]

When ten years earlier, in April 1891, van 't Hoff had described the state of research in physical chemistry at the third Congress of Dutch Scientists and Physicians held in Utrecht, he had compared it to a "new world" which emerges between the great continents of physics and chemistry, clearly reminiscent of Ostwald's metaphor of "new annexed regions." This new world first appeared as clusters of islands, then, on both sides, there emerged mountains which gradually formed bridges, and occasionally swamps. "This is physical chemistry, formerly a colony, now a great, free land" which would yield rich harvest to the scientific community. Van 't Hoff used the metaphor of America as a new world to describe the efforts in support of osmotic pressure studies. The endorsement was eventually given by the physicists: "But the physics gentlemen in our new continent had a kind of American experience: we were, one must confess, quite in a hurry, while our more aged sister had already overcome this stage. And so it happened that, first gradually, after we had already enjoyed some results, the mathematical-physical treatments followed: Max Planck, Duhem, van der Waals, Lespieaux [sic], Lorentz, Boltzmann; but we can rest assured: osmotic pressure is well seen by all of them."[41]

Van 't Hoff had been wondering, prior to his lecture in Utrecht, how to make an attractive presentation of physical chemistry. He had written at the time to Ostwald:

I have read with much interest your lecture in Bremen, especially because I wished to know whether anything understandable can be produced for a mixed audience. You have solved the problem *historically*; I am thinking for Utrecht about a *physiological solution*. [emphasis added][42]

By 1901, the successful and detailed study of osmotic pressure had become one of physical chemistry's "best known and most far-reaching achievements" – largely due to van 't Hoff's elaboration in his *Études de dynamique chimique* (1884) – and thus van 't Hoff constructed much of his Chicago lectures around this problem. In January 1894, in a lecture presented to the German Chemical Society in Berlin, van 't Hoff had introduced the origin of his *Études* as the interest he had had in the differential oxidation rates of organic compounds and the observed phenomenon that oxygen acts as a catalyst [*beschleunigende Wirkung*] in some reactions. Thus he had become involved in reaction velocities. However, he also recounted another, anecdotal version, which related that his *Études* originated in difficulties encountered while attempting to establish stereochemical properties of a "beautiful" succinic acid bromide, which did not yield the expected results. Since the acid reacted in aqueous solution at one hundred degrees at a very easily observable pace, he used it for dynamical experiments, which led to the *Études*.[43]

Indeed, van 't Hoff went on to state that as opposed to the "colossal labors which were spent in the service of atomic conceptions," apparently unsuccessfully, physical chemistry has solved very complex questions, as for example the importance of osmotic pressure to the physiological functions of living organisms.[44] Taking his clues directly from Nernst's *Theoretische Chemie* (1893), van 't Hoff used the example of the law of Avogadro and its relationship to osmotic pressure, which combined in the theory of solutions, to produce "the first principle which has contributed to the recent development" of the new discipline.[45]

What was the meaning of physical chemistry for van 't Hoff in his missionary enterprise? In a general sense it was the "application of physical expedients, methods, and instruments to chemical problems." Like Ostwald and Duhem, he argued that Lavoisier could be the legitimate founder of physical chemistry since he had applied "the balance . . . to the testing and establishing of fundamental laws in the realm of chemistry," together with Robert Bunsen and his use of spectral analysis. On the other hand, van 't Hoff subscribed more fully to Nernst's assertion in the Address at the Opening of the Institute of Physical Chemistry and Electrochemistry, delivered in Göttingen in 1896, that physical chemistry had developed essentially within the last generation, that is, within the last fifteen years of the nineteenth century. Hence van 't Hoff relegated Bunsen's spectroscope and the calorimeter used by Marcellin Berthelot and Julius Thomsen to the status of precursors. Although highly important, their uses were limited to a particular phenomenon such as light emission or heat production.[46] The new physical chemistry, however, had not primarily introduced new instruments or new methods of observation but established "comprehensive principles which fertilized the whole foundation of the science" and established new laws. In his Utrecht lecture, van 't Hoff had discussed Helmholtz's Faraday Lecture of 5 April 1881 in London, where the doyen of German science had urged chemists to deal more extensively with the study of electrical problems. "Since then," writes van 't Hoff, chemists have "striven and have formed a clearly defined view of the ionic dissociation caused by water, based on electrical and chemical research." Wouldn't it be time, van 't Hoff asked, that physicists should express an opinion on the matter, now that ten years have elapsed? He pleads, however, to use restraint with the notorious inkwell: Duhem, for example, had given a formula for osmotic pressure which contained 41 symbols, written "next to each other, above, under, and . . . confused (*durcheinander*)" letters. "But the problem of ionic dissociation is one of insight into living nature, not into symbols."[47]

He accorded the greatest merit in the development of the new science to Ostwald, Arrhenius and Nernst. Their work opened wide prospects, in particular for the field of inorganic chemistry, he argued, basically due to the simpler character of inorganic substances. Physical chemistry had provided inorganic chemistry with a new working method; with a principle which can predict direction and how far a chemical reaction will proceed; and had elucidated the nature of the solutions of electrolytes (bases, acids, and salts.)[48] In his subsequent lectures on the uses of physical chemistry for neighboring fields, van 't Hoff made almost no reference to its applications to organic chemistry. The reasons for such omissions are sometimes evident from his repeated stress on the simplicity of inorganic reactions and substances and the generality of physical chemical laws which need not take into consideration the underlying composition or structure of the substances involved. But a hidden agenda was the revival of inorganic chemistry at the expense of the flourishing organic chemical establishment, which had dominated the German chemical community in particular since the mid-nineteenth century.

Van 't Hoff's allegiance to the dualistic program of atomism and classical thermodynamics led him to stress the application of thermodynamics to chemical investigations, without however providing much insight into the origins and motivations of this approach. The principle of the conservation of energy is simple, but the Carnot–Clausius principle, the second law, is difficult to approach and apply. Therefore two options are available to the investigator who would apply it to chemical problems: "... by carrying out so-called reversible cycles of operations" or by introducing "abstract physical conceptions and mathematical functions, such as entropy, as is done by physicists like Gibbs, Planck, and Duhem." Van 't Hoff thought the first to be better suited, even though it entails defining "reversible cycles," and described the example of phase transitions in ice-vapor-water-ice; the transformations of rhombic and monoclinic sulfur; of cyamelid and cyanuric acid. And in this context, he again stressed the fact that when applying thermodynamic principles to such investigations, the composition of the substances "is of so little importance that atomic and molecular conceptions need not be considered ... "[49]

Van 't Hoff attempted to create space for the discipline by using both the history of scientific ideas and the prospects opened by physical chemistry to a great number of other sciences and practical applications. He established continuity with chemistry, the kinetic theory of gases and with the simpler concepts of heat theory – such as reversible processes. However, he consciously refrained from letting physical chemistry become entangled with the more complex "abstract" and mathematical formulations of the newer thermodynamics, i.e., with problems related to the second law, with which chemists and biologists were poorly acquainted and even opposed to applying. The context in which the lectures were delivered was also of some importance: in order to be an effective messenger of the new science to the United States, van 't Hoff adopted the pervasive view – which persists to our day – that American scientists preferred to operate with experimental evidence and visualizable concepts. Van 't Hoff therefore stressed the advancements which the new science would produce in physiology, mineralogy, metallurgy and traditional fields of chemistry, implicitly appealing to the American style of scientific research and to the needs of a rapidly expanding chemical industry.

The history of physical chemistry as it emerges from the introductory remarks of van 't Hoff in 1901 to the American scientific public is a roller-coaster of success waves. Never does he waver in his conviction that physical chemistry and its brief history embody and are bound for self-evident success of novel ideas. This atmosphere and rhetoric has led scholars, as we have seen, to challenge some aspects of recent scholarship, and to take issue with the "omniscient" characteristics of some historical writing that focuses on only a handful of major personalities, without taking into account the specific research traditions which contributed to the development of physical chemistry.[50] This critique, however, overlooks the functionalist aspect of these contemporary histories in the formation of a disciplinary identity, a process by no means peculiar to physical chemists. Physicists and chemists alike had used scientific and economic arguments in securing state support and status, in particular in those periods when a humanistic education was often furthered to the detriment of the natural sciences.

This need for securing an ancestry and tradition is reflected by the changes brought about in self definition after the advent of quantum theory and the increasing physicalist outlook of modern chemistry after the turn of the century. By 1913, physical chemical textbooks contained what is now the accepted chronology for physical chemistry, one that rendered obsolete the historical postulations of Ostwald, van 't Hoff and Arrhenius, but one that also cast its own tradition, its own space in the expanding world of science.

By the second decade of the twentieth century, this changing self-image within a received tradition is reflected in the work of Karl Jellinek.

A thirty-one year old Privatdozent at the Königliche Technische Hochschule in Danzig on the Baltic Sea, Jellinek completed in 1913 a voluminous textbook on *Physical Chemistry of Homogeneous and Heterogeneous Gas Reactions with Special Consideration of the Theory of Radiation and Quanta and Nernst's Theorem*.[51] He established a chronology according to major theoretical landmarks during the past generation, thus tying the origins of the new science to moments in history at which various "groups of ideas" entered "chemistry from physics." Thus, the first creative decade began with the application of the first two principles of thermodynamics to chemical problems by van 't Hoff (1884), leading to the theory of dilute solutions (van 't Hoff 1885), the electrolytic dissociation theory (Arrhenius 1887) and the osmotic theory of galvanic cells (Nernst 1888/89). This was followed by a second period from 1890 to 1905 in which fundamental theories were further explored and expanded. The discovery of a third principle of thermodynamics (Nernst 1906) initiated a new creative period of physical chemistry, during which Planck's quantum theory (1900) entered rapidly into the new science. The fourth period, whose inception Jellinek placed at the time of publication, was heralded by Nernst's theorem, Planck's quantum theory, the increasingly clearer conception of the atomic structure of matter and electricity, and the budding influence of electron theory and modern radiation theory on physical chemistry.[52]

By the time a new generation of scientists was entering an essentially changed academic environment, the shift in emphasis from chemistry to physics had been complete. Lavoisier, electrochemistry, and heat theories retreated into the distant past. No longer was physics a remote ancestor, but rather a direct and immediate influence at all stages of the history of physical chemistry.[53] The self-definition of physical chemists therefore changes to a large extent after the development of the quantum theory and after Nernst's postulation of the heat theorem at the end of 1905. With several strong physico-chemical institutes in place in all major European and American universities, with the appendage of widely circulating journals, new students were to be lured into the discipline by appealing to the successes already achieved by physical chemists in the recent past.

The power of historical arguments has persisted into the twentieth century. The introductory remarks to be found in many first volumes of new scientific journals attest to the tension inherent in the process of scientific specialization. On the one hand, the desire to be innovative, to participate in high-powered research at the frontiers of knowledge, motivates the editors. On the other hand, the need to justify a new scientific enterprise in terms of a long established historical tradition, of building upon the shoulders of significant predecessors, compels scientists to produce programmatic statements. In 1962, with the publication of the first volume of *Advances in Physical Organic Chemistry*, Sir Christopher K. Ingold wrote in its foreword:

The appearance of this series ... marks the most mature of the steps to ensure continuing progress in an aspect of Organic Chemistry, which has so developed in the last few decades as to lift that subject very largely out of its former characteristically empirical condition .... The beginnings of physical organic chemistry were tentative and gradual, spread out thinly between *two centuries and over at least four decades*. Then during the last four decades, came the uprush, which could at any stage have choked itself ... [54] [emphasis added]

By postulating a centuries' old heritage, Ingold was providing his fledgling journal and colleagues ample justification for a solid identity. In addition, narrow specialization had come to be recognized as a landmark of the new scientific culture. Thus in the first volume of its *Transactions*, Joseph W. Richards of Leigh University all but proclaimed a social Darwinist *raison d'être* for the American Electrochemical Society, established in 1902,

Division of labor in industry has its analogue in specialization in science ... Differentiation and specialization are the watchword, now, of all progress, – industrial, scientific, philosophical ... Evidence accumulates on every hand that the analogue of the specialist in science is the *society which specializes*. Whether for good or ill, whether some of its influences are narrowing in some directions or not, the society which specializes is the necessary corollary of the scientific specialist; the latter came *perforce* into existence, has made the whole world its debtor, and is recognized as the present factor for progress; the former is coming *perforce* into existence, will soon make the world its immeasurable debtor, and will be a wonderfully potent factor in future scientific progress. Such is the force, the necessary condition, which has brought into existence the American Electrochemical Society. *It could not help but come into existence* ... [55]

A generation later, the same rhetoric pervaded the first editorial of the American Physics Institute's new journal for *Chemical Physics*. Its managing editor, the prominent Harold C. Urey, declared the journal to be "a natural result of the recent development of the chemical and physical sciences." In fact, the editorial asserted what few practicing scientists would be willing to accept today, namely, that

> at present, the boundary between the sciences of physics and chemistry has been completely bridged. Men who must be classified as physicists on the basis of training and of relations to departments or institutes of physics are working on the traditional problems of chemistry; and others who must be regarded as chemists on similar grounds are working in fields which must be regarded as physics. These men, regardless of training and affiliations, have a broad knowledge of both sciences and their work is admired and respected by their co-workers in both sciences. The methods of investigation used are, to a large extent, not those of classical chemistry and the field is not of primary interest to the main body of physicists, nor is it the traditional field of physics. It seems proper that a journal devoted to this borderline field should be available to this group.

It is remarkable how little the essay actually tells us about the journal's goals; nor does it seem to address any specific audience except the vague definition of a "borderline field." Urey only asserted that the field is *not* traditional physics, nor is it practiced with classical chemical methods. After reminding his readers that since the beginning of the twentieth century atoms and molecules have increasingly become of interest to physicists and not only chemists, Urey listed the "new and effective methods, experimental and theoretical, for the study of these units" (atoms and molecules), methods which "have developed largely from physical discoveries which at the time did not appear to have the importance to century-old chemical problems that they have since assumed." Urey referred to the new insights gained in the composition of matter, interatomic forces, molecular and atomic structure etc. Radioactivity, mathematical physics, new experimental methods and in particular the work associated with the quantum theory have "made a profound contribution to our knowledge of physics and chemistry. Moreover, the history of these sciences in recent years teaches the effectiveness of applying the exact logic of mathematics to chemical as well as physical problems." Urey concluded by leaving an open door to any contributions in "that narrow boundary between the known and the unknown."[56]

With welcome, self-reflective humor, Gunther Stent, professor of genetics at the University of California, recalled in the preface to his textbook *Molecular Genetics*, that in 1954 he began teaching an undergraduate course:

that was supposed to bring the latter-day gospel of molecular genetics to the Berkeley students. It was an extraordinarily gratifying pedagogical undertaking to face an audience of innocents, who had not yet heard of the DNA double helix, and preach to them *that a new era was dawning for the understanding of heredity*. So enthusiastic were we in those days that we managed to give thirty lectures on what comparatively little was then known about mutation and genetic recombination in bacteria and their viruses.

However, he proceeds immediately to reassure the new generation:

How times have changed! Molecular genetics has since then grown from the esoteric specialty of a small, tightly knit vanguard to an elephantine academic discipline whose basic doctrines today form part of the primary school curriculum.. . . .
The evolutionary origin and essentially pedagogic purpose of this book are reflected in the narrative presentation of the material in the historical sequence in which it actually came to be known . . . Not only did the text simply grow in this way, but also I happen to believe that an understanding of the essentials of molecular genetics can best be taught in an organic (rather than logical) manner.[57]

Here again we witness the pull of historical rhetoric, set within the accepted standards of the twentieth century. Increased volumes of publications, institutions, students continue to give the scientists a feeling of security and progress, of advancement from an initial stage of confusion to one of a full-fledged discipline. What little was known in the 1950s was nonetheless presented as compelling evidence and truth for a solid, yet new, scientific outlook.

Department of History
California Institute of Technology

NOTES

[1] R. G. A. Dolby, 'Thermochemistry versus Thermodynamics: The Nineteenth Century Controversy,' *Hist. Sci.* **22** (1984): 375–400, pp. 379–380.

[2] An outcome of a conference organized by the historical journal *Past and Present*. Eric Hobsbawm and Terence Ranger, eds., *The Invention of Tradition* (Cambridge: Cambridge University Press, 1983).

[3] *Ibid.* See especially the Introduction, and pp. 263ff, 273. But even the establishment of sports clubs, as for example the emergence of soccer as a "mass proletarian cult" exhibits the very rapid acquisition of "institutional and ritual characteristics": between the mid-1870s and late 1880s, professionalism in soccer, the League, the Cup, "the regular attendance at the Saturday match," the triumphal marches in the capital and the fanaticism of supporters had become well established.

⁴ *Ibid.*, p. 293.

⁵ In an important paper on the historicist program embraced by positivists, in particular by Georg Helm, Ernst Mach and Ostwald, Norton Wise has shown that energetics and the ensuing debates concerning the interpretation of the second law of thermodynamics were heavily dependent upon earlier reductionist attempts – such as Hermann von Helmholtz's – to trace the basis of thermodynamics to mechanical principles. Norton Wise, 'On the Relation of Physical Science to History in Late Nineteenth-Century Germany,' in Loren Graham, Wolf Lepenies and Peter Weingart, eds., *Functions and Uses of Disciplinary Histories*, Volume VII, *Sociology of the Sciences Yearbook* (Dordrecht/Boston/Lancaster: D. Reidel, 1983), pp. 3–34.

⁶ David Cahan writes that: "The progress made by [Helmholtz, Kundt, Warburg, Magnus, Kohlrausch] ... encouraged the belief, not uncommon by the 1880s, that the fundamental principles of physics had already been discovered and that, hence, their major task lay in the refinement of physical laws and constants to the highest degree of precision possible." David Cahan, 'The institutional revolution in German physics, 1865–1914," *HSPS* **15** (1985), 1–65, p. 38. This attitude apparently contributed to a large extent to the establishment not only of Physikalische Technische Reichsanstalt, Germany's foremost institution for precision measurements, but also to the enlargement of experimental physics institutes and expanded funds for more updated and sophisticated instrumentation.

⁷ W. Ostwald, *Elektrochemie. Ihre Geschichte und Lehre* (Leipzig: Verlag von Veit & Co., 1896), 1150 pp. Ostwald's *Electrochemistry* clearly set standards for historically significant scientific personalities, papers, and concepts. This seems evident when compared to Partington's *History of Chemistry*. See G. S. Morrison, 'Wilhelm Ostwald's 1896 History of Electrochemistry: Failure or Neglected Paragon,' in *Selected Topics in the History of Electrochemistry*, ed. by G. Dubpernell and J. H. Westbrook, Vol. 78, *Proceedings of the Electrochemical Society* (Princeton, N.J., 1978), pp. 213–225.

⁸ Ostwald, *Elektrochemie*, pp. v–vii.

⁹ Ostwald, *Elektrochemie*, p. 2.

¹⁰ W. Ostwald, *Überwindung des wissenschaftlichen Materialismus*, lecture held at the third general session of the Association of German Scientists and Physicians in Lübeck, 20. September 1895 (Verlag von Veit & Co.: Leipzig, 1895), p. 25. See Jost Weyer, *Chemiegeschichtsschreibung von Wiegleb (1790) bis Partington (1970)* (Hildesheim: Gerstenberg, 1974), Vol. III in *Arbor Scientarum, Beiträge zur Wissenschaftsgeschichte*, ed. by Otto Krätz, Fritz Krafft, Walter Saltzer, Hans-Werner-Schütt and Cristoph J. Scriba, p. 119.

¹¹ Ostwald, *Überwindung*, p. 35.

¹² For discussions see: Erwin Hiebert, 'The Energetics Controversy and the New Thermodynamics,' in Duane H. D. Roller, ed., *Perspectives in the History of Science and Technology* (Norman: University of Oklahoma Press, 1971), pp. 67–86; Robert Scott Root-Bernstein, 'The Ionists: Founding Physical Chemistry, 1872–1890' (Princeton University Ph.D. Thesis, 1980); and Hans Günther Körber, ed., *Aus dem wissenschaftlichen Briefwechsel Wilhelm Ostwalds*, 2 vols. (Berlin: Akademie Verlag, 1969), I, pp. 79–83 and 119–120; II, pp. 138–140; 252, 352. It is to Ostwald's merit that the extremely favorable and detailed analysis of Nernst's work given in his history of *Electrochemistry* seems not to have been altered after Lübeck, where Nernst sided with Boltzmann and Planck in rejecting Ostwald and Helm.

[13] In 1797 Volta put forth the theory that the electromotive forces generated in the battery are produced by the potential difference originated by the contact of dry metals only. In Volta's theory, the solution in which the battery electrodes were immersed (usually water) played only the role of an electric conductor. Opposition to Volta's contact theory and the proposal that chemical reactions are responsible for the electric current arose soon thereafter, and the split over the correct interpretation has persisted into the 20th century. Thus, "on the side of Volta were Davy, Pfaff, Peclet, Marianini, Buff, Fechner, Zamboni, Matteuci, Kohlrausch, Pellat and [Lord] Kelvin," while in opposition stood "Fabbroni, Ritter, Wollaston, Parrot, Oersted, Ritchie, Pouillet, Schoenbein, Becquerel, De La Rive, Faraday, Nernst, Ostwald, and Lodge." Sidney Ross, 'The Story of Volta Potential,' in *Selected Topics in the History of Electrochemistry*, ed. by George Dubpernell and J. H. Westbrook, Princeton: The Electrochemical Society, 1978, pp. 257–271, p. 260.

[14] *Elektrochemie*, pp. 65–66.

[15] *Ibid.*, p. 147.

[16] Ostwald, *Elektrochemie*, p. 1148.

[17] W. Ostwald, *Electrochemistry. History and Theory*, transl. N. P. Date (Washington: Smithsonian Institute and the National Science Foundation, Amerind Publishing Co., 1980), Vol. II, p. 1123.

[18] *Ibid.*, pp. 1147–48. Ostwald's name could therefore be added to the list of scientists "who said explicitly that [their] contribution was revolutionary or revolution-making or part of a revolution" drawn up by I. B. Cohen, and which contains sixteen scientists from Robert Symmer to Benoit Mandelbrot. See I. Bernard Cohen, *Revolution in Science* (Cambridge, Mass.: The Belknap Press, 1985), p. 46.

[19] W. Nernst, *Die Ziele der physikalischen Chemie*, Festrede gehalten am 2. Juni 1896, zur Einweihung des Instituts für physikalische Chemie und Elektrochemie der Georgia Augusta Universität zu Göttingen (Göttingen: Vandenhoek & Ruprecht, 1896), pp. 2–3.

[20] *Ibid.*, p. 9.

[21] When a decade later physical chemistry was no longer battling on the forefronts of scientific recognition, and certain fields seemed to have been exhausted, Nernst's emphasis on thermodynamics and his recent work on the heat theorem led others to accuse him of heralding a "new age" in physical chemistry, based on "his thermodynamics." Nernst's institute in Göttingen had dramatically declined in reputation and scientific output after Nernst's departure to Berlin in 1905. Gustav Tammann to Svante Arrhenius, 21 November 1906, 3 pp.; 2 November 1907; Arrhenius Papers, Royal Swedish Academy of Sciences, KVA, Stockholm.

[22] P. Duhem, 'Une science nouvelle. La chimie physique,' *Revue Philomatique de Bordeaux et du Sud-Ouest*, 1899, pp. 205–219; 260–280.

[23] Duhem was at the time embroiled in a bitter rivalry and dispute over his position within the Faculty at Bordeaux. See M. J. Nye, *Science in the Provinces. Scientific Communities and Provincial Leadership in France, 1860–1930* (Berkeley: University of California Press, 1986), p. 211ff.

[24] Duhem (ref. 22), p. 270.

[25] *Ibid.*, p. 213.

[26] *Ibid.*, p. 215.

[27] *Ibid.*, pp. 264–5.

28 *Ibid.*, pp. 272–3. Duhem here quotes Haller, 'Science et Industrie,' *Bull Soc. Industrielle de l'Est*, 1897, nr. 2.
29 *Ibid.*, pp. 273–4.
30 Duhem was at the time the only chairholder of theoretical physics in France, and despite his fundamental disagreements with Jean Perrin over atomism and mechanical explanation in physics, he forged an alliance within the context of the essay. See Nye, *ibid*.
31 Duhem (ref. 22), pp. 278–280.
32 'Sur quelque propriétés des dissolutions,' *Journ. Physique* **7** (1888), 5–25.
33 Van 't Hoff to Ostwald, 20 January 1888 and 3 February 1888, Körber (ref. 12), Vol. II, pp. 212–214. But van 't Hoff decried Duhem's heavy mathematical formalism: "What a pity that Duhem cannot deformulize himself." Van 't Hoff to Ostwald, 17 July 1888, *ibid.*, p. 218.
34 The lectures, presented extemporaneously on the occasion of the decennium of the university's foundation, were later published from the original stenographic report. They include van 't Hoff's introductory lecture before a general educational conference in the Kent Chemical Laboratory, followed by three others intended for a scientific public. His presentations had been attended by large audiences "which included professors of chemistry and related sciences from many distant states of the Union." J. H. van 't Hoff, *Physical Chemistry in the Service of the Sciences*, transl. by Alexander Smith (Chicago: University of Chicago Press, 1903), p. 3.
35 G. N. Lewis, 'The Development and Application of a General Equation for Free Energy and Physico-chemical Equilibrium,' *Proceedings of the Am. Acad. of Arts and Sciences* **XXXV** (1899), 3–38, p. 3.
36 Albert Ladenburg (1842–1911) had published a series of lectures held in the summer of 1868 as *Vorträge über die Entwicklungsgeschichte der Chemie in den letzten hundert Jahren* (Braunschweig: Vieweg 1869). Ladenburg denied the prevalence of scientific revolutions. ("In the higher spheres of science a revolution is almost unthinkable.") And although he viewed experiments as crucial episodes in the history of science, Ladenburg considered them not to be preeminently determining the history of science: new theories, he thought, have greater experimental generality and are hence "truer." Ladenburg maintained that succeeding hypotheses can be shown to be related, that traces of the past can always be discerned. Quoted in Jost Weyer, *Chemiegeschichtsschreibung von Wiegleb (1790) bis Partington (1970)* (ref. 10), pp. 108–111.
37 Van 't Hoff, *ibid*, p. 4.
38 *Ibid.*, p. 5.
39 Van 't Hoff performed a new synthesis of propionic acid, published as his first work in the *Ber. Deutsch. Chem. Ges.* **6** (1873), 1107. Ernst Cohen, *Jacobus Henricus van 't Hoff. Sein Leben und Wirken* (Leipzig: Akademische Verlagsgesellschaft, 1912). Series *Grosse Männer. Studien zur Biologie des Genies.*, Vol. 3., edited by Wilhelm Ostwald, p. 41.
40 J. H. van 't Hoff, *Vorschlag zur Ausdehnung der gegenwärtig in der Chemie gebrauchten Strukturformeln in den Raum nebst einer damit zusammenhängenden Bemerkung über die Beziehung zwischen dem optischen Drehvermögen und der chemischen Konstitution organischer Verbindungen* (Utrecht: J. Greven, 1874), in Cohen, p. 72.

A USABLE PAST

41 E. Cohen, p. 287, 298.

42 Van 't Hoff to Ostwald, 17 October 1890, Körber, Vol. II, pp. 227–8. Van 't Hoff is referring to Ostwald's lecture at the 63rd Meeting of the German Association of Scientists and Physicians in Bremen in September 1890, on 'Altes und Neues in der Chemie.' In Ostwald, *Abhandlungen und Vorträge* (Leipzig, 1904), p. 13–33; Van 't Hoff, 'De physiologisch beteekenis der jongste Stroomingen op Natuur- en Scheikundig gebied,' *Handl. van het Nederlandsch Natuur- en Geneeskundig Congres*, Utrecht, 1891, in Körber, Vol. II, fn. 3, p. 228.

43 Ernst Cohen considers that the book, which also deals with equilibria, affinity and other subjects, led to a "révolution chimique" around the year 1884. E. Cohen, p. 182.

44 See: a compilation extending to 1900 on what physical chemistry had achieved in this field in the years 1890 to 1900 in Koeppe, *Physikalische Chemie in der Medizin* (Wien, 1900) and Burton E. Livingston, *Role of Diffusion and Osmotic Pressure in Plants* (Chicago: Univ. of Chicago Press, 1903). The most relevant work had been performed by J. Loeb at Chicago. In his well-known report, 'The Physiological Problems of Today,' *American Society of Naturalists*, Ithaca, 1897, he wrote: "At no time since the period immediately following the discovery of the law of conservation of energy has the outlook for the progress of physiology appeared brighter than at present, this largely being due to the application of physical chemistry to the problems of life." Quoted in van 't Hoff, *Physical Chemistry* (ref. 34), p. 11.

45 Van 't Hoff, p. 18.

46 *Ibid.*, p. 16.

47 Quoted by Cohen, p. 299.

48 *Ibid.*, p. 24.

49 Van 't Hoff, p. 23.

50 See in particular R. G. A. Dolby, 'Thermochemistry versus Thermodynamics' (ref. 1); also, R. G. A. Dolby, 'The Case of Physical Chemistry,' in G. Lemaine, R. Macleod, M. Mulkay, and P. Weingart, eds., *Perspectives on the Emergence of Scientific Disciplines* (Mouton/The Hague/Paris: Maison des Sciences de l'Homme, 1976), pp. 63–73; Root-Bernstein, *The Ionists* (ref. 12), pp. 5–7.

51 Karl Jellinek, *Physikalische Chemie der homogenen und heterogenen Gasreaktionen, unter besonderer Berücksichtigung der Strahlungs- und Quantenlehre, sowie des Nernstschen Theorems* (Leipzig: S. Hirzel, 1913).

52 *Ibid.*, preface, np.

53 Evidently, subsuming chemistry into physics is a recurring theme both in the history and the methodology of chemistry. Marx and Engels, for instance, were influenced by H. Roscoe and C. Schorlemmer's view that "Chemistry as the science of atoms is a descendant of physics or of the science of molecules, which is itself based on mechanics or the science of masses. These had therefore to develop to a certain degree before chemistry could appear in its crudest form." Quoted in Mikulas Teich, 'Neue Materialien über Carl Schorlemmer,' *Beiheft zu NTM*, Leipzig, 1964, pp. 108–130. On Roscoe and Schorlemmer's *Ausführliches Lehrbuch der Chemie* (1877–92) and Schorlemmer's *The Rise and Development of Organic Chemistry*, an unpublished manuscript, see K. Heinig, 'Ein unveröffentliches Manuskript Carl Schorlemmers zur Geschichte der Chemie,' in NTM 1 (1960), 62–71. Quoted in Weyer (ref. 10), p. 104.

[54] V. Gold, ed., *Advances in Physical Organic Chemistry* (London/New York: Academic Press, 1963), p. vii.
[55] J. W. Richards, 'The American Electrochemical Society,' *Trans. Am. Electrochem. Soc.* **1** (1902), pp. 1–2.
[56] Harold C. Urey, 'Editorial,' *J. Chem. Phys.* **1** (1933), 1–2, p. 1.
[57] Gunther S. Stent, *Molecular Genetics. An Introductory Narrative* (San Francisco: W. H. Freeman and Company, 1971), p. ix–x.

PART IV

EXPLANATION AND DISCOVERY:
THE CLAIMS OF CHEMISTRY, PHYSICS,
AND FORTRAN

MARY JO NYE*

PHYSICS AND CHEMISTRY: COMMENSURATE OR INCOMMENSURATE SCIENCES?

In the preface to this *Introduction to Chemical Physics* (1939), the American physicist and co-founder of quantum chemistry John C. Slater wrote the following:

> It is probably unfortunate that physics and chemistry ever were separated. Chemistry is the science of atoms and of the way they combine. Physics deals with the interatomic forces and with the large-scale properties of matter resulting from those forces. So long as chemistry was largely empirical and non-mathematical, and physics had not learned how to treat small-scale atomic forces, the two sciences seemed widely separated ... Now that statistical mechanics has led to quantum theory and wave mechanics, with its explanations of atomic interactions, there is really nothing separating them any more .... [However,] for want of a better name, since Physical Chemistry is already preempted, we may call this common field Chemical Physics.[1]

Slater's chemical physics is a realization in many respects of the program envisioned a decade or two earlier by the physicists Max Born and Paul Dirac. Reflecting upon the initial successes of the Bohr–Sommerfeld atom, Born said in a lecture published in 1920:

> When we contemplate the path by which we have come, we realize that we have not penetrated far into the vast territory of chemistry, yet we have travelled far enough to see before us in the distance the passes which must be traversed *before physics can impose her laws upon her sister science*.[2]

Almost a decade later, in 1929, the young Dirac claimed that, by pursuing the application of quantum wave mechanics, "The underlying part of physics and the whole of chemistry are thus completely known."[3]

Nonetheless, most chemists in the 1920s and 1930s were not convinced that their aims were being met by the new physics, nor that their disciplinary field, even in the domain of physical chemistry, could be reduced to the methods and laws of physics. Fifty years later, the editors of the *Annual Review of Physical Chemistry* rejected the proposal that "Chemical Physics" be added to the title of the journal, even while recognizing, they said, that the difference between the practice of physical chemistry and chemical physics is "small indeed."[4]

M. J. Nye et al. (eds.), The Invention of Physical Science, 205–224.
© 1992 *Kluwer Academic Publishers. Printed in the Netherlands.*

What, then, we might be inclined to ask, *is* the difference between physics and chemistry, given the overlap between the two domains? Or, changing the perspective from present to past, and from essences to contingencies, how have the two domains been distinguished by their practitioners? How have these distinctions changed in time? How have the two sciences interacted and influenced the nature and the practice of the other? And what can be said about the degree of commensurability between the two?

In addressing these questions here, the approach taken is primarily that of an explanation-sketch from a conceptual and intellectual perspective, without account of the personal, institutional, and sociological aspects of discipline formation. These latter factors are treated elsewhere, although they are not absent entirely from this analysis.[5]

A guiding theme in this present study is the ubiquitous influence on scientific practitioners of the tradition of the hierarchy of the sciences, especially the privileged place given mathematics among canons of knowledge. Another theme is the very real conceptual difference between physicists' and chemists' ways of thinking. This difference is, in fact, illustrated above in Slater's somewhat misleading definition of chemistry as "the science of atoms and of the way they combine." Most chemists since the nineteenth century have defined chemistry as the science of *molecules* and the way they behave.

A statement from Ernest Rutherford further illustrates frequent incomprehension between physicists and chemists, this, in a letter to George Ellery Hale in 1918:

I think it is highly important that physics and chemistry labs should be in the same building or in buildings close by, and that there should be the closest liaison between the physical and chemical researchers. Both will benefit ... At present, chemistry lags in ideas behind physics and they want a broader philosophy of science, while the physicist is *woefully ignorant of the reasons why a chemist's mind works in such particular grooves and why he has acquired such an impenetrability to the flying colours of the physicists.*[6]

ESTABLISHING THE AUTONOMY OF CHEMISTRY FROM PHYSICS

To begin, what conceptual distinctions historically defined the subject matter of physics and chemistry? The French word *physique*, like its counterparts, is rooted in the Greek word for [N]nature, in the sense of the nature of things around us. Aristotle defined physics as theoretical knowledge. Aristotle's "physics" included the study of those changes which we now identify with "chemistry," a word rooted in Egyptian and Arabic words which came into western parlance only in the middle ages.

In its seventeenth and eighteenth-century usage, "physics" still was theoretical by definition, whereas "chemistry" had both theoretical and practical parts, the practical part having to do with instruments of heat and means of solution, the theoretical part being "philosophical," sometimes historical, and always systematic.[7] The broad field of "Natural Philosophy" or "physique générale" popularly included mechanics ("mathematical physics") and almost everything else that has to do with the nonliving world, thus including both "experimental physics" ("experimental philosophy") and "chemistry."[8] In a successful and typical popular text of natural philosophy, which went through twenty editions from 1766 to 1827, Adam Walker arranged the book's contents in sections of mechanics, astronomy, light (including optics), fluids (including air), chemistry (including heat), magnetism, and electricity.[9]

One of the founders of modern chemistry, who aimed to establish its clear autonomy from physics, was Georg Stahl. In 1723 he wrote, "Those are called physical principles whereof a mixt is really composed, but they are not hitherto settled ... And those are usually termed chemical principles, into which all bodies are found reducible by the chemical operations hitherto known."[10] Thus, chemical operations give us chemical principles.

This operationalist definition of chemical knowledge similarly can be found in G. F. Venel's 1753 *Encyclopédie* article "Chymie": "Chemistry by its visible operations resolves bodies into certain gross and palpable principles, salts, and sulfurs, etc; but Physics, by delicate speculations, acts on principles as chemistry has done on bodies; it resolves them into simpler principles, into smaller bodies propelled and shaped in an infinity of fashions."[11] Thus, in a somewhat tendentious definition much cited from then on, chemists operate, physicists speculate on the underlying principles of matter.

In the eighteenth century, Stahl, Venel and French Stahlians like G. F. Rouelle and P. J. Macquer were mounting a campaign against earlier, seventeenth-century attempts to treat chemistry as a branch of mechanical philosophy. In particular, they resisted the reading of Newton's natural and experimental philosophy which prescribed the reduction of chemical transformations to nothing but mechanical motions and interatomic forces. This was a mechanical reductionist program seated in a precise tradition associated with the venerable Newton's injunction (Query #31) that:

> To tell us that every Species of Things is endow'd with an occult specifick Quality by which it acts and produces manifest Effects, is to tell us nothing: But to derive two or three general Principles of Motion from Phaenomena, and afterwards to tell us how the Properties and Actions of all corporeal Things follow from those manifest Principles, would be a very great step in Philosophy.[12]

However, instead of defining or deriving mechanical principles for chemistry, the French Stahlians stressed that chemistry was concerned with the *intimate* nature of matter, namely the different *parties intégrantes* into which a homogeneous substance might be divided without decomposing it (i.e., into what later was codified as "molecules"). In contrast, physics was concerned with the superficial, outward appearance of bodies. Thus, two bodies might appear identical and yet differ in their interior (white powders), or two bodies might be physically distinct and yet be fundamentally, or *essentially*, the same (coal and diamond).[13]

In Venel's view, the two sciences of physics (*physique générale*) and chemistry (*la chimie*) were to have the status of co-fraternal siblings. But while *physiciens* in the eighteenth century were already proceeding in a well-order systematic manner, Venel thought that eighteenth-century chemists still were in need of a "new Paracelsus" to put chemistry securely at the side of "la Physique calculée."[14] Lavoisier and his colleagues were to bring to fruition this program for making chemistry an independent and autonomous science, clearly distinct from *physique générale*.

Lavoisier did this in part by claiming to have established a chemistry which could arrive at "results as certain as one can hope for in physics."[15] Lavoisier's claim here was not based on the working out of fundamental mathematical principles for chemical laws in the tradition of rigorous geometrical reasoning, but on the use of quantification and experimental method as complementary methods to the method of natural classification.[16] To be sure, Lavoisier expected that fundamental chemical problems eventually would be accessible to mathematical solution, meaning not just quantification or geometrical reasoning, but algebraic formulation on mechanical principles.[17] However, he did not think that the present autonomy of chemistry was imperilled by its lack of mathematical or "philosophical" foundation,[18] and as F. L. Holmes had demonstrated in a recent history of eighteenth-century chemistry, Lavoisier was right in recognizing that the discipline was well-established well before century's end.[19]

Evan Melhado and other historians frequently have emphasized common aims and methods in physics and chemistry by referring to the "physicalist" tradition in Lavoisier's chemistry.[20] This physicalism included use of instruments, like the thermometer, which still were controversial among chemists elsewhere in the mid-eighteenth century.[21] In Lavoisier's chemistry and in Dalton's chemistry a generation later, physicalist quantification of heat, weight, volume, pressure (and by the mid-nineteenth century, optical properties) became standard practice. Electrical effects, too, were examined, as in the decomposition of water under the influence of an electric spark, and the quantities of electricity required to deposit chemically equivalent weights of elementary substances at electrodes. Indeed, at the turn of the nineteenth century, electricity, magnetism, and galvanism all were classified as "mechanico-chemical" sciences, not "mechanico-physical sciences" on the basis of their origin.[22]

By mid-nineteenth century, after the failure of Jöns Jacob Berzelius' electrical theory of chemical affinity and the success of Michael Faraday's theory of the electromagnetic field, electricity and magnetism ceased to play a central role in the chemical part of most chemistry instruction and textbooks, becoming, instead, instances of "chemical physics" preliminary to the study of the core material of chemistry.[23] Thus, as the Connecticut chemist Thomas Pynchon wrote in his *Introduction to Chemical Physics*,

Chemistry is usually divided into two portions. The first treats of the Chemical Agents, Heat, Light, and Electricity, and is commonly called Chemical Physics; the second, of the chemical properties and relations of the various kinds of matter, inorganic and organic.[24]

The absolutely central meeting ground of physics and chemistry at the turn of the nineteenth century lay in the concepts of "Force" and "Affinity" and the conflation between their meanings. Newton's queries (#23 in 1706, #31 in 1717) posed the problem of the nature of chemical "affinity," a kind of "force" which Robert Boyle, for example understood to be specific to the *kind* of substance, as in the precipitation of metals from a solution of their salts. The word "affinity" originally meant the attraction of like things for each other; it came to mean, in addition, the attraction of unlike things. But crucially, "affinity" was "elective" power which depended on the *kind* of substance, not its mass; and thus affinity differed decisively from the mechanical "force" studied by Newton and Newtonians in astronomy, hydrodynamics, and mechanics generally.

In the physicalist tradition, Lavoisier's colleague Louis B. Guyton de Morveau attempted to quantify short-range forces of chemical affinity by studying the force necessary to lift different kinds of metal disks from the surface of mercury.[25] Later, Berzelius, in his ill-fated analogy between electrical force and chemical affinity, reasoned that the exhibition of inverse-square force would necessitate that all chemical atoms be spherically-shaped. But these physicalist and mechanist approaches in late eighteenth-century and early nineteenth-century were unsuccessful. Arnold Thackray distinguishes John Dalton's new chemistry in the early 1800s by its abandonment of Newtonian affinity-forces, whether conceived on a gravitational or an electrical analogy.[26]

What we emphasize here is chemists' recognition of the very different character of chemical affinity from gravitational-force or electrical-force models. Newtonian force depends on quantity (of matter, charge, etc.) and distance, whereas affinity depends on quantity, distance, and *kind*. It is the problem of *kind* that distinguished eighteenth and nineteenth-century chemistry from physics, even though chemists often felt awkward about making this distinction.

The reason for the embarrassment, of course, is that mathematical and experimental philosophers of the seventeenth and eighteenth centuries were seeking to distinguish themselves as offering a *new science* by repudiating Aristotelian "physical" explanation in terms of categories, kinds, and essences. Though chemists were also intent on establishing a scientific domain which was experimental and non-Aristotelian, they still used explanation systems employing the classification and description of natural kinds.[27] This practice seemed to place them in a backward and inferior position in comparison to natural philosophers and physicists.

The clearest distinction between Lavoisier's chemistry and eighteenth-century mechanical philosophy lay in the former's construction of generic groups. Jean Baptiste Dumas, for all the fact that he aspired to Newtonian chemistry in his youth, abandoned a chemistry of forces in the 1830s and adopted a classification of chemical types, modelled to some extent on the scheme of A. P. De Candolle, Henry Milne Edwards, and Georges Cuvier.[28] Within a couple of decades, Henri Sainte-Claire Deville defined chemistry as a natural science in which chemical theories are methods of classification.[29] This emphasis on systems of classification modelled on natural history made all the more sense in nineteenth-century chemistry, given the increasing emphasis in chemical instruction and investigation upon organic substances of biological origin or biological significance.

Functional and structural chemistry, not affinity or force chemistry, was the great achievement of nineteenth-century chemists, and for most of the century, synthetic organic chemistry dominated the discipline. Liebig's *Annalen* was a journal of chemistry, but one in which the dominance of its pages by organic chemistry reflected the state of the discipline as a whole. These triumphs were such that chemistry became the first science, as Marcellin Berthelot noted, to create its own object of study.[30] For chemists, it can be said, there are two types of empirical laws: a theoretical part concerned with function and composition, which states functional relations between variable properties of a system; and a descriptive and classificatory part, with laws stating that there are kinds of material with reproducible properties, i.e., "natural kinds."[31]

Indeed, during the mid-decades of the nineteenth century, chemistry appeared to be developing into an increasingly nonmathematical science, during a period when physics was becoming increasingly more mathematical. In the late eighteenth century, Immanuel Kant had taught that theoretical reason includes mathematics and physics [Physik], the principles of which are *à priori* synthetic judgments, and that one encounters "genuine science only to the extent that one encounters mathematics in it."[32] This Kantian definition excluded late eighteenth-century chemistry from genuine science. By the late nineteenth century, James Clerk Maxwell could be found characterizing chemistry as a science lacking clarity and rigor, i.e., insufficiently abstract.[33] About the same time Hermann von Helmholtz expressed the opinion that chemistry as a science progresses "not quite rationally."[34]

A RECONVERGENCE OF CHEMISTRY AND PHYSICS

Nineteenth-century organic chemists largely left behind the physicalist and mathematical program of their predecessors or of their own youth. R. Willstätter commented on the very absence of any physical instruments in Adolf von Baeyer's Munich laboratory for organic chemistry.[35] There was resentment among some chemists in the 1860s of Lord Kelvin's optimistic claim that his physical vortex atom could explain chemical affinities, and there was the ennui of "*déjà-vu*" in response to Ludwig Boltzmann's essay in 1901 boasting that a strength of the kinetic-atomic theory was in its being a key to chemical combination and to chemical isomerism.[36]

Still, some chemists continued to pursue the conundrum of the "affinity" problem through studies of electrolysis, physical properties in relation to atomic and molecular constants, and thermochemistry. Julius Thomsen and Berthelot each tried to define and measure chemical affinity inferentially by the force required to overcome the affinity uniting the parts of a compound. This force, they argued, was measurable by the heat evolved in spontaneous exothermic reactions.[37] But their programs foundered on Thomsen's conflict with organic chemists over the energy content of single and double carbon bonds[38] and on the demonstration by Pierre Duhem and others that the law of maximum work was wrong.[39]

More successfully, the physicists or physical chemists J. W. Gibbs, Duhem and J. H. van 't Hoff developed mathematical equations to describe states of chemical and physical equilibria. In their application of these techniques to chemical reactions, the term "affinity" disappeared. "Arbeit," then "Energie" became the fundamental solvable parameters. Thus, in the new thermodynamics and the new physical chemistry associated with the so-called "Ionist School" of van 't Hoff, Svante Arrhenius, and Wilhelm Ostwald, the energy term, later "free energy," was to replace chemical "affinity" as the driving force of chemical reactions.[40] As G. N. Lewis noted in his popular book *The Anatomy of Science* (1926), the concept of "force" gradually fell out of use among chemists.[41]

Physicists and chemists concentrated in their application of mathematics to chemistry on studies of osmotic pressure, colloid and surface chemistry, the phase rule, kinetics and equilibrium, setting up physical and mathematical relationships in relation to energy or entropy. These were important, revolutionary reintroductions of physicalist and mathematical methods into chemistry in order to explain the bulk properties of chemical systems.

The new physical chemistry was conceived by contemporaries both as a bridging science and as a fundamental and autonomous theoretical chemistry.[42] Walther Nernst used the terms physical chemistry and theoretical chemistry interchangeably, conceiving it "not so much . . . a new science, but rather the co-operation of two sciences which hitherto have been, on the whole, quite independent of each other."[43] However, many chemists viewed the new physical chemists as interlopers in a discipline in which they were not fully members. Ostwald, it sometimes was noted, had never synthesized a single new compound.[44]

Yet, chemical practitioners both young and old acknowledged that the new theoretical chemistry left unresolved the conundrum of affinity. In a lecture in Paris in 1912, Arrhenius emphasized this point, saying that the central problem of chemistry remained affinity or the cause of chemical reactions.[45] Among those who heard this lecture was the leading proponent of physical chemistry in France, Jean Perrin, the first scientist to hold a chair of physical chemistry at the Sorbonne. Influenced by Arrhenius' earlier work on "activation" energies and coefficients, and influenced as well by the early quantum and radiation theory, Perrin sought to solve the theoretical problem of the origin of chemical activation, i.e., the reason a molecule becomes unstable. To this end, Perrin became one of several physical chemists and physicists who pursued a radiation theory of chemical activation.

In 1921 Perrin and Thomas Lowry, the first chairholder in physical chemistry at Cambridge University, debated one another at a Faraday Society Symposium in which Perrin presented his theory that radiation energy is responsible for all chemical reactions, that reaction velocities are independent of kinetic effects, and that monomolecular reactions are confirming instances of this theory. Lowry, familiar with both organic and inorganic results, disagreed on these points and argued that all apparent molecular dissociations, including tautomerism in conjugated hydrocarbons, are triggered by a catalyst. The only example of a true monomolecular reaction, Lowry asserted, is the very different phenomenon of radioactivity.[46]

The argument here is fundamentally that a *chemical* reaction involves not one, but *two* particles. This may seem a trivial point, but it was historically fundamental to definitions of how chemistry differs from physics. Thus the British chemist Alexander Williamson, who in large part originated interest among chemists in the kinetic motions of atoms *within* a molecular, wrote in the *Chemical Gazette* in 1851:

... the study of the properties of matter ... as long as they undergo no change, belongs to physics .... . when we study a molecule by itself, we study it physically; chemistry considers the *change* effected by its reaction upon another molecule.[47]

Perrin's radiation hypothesis was rejected by the late 1920s. The arguments which refuted it came from a revival of kinetic studies[48] and from the application of new electron theories to the problem of reaction mechanisms in chemistry. The most important of these electron theories were the static Lewis–Langmuir electron-bond theory and the dynamic Bohr electron-orbital theory.

While organic chemists might have been expected to be recalcitrant to the revival of electrical theories, and indeed German organic chemists showed little interest in them, the English chemists Arthur Lapworth, Robert Robinson, and Christopher Ingold were leaders in applying ionic hypotheses and electron imagery to the prediction of substitution and addition reactions in aromatic hydrocarbon chemistry. The young Ingold, initially under the sway of Alfred Werner's and Bernard Flürscheim's continuous and spatial models of "chemical affinity," adopted the approach of Robinson and Lapworth in the mid-1920s, subsequently working out a language and nomenclature of reaction mechanisms in which "chemical affinity" became "electron affinity." Ingold's chemical hypothesis of "mesomerism" or resonance structure for benzene compounds proved consistent with the mathematical interpretation worked out independently by others from the principles of quantum wave mechanics.[49]

On this different track, physicists worked from Bohr's electron-orbital theory, the spectroscopy of the elements, and the new quantum wave mechanics to tackle the problem of the energy constitution of atom and the molecule. It was in this spirit that Born and Dirac forecast the prospect of solving the fundamental problems of chemistry through basic laws of physics. Theoretical chemistry and physics were not to be cofraternal siblings or even fraternal twins, but commensurate sciences in which chemistry was reduced to physics, the old ideal of seventeenth-century mechanical philosophy.

Thus, there grew a perception among many chemists, and perhaps among physicists as well, that theoretical physicists regarded chemistry as an applied field of quantum mechanics.[50] As Erwin Hiebert has noted, Nernst preferred to be recognized as a physicist doing chemistry rather than as a physical chemist.[51] Many chemists looked with suspicion upon the hubris of Nernst, then of Dirac, and found irrelevant to truly chemical problems the nonetheless admirable achievement of W. Heitler and F. London in demonstrating that two helium atoms will not form a molecule, but two hydrogen atoms do.[52]

A different quantum mechanical approach, treating the molecule as a unit, rather than as a combination of two or more atoms, was Friedrich Hund's molecular orbital method, which took into account the influence of atomic nuclei and average charge density of nearby electrons on the motions of an individual electron. Hund calculated energies fairly successfully for the ethylene double bond and for benzene. Soon John Van Vleck treated the more complicated molecules of methane and ammonia.[53]

The allure of physical and mechanical solutions to fundamental chemical problems always had been strong. The new wave mechanics was even more promising than Newtonian force analogies or Berzelian electrical dualities, perhaps because it was even more abstract than they. A new generation of theoreticians began to emerge which was to professionalize a "theoretical chemistry" distinct from thermodynamics-based physical chemistry.

Charles Alfred Coulson, for example, was a mathematical student at Cambridge in the late 1920s when Dirac gave what became a famous talk at a student mathematical society meeting on commutation relations in quantum mechanics. Someone asked him the question when this sort work was done. When Dirac replied that it was done the previous night, Coulson was hooked.[54]

Coulson finished degrees in mathematics and physics, and then he took a D.Phil. degree in chemistry at Cambridge in 1934. He recalled a lecture there in the early 1930s by the American theoretician Robert Mulliken. Mulliken described how the double bond in ethylene is formed by bringing two CH_2 groups together in terms of sigma and pi orbitals. Coulson later remarked that he found this conception clever and satisfying.[55]

Exactly how was this satisfying? The new wave mechanics seemed to explain why no more than two electrons make a double bond.[56] The chemist Linus Pauling and the physicist John Slater developed equations for electron orbits with patterns pointing strongly in two directions at right angles to each other; thus, according to Coulson, chemists now had a simple explanation of the fact that in water the angle HOH is nearly 90°. Coulson's calculations of fractional bond lengths in 1937 showed that Johannes Thiele's old notion of partial valence could be put on a firm and rational basis.[57] The quantum mechanical resonance theory developed by Pauling could explain why aromatic compounds are so stable. Following her reading of Pauling's *Nature of the Chemical Bond* (1939) in the mid-1940s, Alberte Pullman could put away forever her hundreds of tracing-paper formulae for naphthacene ($C_{18}H_{12}$).[58]

MATHEMATICS AND MOLECULES: IS CHEMISTRY PHYSICS?

But were these results identical to physical explanation? Was the mathematical physics of wave mechanics now the basis for chemistry? We would not be surprised that some synthetic organic chemists denied in the 1930s and 1940s that this was true. But statements by some theoretical chemists also led to the same conclusion. Coulson, for example, described the first ten years of work, from roughly 1928–1938, on the electron valence bond as work spent escaping the thought-forms of the physicist, which were dominated by the idea of a centre of symmetry. Now the chemical notions of directional bonding and localization could be developed. [59]

Quantum mechanics gave chemistry a new "understanding," but an understanding absolutely dependent on purely chemical facts already known. What enabled the theoretician to get the right answer the first time in a set of calculations was a set of experimental facts of chemistry, which in Coulson's view implied the properties needed for solving the wave equations. Chemistry could be said to be solving the mathematicians' problems rather than vice-versa.[60] For Coulson, confidence was warranted that the model was essentially the right one if choosing one parameter resulted in a first right answer, followed by a succession of good fits in the series.[61]

Indeed, Coulson's generation of quantum theoretical chemists did not claim that the mathematical physics of wave mechanics led to fundamental breakthroughs or discoveries in chemistry. Alberte Pullman commented in 1970:

While it is certainly indispensable that theoretical chemists constantly try to improve the values of the sizes they calculate and more and more approach exact energy values ... quantum chemistry risks giving the impression that its essential goal is reproducing by uncertain methods known results, in contrast to all other sciences whose goal is to use well-defined methods for the research of unknown truths.[62]

In the decades following the Second World War, the work of Alberte and Bernard Pullman focussed on the application of quantum mechanics to conjugated molecules and especially to biologically active molecules and carcinogens. Christopher Longuet-Higgins and Coulson collaborated with the Pullmans and Raymond Daudel in some of this work. Similarly, Pauling concentrated after the late 1930s on the study of biochemical substances and especially on relations between molecular structures and properties of compounds.[63] As chemists focussed on larger and larger molecules, physicists were studying smaller and smaller particles. For chemists, the calculations of quantum chemistry were one tool in a network of multiple explanation; for particle physicists, the calculations tended to be the end of explanation.

When, in the 1960s, Mulliken commented on the difference between chemists and physicists, he suggested that, at least in chemistry until then, the difference is that "chemists love molecules, and get to know them individually, in the same way that politicians love people." In contrast, he suggested, "physicists are more concerned with fields of force and waves than with the individual personalities of molecules or matter."[64] In an interview with S. S. Schweber, the theoretical chemist E. Bright Wilson corroborated Mulliken's point of view, saying that for his part, "I love my molecules."[65] The theoretical chemical Roald Hoffman has likened the chemists' world of molecules to the straphanger's world of the subway.

I think that in their richness and variety molecules are to be compared with people. This is what I like about riding the subway in New York – the incredible range of ethnic type, physiognomy, clothes, and emotions. I see tired swarthy men, women with henna-dyed hair, people reading Korean and Russian newspapers, Caribbean blacks, a sleepy Indian girl. Angelic or rough, they're alive, and in their lives are a million novels. When I open a page of *Chemical Communications* or *Angewandte Chemie*, I get a similar feeling. I recognize the molecule types (my prejudices and education determine that), but in these pages someone has pulled off something new – here a cluster of nine nickel atoms, one inside a cube of eight, there someone else has found out the curious way an NO molecule tumbles as it jumps off a metal surface to which it has been stuck. I'm a voyeur of molecules.[66]

The molecule, or the eighteenth-century *partie intégrante*, has been at the heart of modern chemistry. The affinity problem is a molecular problem and it manifests itself when one molecule encounters another of a different kind. There seems to have been a consensus among most chemists from the eighteenth through the twentieth centuries that chemistry ordinarily involves two molecules, or a molecule and a catalyst. There also has been a consensus, at least among theoretically-minded chemists since the mid-nineteenth century, that if there is a molecule-at-rest, which can be described in a "rational" formula or equation, it must have a different internal structure than the molecule-in-reaction. Auguste Kekulé, Williamson, and Lowry, for example, all recognized that the molecule-at-rest must have a dynamic internal structure not amenable to ordinary graphic or visual imagery, as I have argued elsewhere.[67]

Stephen Weininger has brought to the attention of historians of chemistry a recent controversy in quantum chemistry regarding the irreducibility of the chemical molecule to an isolated physical molecule defined by the equations of quantum mechanics. Thus the molecular physicist R. G. Woolley (Cavendish Laboratory at Cambridge University) and the physical chemist Hans Primas (E.T.H., Zurich) reminded theoretical chemists and physicists that molecular structure is a classical idea, foreign to the principles of pioneer quantum mechanics which "neither gives a correct nor a consistent description of molecules ... quantum mechanics gives perfect predictions for all spectroscopic experiments. However, chemistry is not spectroscopy."[68]

The problem according to Woolley, is that quantum mechanical calculations employ the fixed, or "clamped," nucleus approximation in which nuclei are treated as classical particles confined to "equilibrium" positions. Woolley has claimed that a quantum mechanical calculation carried out completely from first principles, without such an approximation, yields no recognizable molecular structure and that the maintenance of "molecular structure" must therefore be a product not of an isolated molecule, but of the action of the molecule functioning in its environment.[69]

On a different tack, but not a completely dissimilar one, Hoffman has noted the irreducibility of many molecular concepts to physical or mathematical thinking: "Most of the useful concepts of chemistry (for the chemist: aromaticity, the concept of functional group, steric effects, and strain) are imprecise. When reduced to physics they tend to disappear."[70] Or, returning to the more distant past, as Robert Lespieau put it at the beginning of the twentieth century, the method of chemical science is not at all like that of mathematical science, because a chemical formula cannot be demonstrated like a theorem.[71]

By way of conclusion, we may recall that there are many who have believed that physics is the "basic" science and that all sciences fall short of an ideal form of knowledge to the extent that they elude mechanical first principles.[72] Modern chemistry began to take its characteristic form in the mid-1730s, evolving out of the Aristotelian tradition of *physics*. Chemistry thus has been a "physical" science, but it has not been physics.

Other sciences, too, have labored under the shadow of the physics ideal. As David B. Kitts outlined in his work on the epistemological structure of geology, and as Rachel Laudan similarly has argued, geology is an immensely complex science of interdependent retrodictive inferences. In pursuing geological explanation, which is historical explanation, the geologist employs the conceptual framework of contemporary physical theory.[73] Yet physics, as in historical cases like the Darwinian time-scale and Wegener's theory of continental drift, has sometimes led geologists seriously astray from superior kinds of explanations based on geology's own methods and concerns.

Historically, chemistry and physics have been regarded as complementary, but not commensurate sciences. The conceptual distinctions which led to the definition of an autonomous chemical science in the eighteenth century remained intact in the nineteenth and twentieth centuries, despite the two sciences' considerable conceptual and institutional overlap both in the past in more recent decades. Chemists have waxed and waned in their enthusiasm for applying physicalist programs in chemical science, but the reductionist program appears not to have been achieved by the founders of quantum chemistry.

Department of the History of Science
University of Oklahoma

NOTES

* It is a pleasure to acknowledge research support for this study from the National Endowment for the Humanities and the University of Oklahoma. I am indebted, too, to students and colleagues in my seminar at Harvard University during the spring of 1988; to the Bodleian Library at Oxford University; and to comments on an earlier draft from Robert Nye. The basic problem-set of this essay is one to which I was first introduced by Erwin Hiebert.

[1] J. C. Slater, *Introduction to Chemical Physics* (New York: McGraw-Hill, 1939), pp. v–vi. On Slater, see S. S. Schweber, 'The Young John Clarke Slater and the Development of Quantum Chemistry,' *Historical Studies in the Physical and Biological Sciences* **20** (1990), 339–406.

[2] My emphasis. Max Born, *The Constitution of Matter. Modern Atomic and Electron*

Theories, trans. from 2nd German ed., E. W. Blair and T. S. Wheeler (London: Methuen, 1923, 1st ed., 1920), on p. 78.

[3] P. A. M. Dirac, 'Quantum Mechanics of Many-Electron Systems,' *Proceedings of the Royal Society of London*, Series A **123** (1929), 714–733, on p. 714.

[4] *Annual Review of Physical Chemistry* **39** (1988), Preface.

[5] See M. J. Nye, *From Chemical Philosophy to Theoretical Chemistry: Dynamics of Matter and Dynamics of Disciplines, 1800–1950* (Berkeley: University of California Press, in progress).

[6] My emphasis. Ernest Rutherford to G. E. Hale, 15 November 1918, Rockefeller Archives Center, Rockefeller Foundation, RG1.1, Series 200, Box 37, Folder 417, quoted in Alexi Assmus, 'The Creation of Postdoctoral Education and the Siting of American Scientific Research,' ms., p. 14.

[7] See Antoine de Fourcroy, *Système des connaissances chimiques, et de leurs applications aux phénomènes de la nature et de l'art*, 11 vols. (Paris: Baudoin, 1800), Vol. I., pp. xxx–xxxi.

[8] See Adam Walker, *Analysis of a Course of Lectures on Natural and Experimental Philosophy*, 1st ed., 1766, 20th ed., 1827, discussed below. Also see John Elliot, *Elements of the Branches of Natural Philosophy Connected with Medicine* (London: J. Johnson, 1782); William Nicolson, *Introduction to Natural Philosophy*, 2 vols, 2nd ed. (London: J. Johnson, 1787); and F. A. C. Gren, *Grundriss der Naturlehre in seinem mathematischen und chemischen Theile* (Halle, 1793; 1st pubd. 1788).

[9] Adam Walker (ref. 8). For an early nineteenth-century critique of this tradition, see E. G. Fischer, *Lehrbuch der mechanischen Naturlehre*, 3rd ed., 2 vols. (Berlin, 1826–1827), Vol. I, pp. vii–viii, 4–5, who argued that what belonged to Physik proper were investigations of natural phenomena on the basis of the laws of mechanics, including heat, light, and electricity, but not chemistry. Discussed in Christa Jungnickel and Russell McCormmach, *Intellectual Mastery of Nature. Theoretical Physics from Ohm to Einstein. Volume I. The Torch of Mathematics. 1800–1870* (Chicago: University of Chicago Press, 1986), p. 31.

[10] Georg Stahl, *Philosophical Principles of Universal Chemistry* (1723), trans. 1730, quoted in Arnold Thackray, *Atoms and Powers. An Essay on Newtonian Matter Theory and the Development of Chemistry* (Cambridge: Harvard University Press, 1970), p. 175.

[11] G. F. Venel, 'Chymie,' pp. 408–437, in facsimile edition of *Encyclopédie ou Dictionnaire raisonné des arts et des métiers*, Vol. 3 (Stuttgart: Frederick Frommann, 1966; from 1753 ed.), on p. 409.

[12] Isaac Newton, *Opticks*, pref. I. B. Cohen (New York: Dover, 1952; from 4th ed., 1730), pp. 401–402.

[13] See Hélène Metzger, *Newton, Stahl, Boerhaave, et la doctrine chimique* (Paris: Albert Blanchard, 1930); and J. B. Gough, 'Lavoisier and the Fulfillment of the Stahlian Revolution,' *Osiris*, 2nd series, **4** (1988), 15–33, esp. 21–29.

[14] Metzger, 409–410.

[15] In Antoine Lavoisier, *Opuscules physiques et chimiques*, Vol. I, *Oeuvres de Lavoisier* (Paris: Imprimerie Impériale, 1864), p. 446.

[16] On this point, see Arthur Donovan, 'Lavoisier and the Origins of Modern Chemistry,' *Osiris*, 2nd series, **4** (1988), 214–231, esp. pp. 219–228.

[17] One day, Lavoisier wrote, it would be possible to "know the energy [connaître

l'energie] of all these forces, to succeed in giving them a numerical value, to calculate them – this is the aim which chemistry must have." Lavoisier, *Oeuvres de Lavoisier*, Volume II (Paris: Imprimerie Impériale, 1862), p. 525; quoted in Maurice Crosland, 'The Development of Chemistry in the Eighteenth Century,' *Studies on Voltaire and the Eighteenth Century* **24** (1963), 369–441, on p. 407.

[18] See Henry Guerlac, 'Chemistry as a Branch of Physics: Laplace's Collaboration with Lavoisier,' *Historical Studies in the Physical Sciences* **7** (1976), 193–276; Evan Melhado, 'Chemistry, Physics, and the Chemical Revolution,' *Isis* **76** (1985), 195–211.

[19] Frederic Lawrence Holmes, *Eighteenth-Century Chemistry as an Investigative Enterprise* (Berkeley: Office for History of Science and Technology, University of California at Berkeley, 1989).

[20] Melhado (ref. 18), pp. 209–210.

[21] Maurice Crosland, 'The Development of Chemistry in the Eighteenth Century,' *Studies on Voltaire and the Eighteenth Century* **24** (1963), 369–441, on p. 409.

[22] See classification in William Whewell, *Philosophy of the Inductive Sciences* (1840, new ed., 1847) in *Selected Writings*, ed. Y. Elkana (Chicago: University of Chicago Press, 1984), p. 187.

[23] On "physics," "experimental physics," and "mathematical physics" in the eighteenth and nineteenth centuries, Thomas Kuhn, 'Mathematical vs. Experimental Traditions in the Development of Physical Sciences,' in *The Essential Tension: Selected Studies in Scientific Tradition and Change* (Chicago: University of Chicago Press, 1977), pp. 31–65.

[24] Thomas P. Pynchon, *Introduction to Chemical Physics*, 3rd rev. ed. (1881).

[25] Crosland (ref. 21), p. 387.

[26] Arnold Thackray, (ref. 10); Melhado argues that Lavoisier, even before Dalton, rejected the Newtonian program, although Lavoisier's colleagues Guyton de Morveau and C. L. Berthollet continued to pursue it (ref. 18).

[27] Crosland (ref. 21), p. 401.

[28] See J. B. Dumas, 'Mémoire sur la constitution de quelques corps organiques et sur la théorie des substitutions,' *Comptes Rendus Hebdomaires de l'Académie des Sciences*, 1839, 609–622; and Mi Gyung Kim, 'Practice and Representation: Investigative Programs of Chemical Affinity in the Nineteenth Century' (UCLA Ph.D. Thesis, 1990), p. 75, n. 37.

[29] Henri Sainte-Claire Deville, *Leçons sur l'affinité* (Paris, 1867), quoted by Pierre Duhem, *The Aim and Structure of Physical Theory* trans. Philip P. Wiener (Princeton: Princeton University Press, 1954), p. 125.

[30] See Marcellin Berthelot, *La synthèse chimique* (Paris: Baillière, 1876), p. 275: 'La chimie crée son objet.'

[31] E. F. Caldin, *The Structure of Chemistry in Relation to the Philosophy of Science* (London: Sheed and Ward, 1961), pp. 18–19; also see comments by Ida Freund at the turn of the century, in *The Study of Chemical Composition* (Cambridge: Cambridge University Press, 1904), pp. 2–3.

[32] Aristotle, of course, had considered "physics" to be a very different kind of knowledge from "mathematics" and "theology" since physics, in contrast to the other two disciplines, concerned itself with imperfect, changing objects. See Immanuel Kant, *Critique of Pure Reason* (Norton edition), p. 54; and preface to *Metaphysical Foundations of Natural Science*, quoted in Frederick Gregory, 'Romantic Kantianism and the End of the

Newtonian Dream in Chemistry,' *Archives Internationales d'Histoire des Sciences* **34** (1984), 108–123, on p. 109.

[33] See Keith Nier's characterization of Maxwell's attitude: "Maxwell admitted, almost grudgingly, that chemistry is a physical science. But he could find nothing agreeable to say about it. He passed over it with a perfunctory acknowledgment of high rank and an implied slur regarding lack of clarity, organization, and so forth." In Keith Alfred Nier, 'The Emergence of Physics in 19th-Century Britain as a Socially Organized Category of Knowledge: Preliminary Studies' (Harvard University Ph.D. thesis, 1975), pp. 102–105. James Clerk Maxwell, 'Physical Science,' *Encyclopedia Britannica*, 9th ed., 1875–1889, Vol. 19, pp. 1–3.

[34] Hermann von Helmholtz, quoted in Henry Edward Armstrong, 'Presidential Address of the Chemical Section,' *Reports of the British Association for the Advancement of Science*, Winnipeg, 1909, 420–454, p. 423.

[35] Quoted in Jeffrey Johnson, 'Academic Chemistry in Imperial Germany,' *Isis* **76** (1985), 500–524, p. 510.

[36] William Thomson [Lord Kelvin], 'On Vortex Atoms,' *Philosophical Magazine* (4), **34** (1867), 15–24, pp. 16–17; and Ludwig Boltzmann, 'On the Necessity of Atomic Theories in Physics,' *The Monist* **12** (1901), 65–79, pp. 73–74.

[37] See Helge Kragh, 'Julius Thomsen and Classical Thermochemistry,' *British Journal for the History of Science* **17** (1984), 255–272; Marcellin Berthelot, *Essai de mécanique chimique* (Paris, 1879), p. 259.

[38] See Kragh, pp. 264–266.

[39] Pierre Duhem, *Le potentiel thermodynamique et ses applications à la mécanique chimique et à la théorie des phénomènes électriques* (Paris, 1886).

[40] *Ibid.* Crosbie Smith argues that in physics the concept of energy was substituted for force, beginning around 1850 with the work of William Rankine and William Thomson (Lord Kelvin). See Crosbie Smith, 'Mechanical Philosophy and the Emergence of Physics in Britain, 1800–1850,' *Annals of Science* **33** (1976), 3–29; and 'A New Chart for British Natural Philosophy: The Development of Energy Physics in the Nineteenth Century,' *History of Science* **16** (1978), 231–279. I am grateful to Richard Beyler for pointing me originally to these references.

[41] Gilbert N. Lewis, *The Anatomy of Science* (Washington D.C.: American Chemical Society, 1926; reprint, Books for Libraries Press, 1971), pp. 99–102.

[42] See John Servos, *Physical Chemistry from Ostwald to Pauling. The Making of a Science in America* (Baltimore: The Johns Hopkins University Press, 1990).

[43] Walther Nernst, pref. to the German ed., *Theoretical Chemistry. From the Standpoint of Avogadro's Rule and Thermodynamics*, trans. Charles S. Palmer (London: Macmillan, 1895), p. xii.

[44] There was considerable resentment of Ostwald's criticism of organic and structural chemistry. See Richard Willstätter, *Aus meinem Leben; von Arbeit, Musse, und Freunden* (Weinheim: Verlag Chemie, 1949; 2nd ed. 1958), pp. 89–90.

[45] Discussed in Charles Brunold, *Le problème de l'affinité chimique et l'atomistique. Etude du rapprochement actuel de la physique et de la chimie* (Paris: Masson, 1930), p. 3.

[46] Thomas Lowry, 'Is a True Monomolecular Action Possible?' *Transactions of the Faraday Society* **17** (1921–1922), 596–597.

[47] Alexander Williamson, 'On the Constitution of Salts,' *Chemical Gazette* **9** (1851), 334–339, on p. 334; quoted in O. T. Benfey, 'Concepts of Time in Chemistry,' *Journal of Chemical Education* **40** (1963), 547–577, on p. 574.

[48] See Kenneth J. Laidler, 'Chemical Kinetics and the Origins of Physical Chemistry,' *Archives for History of Exact Sciences* **32** (1985), 43–75.

[49] For a detailed study of the radiation and the ionic and electronic theories of chemical activation, see M. J. Nye, 'Chemical Explanation and Physical Dynamics: Two Research Schools at the First Solvay Chemistry Conferences, 1922–1928,' *Annals of Science* **46** (1989), 461–480.

[50] See Yuko Abe, 'Pauling's Revolutionary Role in the Development of Quantum Chemistry,' *Historia Scientiarum* **20** (1981), 107–124.

[51] Erwin Hiebert, 'Nernst and Electrochemistry,' in George Dubpernell and J. H. Westbrook, eds., *Selected Topics in the History of Electrochemistry* (Princeton: The Electrochemistry Society, 1978), 180–200, on p. 188.

[52] W. Heitler and F. London, 'Wechselwirkung neutraler Atome und homopolare Bindung nach der Quantenmechanik,' *Zeitschrift für Physik* **44** (1927), 455–472.

[53] F. Hund, 'Zur Deutung der Molekelspektren,' especially Part V. "Die angeregten Elektronenterme von Molekeln mit zwei gleichen Kernen (H_2, He_2, Li_2, N_2^+, N_2 ...)," *Zeitschrift für Physik* **63** (1930), 719–751; J. H. Van Vleck, 'On the Theory of the Structure of CH_4 and Related Molecules,' *Journal of Chemical Physics* **1** (1933), 177–182, 219–238; **2** (1934), 20–30; and review article by J. H. Van Vleck and Albert Sherman, 'The Quantum Theory of Valence,' in *Reviews in Modern Physics* **7** (1935), 167–228.

[54] C. A. Coulson, 'After-Dinner Speech,' 16 August 1971, at Fourth Canadian Symposium on Theoretical Chemistry in British Columbia, Coulson Papers, #40, Bodleian Library (Duke Humfrey's Library), Oxford University.

[55] C. A. Coulson, Papers, 'Recent Developments in Valence Theory,' paper delivered at Australian symposium 'Fifty Years of Valence Theory,' page proofs, p. 2; Coulson Papers, #41.10, Oxford University.

[56] This no longer is thought to be explained by quantum mechanics. See Stephen J. Weininger, 'The Molecular Structure Conundrum: Can Classical Chemistry be Reduced to Quantum Chemistry?' *Journal of Chemical Education* **62** (1984), 939–944.

[57] On this point, see Coulson's 'Inaugural Lecture' for the chair of theoretical physics at King's College, London, 1948, typescript, Coulson Papers, #21, Oxford University.

[58] Alberte Pullman, Introduction, in Raymond Daudel and Alberte Pullman, eds., *Aspects de la chimie quantique contemporaine*, Colloques Internationaux de CNRS, #195 (Editions du CNRS, 1971), p. 10.

[59] Coulson, 'Recent Developments' (ref. 55), p. 3.

[60] Coulson, 'Inaugural Lecture,' 1948 (ref. 57).

[61] *Ibid.*

[62] Pullman (ref. 59), p. 13.

[63] See Yuko Abe (ref. 50), 109–110. Also, see Anthony Serafini, *Linus Pauling. A Man and His Science* (New York: Paragon House, 1989).

[64] R. S. Mulliken, 'Spectroscopy, Quantum Chemistry, and Molecular Physics,' *Physics Today* **21** (1968), 52–57, on p. 54. Mulliken here advised against the trend in introductory chemistry and physics courses to treat theory before "a generous confrontation with facts" (p. 55).

65 Quoted in S. S. Schweber, 'The Young Slater' (ref. 1), pp. 403–404.
66 Roald Hoffmann, 'The Grail,' in Vivian Torrence and Roald Hoffmann, *Chemistry Imagined. An Art/Science/Literature Collaboration*, and exhibit, book ms., to appear in 1992.
67 See M. J. Nye, 'Explanation and Convention in Nineteenth-Century Chemistry,' pp. 171–186 in R. Visser *et al.*, eds., *New Trends in the History of Science* (Amsterdam: Rodolpi, 1989).
68 See the discussion in Weininger (ref. 56), citing Hans Primas, 'Foundations of Theoretical Chemistry,' pp. 39–113 in R. G. Woolley, ed., *Quantum Dynamics of Molecules: The New Experimental Challenge to Theorists* (New York: Plenum Press, 1980); quotation on p. 105; and Primas, *Chemistry, Quantum Mechanics, and Reductionism* (New York: Springer Verlag, 1981).
69 R. G. Woolley, 'Must a Molecule Have a Shape?' *Journal of the Chemical Society (London)* **100** (1978), 1073–1078; and 'Further Remarks on Molecular Structure in Quantum Theory,' *Chemical Physics Letters* **55** (1978), 443–446.
70 Roald Hoffmann, 'Under the Surface of the Chemical Article,' *Angewandte Chemie. International Edition in English* **27** (December 1988), 1593–1602, on p. 1597.
71 Robert Lespieau, 'Poids moléculaires et formules développés,' 8-page extract from the *Journal de Physique*, June 1901, pp. 6–8.
72 See D. W. Theobald, *Chemical Society Reviews* **5** (1976), 203, regarding chemists and this view.
73 See David B. Kitts, *The Structure of Geology* (Dallas: Southern Methodist University Press, 1977); Rachel Laudan, *From Mineralogy to Geology. The Foundations of a Science, 1650–1830* (Chicago: University of Chicago Press, 1987).

PETER GALISON

FORTRAN, PHYSICS, AND HUMAN NATURE

1. INTRODUCTION: THE READING OF MACHINES

During the late 1950s and 1960s, bubble chambers buried physicists. At first thousands, soon tens of thousands, then hundreds of thousands, and at last millions of photographs issued from the cameras at the massive chambers of the largest particle physics laboratories in the world: Lawrence Radiation Laboratory (LRL), Brookhaven National Laboratory (BNL), the Center for European Nuclear Research (CERN) along with a host of other laboratories including Dubna (in the Soviet Union), and Rutherford Laboratory (in the U.K.). Together, these sites produced data on new particles that transformed the discipline intellectually through a much-deepened understanding of symmetry principles and nuclear forces. It did so at a cost. Already inflated by construction outlays, the price of doing particle physics surged further as "armies" of technicians and scanners joined physicists in sorting through the endless stream of 70 mm negatives. But the burden of photographic plenitude was not simply economic: physicists struggled to define the right relation between scientist and data, between physicist and technician, and between human and machine. At stake, the physicists argued, was not simply a matter of new technologies – each solution to the picture problem was at one and the same time an argument about the boundary of what it meant to be a physicist, a scientist, and a human being.[1]

All of these concerns were embodied in a novel class of devices designed to "read" photographs, often bearing names like "Rapid Reader" or "Spiral Reader." It is to these "reading machines" that this chapter is devoted. The nomenclature itself forces a twist on Michael Mahoney's important study "Reading Machines."[2] For in the case at hand we must work in two directions simultaneously: we must "read" the machines for the cultural assumptions built into them, and simultaneously reconsider the nature of reading implied by the textualization of photographs. In the gerundive usage of "reading," we need to parse the use of the machines, the hardware, and the software that compose it. In the adjectival sense of "reading" machines we must ask what stands behind the project of treating photographic evidence as text. What do the builders (are they authors?) of these devices have in mind when they decompose "reading" into constituent parts (such as scanning, measuring)? Reading machines (machines that read) offer us an opportunity to approach the vexed problem of interpretation from the ground up, as it were. Instead of beginning with a high-level theory about reading, rhetoric, and writing and then "applying" it to the analysis of an instrument, I want here to begin with the object and the day-to-day operations of economics, work, gender, and science in which the instrument participated. The difficulties of reading, interpretation, and discovery then emerge embedded in the machine design and use. To see how these broader issues are entangled with the practice of data reduction, we must enter the fray of a struggle between two strategies of reading.

At one pole was Luis Alvarez, leader of the crash bubble chamber program at LBL. In a long series of hard- and software developments, Alvarez's group defended what I will call an "interactionist" view; they held fast to the position that human beings, by virtue of their peculiar capacities, had to remain central to the processing of track pictures. Machines would *aid* the human, but the technology would revolve around "good enough" engineering that eschewed any attempt to supplant the intrinsically human gift of recognizing patterns and seizing upon the striking or unusual.

At the other pole was Lew Kowarski, an early participant in, propagandist for, and organizer of CERN. From the beginning of the "reading machine" industry, Kowarski defended what might be called the "segregationist" view: that, provisionally, humans would do what preparatory work they had to do before the machines would take over – but ultimately machines would put people out of the photo-reading business. The machine which he hoped would usher in the posthuman age was invented by Paul Hough, a Michigan physicist, and Brian Powell, a staff physicist at CERN. From their first publication, Hough and Powell saw their partially automated scanner as but a first step in the complete elimination of human beings from the process of reading photographs. In Kowarski's later thinking, the ultimate goal transcended not only high-energy physics but physics more generally. The reading machine was to "give the computer an eye," by solving the pattern recognition problem and eventually transforming other scientific technologies, from the tools of cellular biology to those of aerial reconnaissance.

Each side imposed what I will call a *reading regimen* – a specification of who would see the pictures, what they would search for, how the information would be recorded. In the process of exploring the dynamics of each regimen, we can see more than mere data for particle resonances. Embodied in the technology of reading is a social order of the workplace, an epistemological stance towards discovery, and a vision of the relation between physics and the engineering arts. The task of this essay is to trace the development of these competing strategies, and to situate them in a wider universe of beliefs about human nature.

2. FRANCKENSTEIN READS

Bubble chambers were introduced into particle physics as a major tool of inquiry by Alvarez's group at Berkeley during the mid-1950s. Their schematic function is easy to describe. A volume of superheated liquid hydrogen (for example) is held under sufficient pressure to keep it from boiling. Just before a beam of particles is blasted into it, that pressure is released, leaving the liquid unstable against eruption into bubbles. When the beam enters the chamber, the charged beam particles leave behind a spike of heat in the liquid, and these hot spikes precipitate boiling. Illuminated by huge strobes, cameras record the tracks of bubbles, and the chamber is recompressed in a matter of milliseconds to prepare for the next pulse of the beam.

From Alvarez's first plunge into bubble chambers, he set three principal constraints on future work. To bring the device into the center of particle physics, the chambers had to exploit liquid hydrogen, with its ultimately simple nucleus of a single proton, not the easier-to-manipulate hydrocarbons that the originator of the method (and most other groups) had chosen. Second, Alvarez wanted his team to scale up to an industrial size as quickly as possible. This would insure more data, and a larger useful volume in which decays, scatters, and interactions could be observed. Third, and most importantly for our purposes, from the mid-1950s forward, Alvarez made it clear that only a fully integrated cycle of data production and data reduction would constitute a useful innovation.[3] Unlike CERN, from the earliest days of bubble chamber work the analysis of photographs was considered equally (both institutionally and conceptually) with the mechanical and cryogenic problems of big chambers. Alvarez was fond of saying that bubble chamber physics could only function as an *integrated system* – like the weapons systems to which he had devoted so much of his career.

The first of the reading machines drew its name from one of its inventors, the LBL engineer, Jack V. Franck. On seeing it, the physicist Arthur Rosenfeld was sufficiently horrified by its jerry-built illumination stand, microscope measuring engine and other features to label Franck's contraption a monstrosity – Franckenstein.[4] Despite its appearance, the machine clearly accelerated the measuring process by allowing an operator to "drive" a cursor along the projected bubble chamber track. By pushing a button, the machine would automatically register the precise coordinates of a given spot on the track. These coordinates then were taken up into a series of computer programs that first fit the points into a geometrical curve, then performed a kinematical analysis on the curve to check the relation of energy and momentum, and in later years executed a variety of more elaborate tasks associated with data reduction.[5]

From the start, the Berkeley team integrated physicists and engineers, experimentalists and computer programmers: Alvarez, for example, was involved in many ways in work on scanning and measuring devices, while many of the early bubble chamber articles were authored by engineers. In constant interaction with plans for new bubble chambers, the scanning tables and measuring machines went through a long series of new models: Measuring Projector (MP) I, MPIa, b, c, d, e, and f, followed by larger-scale devices for the 46 mm film needed for the bigger chambers. By contrast, at CERN efforts to produce a Franckenstein-like device (Instrument for the Evaluation of Photographs, IEP) were relegated to the Scientific and Technical Services (STS) division, renamed *Données et Documents* in 1961, and headed by Kowarski until 1963.[6] I will argue below that this institutional contrast between the Alvarez group and the CERN approach went hand in hand with differences in the design of hardware, and with much wider differences in the relation of physics to engineering. Indeed, in many respects the reading machines captured, and then helped fashion, the work life that went with experimentation.

For example, the practice of nuclear physics in the two decades after World War II was largely a male preserve, and this included the design of reading machines. But the actual reading of photographs was, from the start, women's work. For years, women had been poring over astronomical star plates[7] and scrutinizing scintillation screens.[8] Since the

Fig. 1. Named after one of its inventors, Jack V. Franck, the Franckenstein became the prototype for a generation of track measuring 'reading machines,' not only in the United States, but in Europe and the Soviet Union as well. *Source*: UCRL BC 764.

1940s, women had been the first to examine and record the nuclear and particle tracks on photographic emulsions.[9] At the same time women had served as "computers" (as they were called) who calculated numerical solutions to the tangle of differential and integro-differential equations that arose in various war projects, most prominently in the nuclear weapons work at Los Alamos during and after World War II. (When the electronic calculators began to execute this work more quickly, they took on the name "computer" and women were rapidly assimilated into the job of programming their functional namesakes). Thus when it came time to define who was likely to fit the job of measuring tracks on bubble chamber film using a computer-aided device, it was overdetermined that the occupation should appear naturally as women's work.

This gendered division of labor in the bubble chamber laboratory seemed obvious to the participants, quite independently of the scanning system. The self-evidence of this appears in a myriad of places. Here is one: at the end of the first track analysis conference at CERN in 1962, Kowarski rose to summarize the fundamentally different approaches of the Alvarez system and the Hough Powell Device (HPD); in so doing, he pointed to many axes of disagreement. But ultimately "both [approaches] pursue the same aim – to solve the problem of man vs machine (or, rather, the scanning and measuring girl vs machine.)"[10] In a similar vein, Alvarez reported in 1966 that, absent innovations, Franckensteins would have demanded a staff of a thousand to measure a million events per year – then the current production quota for a single bubble chamber. By Alvarez's lights, it was a situation not unlike the one that faced the telephone company half a century before; the industry had predicted that "if everyone was going to own a telephone, about half of all the women [in the United States] would be required as telephone operators. [The company] concluded that the efficiency of each operator had to be increased enormously; the dial system is the result of that engineering analysis."[11] A similar engineering analysis was the order of the day in the 1960s: how to build on and augment the productivity of everyone within the system, from the "scanning girl" to the physicist.

I emphasize the separation of the "scanning girl" from the physicist because as the task of scanning was removed from the experimentalist's domain, the definition of experimental physics itself shifted. The point here is not so much that the scanning of bubble chamber images was intrinsically unskilled labor and was cast to women. Rather, while scanning had routine aspects, it was in fact intensively *skilled* labor – and its separation from the experimentalist's domain altered the meaning of experimentation. Observation and the first confrontation with data moved away from the laboratory to the "shop floor." But to characterize the practices involved in scanning and measuring, we need to move beyond generalities; we need to articulate the specific technical operations required at each site of the contested reading regimens.

We would not know the details of the work of reading, were it not for an extraordinary set of training documents that were written at Berkeley, and used for years in laboratories all over the world. Track reading skills were passed orally from physicist to division leader to scanning supervisor to scanner. As the procedures expanded and became routinized in the early 1960s, however, the Alvarez team began to prepare training documents.[12] Starting in 1961, and established firmly by the mid-1960s, a rather formal "course" was in place to instruct the new scanner. In the most popular of these, the scanner learned while seated at the scanning table, working through the instructions frame by frame. Projected onto the table, just as bubble chamber film would be, the training film covered a distinctive admixture of physics, scanning tricks and procedures, and the broader culture of the Radiation Laboratory. After training, she could become a routine scanner, able to follow the classificatory demands of the physicists, or rise to the level of advanced scanner, capable of disambiguating complex events that had baffled the computer, the physicists, or both.

The Alvarez Group Scanning Training Memo (written in 1961, revised in 1964, and again in 1968) opens with a presentation of the basics of accelerator design: how the bending magnets isolate particles with a chosen momentum, how velocity spectrometers separate just those particles with a specific velocity (and therefore mass), and how focussing magnets collimate the beam. At the same time, the apprentice scanner scrolled through pictures not only of the chamber and its photographs, but of the wider accelerator laboratory itself. If the environment appeared chaotic, this was no mere accident – rather it was presented as a part of the world the scanner was entering: "The elements of the beam described above are sometimes sophisticated. Designing such a beam and putting it together in working order requires great skill. Nevertheless, as is so often the case with brilliant scientists, their contraptions are rarely beautiful. Consider the photo below . . . "[13]

The vision of physics passed to the scanners by the physicists is both immensely complex and highly simplified. It is simplified insofar as a large fraction of the entities and the laws governing them are deleted in the presentation. For example, of the neutrino the guide has this to say: "The neutrino . . . can pass thru[sic] a block of concrete the size of the earth without realizing it was there. You will never see one or suspect that one was involved in the interactions you find while scanning. For our purposes, neutrinos may be forgotten."[14] Or, the manual informs the scanners that the η will not be discussed because, like other resonances, it decays too quickly to leave a track. "None [of these resonances] are significant to scanners because, like the η, there is no way to tell whether or not they are there." No way, that is, for the *scanner* to tell. And again (now referring to the classification of forces): "The scanning instructions on the experiment for which you'll be scanning will very probably say nothing about strong, electromagnetic and weak interactions. Scanning is mostly a function of topology, and a knowledge of what is occurring inside the particular nucleus involved will rarely affect the proper identification of the event."[15]

But if the physics is truncated in some places, in others (where it leaves the visible tracks projected on the scanning table), it is subtle, difficult, and anything but routinized. For a start, the scanner acquired five basic tricks, and a myriad of delicate variations upon them:

1. Scanning two views at a time (superimposing)
2. Reconstructing invisible paths of neutral particles (deducing lines of flight)
3. Matching paths of tracks to known trajectories (deploying stopping templates)
4. Measuring curvatures of track (using curvature templates)
5. Counting bubble density (consulting ionization charts).

At the simplest level, superimposing meant taking two of the three stereo views, and positioning a known point seen on both images one on top of the other. For example, if a particle stopped in the interior of the chamber, the end point of the track (such as point p) could easily be identified and the two views superposed – as in Figure 2.

As an aid to the determination of points inside the liquid hydrogen fiducial points (surveying marks) are inscribed on the interior of the top bubble glass and on the bottom of the chamber. If the point p is near the top of the chamber, when the two views of p are made to sit on top of one another, the top fiducial points will also appear to be superposed. If p is deep in the chamber the top fiducial marks will appear to be far apart. A much more difficult exercise can be undertaken as the scanner becomes more experienced; she can take two views of the track and "run" along it, superposing points further and further along the track. By watching the relative position of the two pictures of the fiducial, she can determine if the track is dipping or rising in three-dimensional space. What begins as an intellectual exercise becomes a three-dimensional visual intuition upon which many of the other operations will build.

In particular, many of the scanner's tricks only work for an event that occurs roughly in the horizontal plane; superposition could help the scanner establish that the event neither dipped nor rose. For example,

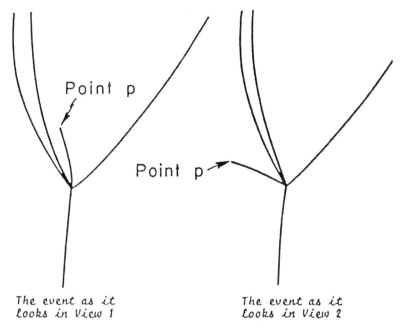

Fig. 2. Superimposition. Bubble Chamber photographs (like cloud chamber pictures) were taken in stereo, and these different perspectives were combined to deduce the full spatial trajectory. Though the final cartesian coordinates were calculated by computer, the initial classification and identification had to be executed by 'scanners' whose job it was to locate 'interesting' phenomena. This skilled task demanded the ability to slide the stereo projections over one another in order to create an abstract mental 'movie' of the unfolding microphysical interactions. *Source*: Alvarez Group Scanning Training Film (1968).

one trick for reconstructing the path of a neutral particle is to mark off tangents to the visible particles of a "vee" – an event with two visible prongs. Using curvature templates, the scanner measures the approximate radii of the left moving particle and marks off a length on the tangent line proportional to that radius; the same is done for the right-moving particle (see Figure 3).

The two marked-off line segments along the tangents then form half a parallelogram, whose diagonal gives the line of flight of the unseen neutral. If the reconstructed neutral has a path leading back to a plau-

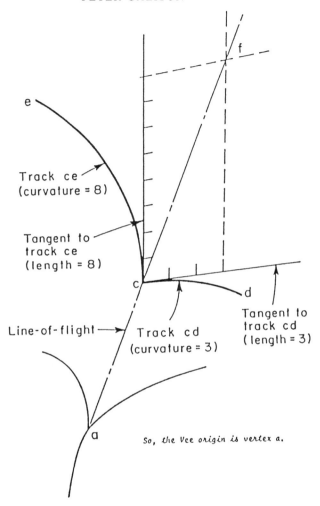

Fig. 3. Tangent Method. Scanners learned to 'see' the unseen by reconstructing the paths of trackless neutral particles. Tangent lines to a 'vee' (two diverging tracks from a vertex) gave their separate momenta. By completing the parallelogram and drawing the diagonal, the scanner could see if there was a candidate source for a neutral particle that could have given rise to the charged (visible) particles of the vee. Nothing *guarantees* the inference here. There is no way, for example, to be absolutely sure that the decay is truly only a two-body process. Here, as elsewhere, scanners were expected to exercise judgment where algorithmic reasoning could not tread. *Source*: Alvarez Group Scanning Training Film (1968).

sible event where it might have originated, the scanner will often conclude that she has completed a valid reconstruction. Back of the method, but unsaid, is the conservation of momentum: by classical electromagnetism, the radius of the track is proportional to the momentum, and assuming the neutral decays into two (and only two) particles, the vectorial sum of these momenta must equal the momentum of the invisible neutral.

For the Alvarez group it was patently clear that these methods could not be completely routinized; this belief was iterated over and again in the scanners' instructions, as for example, in the discussion of neutral particle reconstruction. "Remember," the scanners would read, "that these methods are only approximate. They are not foolproof." From experiment to experiment, the protocol changed: "You should always follow the scanning instructions for your experiment to *the letter*. The mis-use of the above technique could be detrimental to the final data."[16] Even within a single experiment, judgment was required – this was not an algorithmic activity, an assembly-line procedure in which action could be specified fully by rules. For example, if the V dipped or rose, if there was a third (neutral) particle emitted, or if any of a number of other conditions held, the method would fail. Measuring errors were far less dangerous than misidentifications, and the waiting computers were nearly immobilized by wrong qualitative judgments made by the scanners.

Tricks multiplied. The scanner learned to eliminate certain particles as candidates. The track of a particle that stopped in the visible volume of the chamber left a distinctive curve. No particle of greater mass could ever curve more than that. Take, for example, a curvature template for a stopping proton; the guide explained that if a positive track of unknown origin curves more than that, it cannot possibly be a proton, and is likely to be (by an eliminative process) a stopping positive pion.

Further aiding particle identification was the momentum of a particle, which could be determined from its curvature, which the scanner fixed by trying out different curvature templates. This would work, however, only if the track lies on a relatively level plane. In this circumstance, the scanner could use the curvature and the magnetic field (printed on the film) to determine the momentum of the particle as measured in millions of electron-volts (MeV). As in all of the techniques, the scanner spent equal time being instructed in when *not* to use techniques at her disposal. If, for example, the three stereoscopic views gave curvatures differing by 20 degrees or more, it meant that the orientation of the track curved in space in a way that the scanner simply could not determine its curvature; only later, would the computer make this reckoning.[17]

Alvarez frequently expounded on the virtues of human intervention as a methodological precept. But the force of his view was inscribed not in these pronouncements, but on the scanning tables. For however deft the computers became, there were many points where the scanner could advance where the computer could not. One was the density of ions left by a particle, a quantity measured by the density of bubbles. No technical means existed for the computer to gain access to the density of bubbles on a photographic track; it was at best a qualitative guess. But by making such an "eyeball" estimate, and using superposition to find the dip angle, the scanner could examine a chart on which was printed the curve of ionization as a function of dip angle for each particle. Sometimes even a rough estimate could eliminate a candidate particle or reaction, even where a precision measurement could not be made.[18] For example, the track might be so dark or light that an interpretation had to be thrown out: a very heavily ionizing track thought to be a proton for other reasons probably is; its best impersonator, the positive pion, is lightly ionizing. As the Alvarez group emphasized in every section, ionization estimates, indeed scanning in general, was explicitly taught as defying a purely rule-bound activity:

As you have seen, ionization, or track density, can help you to identify particles. As with the other scanning techniques, it is approximate and can only be relied upon as such. Experienced scanners will rarely, if ever, say "I *know* that track was made by a π". What they will more likely say is "I bet it is a π", or "it is most likely a π". One should always use track density information with the awareness that it is not foolproof.[19]

How does this work in practice? After an initial scan, measurement, and computer run, certain events are rejected because they fail to represent "the expected hypotheses." These are then sent back to the scanning table "for further 'eyeballing'". Though the computer does not itself have any ionization information from the photograph, it can use its reconstruction of energy and momentum and particle type to tell "him" [*sic*]:

If this event is, as you suppose, [$K^- + p \rightarrow \Lambda + \pi^- + \pi^+$, with the Λ going to $\pi^- + p$] and track 3 is a π^+, then track 3 should be 1.2 times minimum. It will give similar information for the other tracks. The experienced scanner can then look at track 3 and if it is quite dark, and clearly not 1.2 times minimum, conclude that it is not a π^+ as supposed .... He is in a better position to solve the problem than the computer because he can see how dark the tracks are on the scan table. He also has an advantage over the scanner who originally found the event because he has additional information from the computer.[20]

From these procedures themselves – not from grand methodological pronouncements – emerge deeply embedded assumptions about what goes into the reading of a photograph. First, the marks (the tracks) are allusive: they point elsewhere. A delta ray signals the passage of a heavy charged particle, the e+ e– pair with no opening angle heralds a photon, the deeply ionizing track points against a pion, and the list goes on. Second, the trained scanner has acquired a set of conventions. There are conventions conveyed by instantiation – as in the cornucopia of exemplary photographs used to illustrate K's, electrons, Dalitz pairs, conversion pairs, protons, pions, sigmas, π-μ-e chains. And there are conventions of procedure: rules about throwing out angle measurements when views disagree by more than 20 degrees; rules about when the tangent method can be applied to find a neutral's line of flight. Finally, and perhaps most importantly, there is caution before the text, a kind of hermeneutical hesitancy that explicitly forbids the reader to go too far towards the claim of an exclusive interpretation. Judgment was ineliminable in the Alvarez reading regime: the thoroughgoing integration of "scanning girl," physicist, and computer *was* the system, not a prelude to it.

At CERN (and its allied groups in the United States), the HPD gestured to the arrival of the age of computer-driven artificial intelligence as a goal. Engineering would eliminate judgment, and "scanning girls" would tackle increasingly regularized tasks until every last one was absorbed into automation. It was to this alternative vision of data analysis that we now turn.

3. THE COMPETITION: HPD *VERSUS* SMP

In August 1960 Paul Hough, then visiting CERN from the University of Michigan, collaborated with the British physicist Brian Powell in the submission of a radically new automated method of reading bubble chamber photographs, intended to replace the human entirely. Their goal was to develop a track analysis system that would stand in a natural sequence of devices leading, ultimately, to a fully automatic pattern-recognizing instrument. With this *telos* their scheme held four parts. First, they would eliminate mechanical track-following systems, instead building hardware that would cast a spot of light smaller than the mean bubble diameter, and flash this spot over the film. When the spot hit a darkened bubble image, the intensity of transmitted light dropped, and a photomultiplier recorded this event as a halt in its electrical output. By shining their beam over several thousand parallel lines, they could digitize the positions of individual bubble images to within 2 microns; in only ten seconds, the device could reduce a 70 mm photographic image from the 72-inch bubble chamber to numerical information.

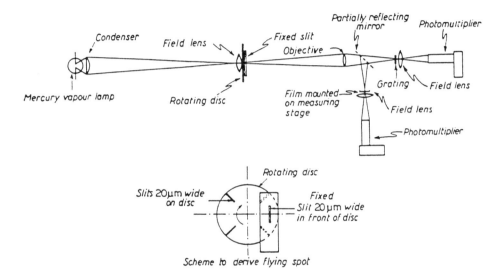

Fig. 4. Schematic HPD. A mercury vapor lamp illuminates a mechanical flying spot generator. The generator works (as indicated in the blown-up insert) when the rotating slit crosses the stationary vertical slit. This spot then passes through the objective lens and is split by a half-silvered mirror: one image continues through a field lens and is captured by the photomultiplier at the upper right-hand portion of the schematic diagram. The other image is reflected down through the bubble chamber film and onto a second photomultiplier. By subtracting the signal from the first photomultiplier from that of the second, only the effects of the film image remain. A dark spot (a bubble) then would cause a spike in the electronic output. These spikes serve as the basis for the computer analysis, which Hough and Powell, hoped would eventually eliminate humans altogether. *Source*: Hough and Powell, 'Faster Analysis,' *Nuovo Cimento* **18** (1960): 1184–1191 on page 1186.

Hardware was, however, only part of the reading regimen. For as they unveiled the device, Hough and Powell introduced a second idea, "parallel human guidance," which specified the precise position people would occupy in reading. In particular, as a necessary (but all hoped provisional) concession, a *person* would intervene before the machine began its work, help it begin the task, then cede control entirely to the technology. Human and machine were therefore segregated, not integrated. The HPD philosophy: where humans were, there automation would be. Reading à la HPD obviated the need to insert human beings serially between each measurement. Instead, humans would apply their judgment in parallel to the whole lot by laying out "roads," crude demarcations of the rough trajectories in which the actual particle track was to be found. At first, Hough and Powell imagined the operator marking out masks around the track to keep the computer from needlessly taking time on irrelevancies or succumbing to the ambiguities of which tracks were relevant, or even what constituted a track. Soon, at Kowarski's suggestion, Hough and Powell simplified road engineering by having the operator merely indicate the location and identity of a track by a few rough designations on or near it.[21] Specifying roads in this way gave the computer a jump start on image processing, since tracks were now unambiguously identified and vast areas of the film were stricken from computer consideration as "uninteresting."

Third, the inventors determined that they needed to extract far less information from a bubble chamber photograph than others had expected. Instead of the order of 10^8 or 10^9 bits, they would only need to register about 10^6 bits. Fourth (and finally) the system involved a transfer mechanism from the flying spot device itself to a computer. "A measuring system of this kind," the authors emphasized over and again, "has the advantage that it can be developed into one capable of varying degrees of pattern recognition."[22] For the Franckenstein designers, such ambitions were irrelevant; their appointed task was the design of engineering solutions to facilitate image processing. They sought to aid the human scanner – not to replace her.

What was the appeal of full automation? Answers varied. One participant on the HPD project, the Brookhaven mathematician J. W. Calkin, conceded that the goals were not purely practical, and spotlighted his fascination with the general problem of automatic pattern recognition:

The great argument for [a full-fledged flying spot device] is of course the elimination of human intervention. The importance of this varies from place to place and is, in effect, a function of the available labor market. I take it, nevertheless, that on the whole we regard it as desirable, uncertainties in the labor supply for such work being what they are. In fairness, I should add that our motivations in pursuing this work at Brookhaven were mixed – certainly, anyway, insofar as I and others in the Mathematics Department at Brookhaven are concerned.[23]

Calkin and his colleagues wanted what they called "blue sky" projects, endeavors leading far beyond the physicists' merely terrestrial concerns with controlling costs and "scanning girls" in a difficult labor market. The framework for such open-ended pattern recognition work included four stages beginning with *acquisition* – the accumulation of raw data from the photograph in numerical form. Second, Calkin stressed, was *erasure* – the striking out of information deemed unworthy ("irrelevant" or "uninteresting") now and forever more. From the surviving information, the goal was to extract "physics," which for Calkin meant effecting a *translation* from information to what he dubbed the "Cartesian–Newtonian–Einsteinian language [that] we speak scientifically."[24] Then, once in the language of physics, experimental results would be sorted either into the domain of understood theory or into the challenging category of new physics.

Success with the HPD was striking in the acquisition stage – where rough digitization of data could be effected at an extraordinarily rapid rate. It was therefore clear almost immediately that the Franckenstein held out little hope against the new device. Scanning at a rate of merely tens of thousands events per year, the Alvarez group had to invent or capitulate. On 8 November 1960, Alvarez finished a 27-page "Physics Note" to be distributed inside the Radiation Laboratory. It began by setting out (by analogy) what Alvarez called a "basic design philosophy" for the new machine: "If one wishes to know the coordinates of his home relative to the appropriate fiducial mark, which is just outside the backyard of the White House, in Washington, he doesn't hire a party of surveyors to run a survey line between his house and the fiducial mark; he makes use of an extensive grid of 'bench marks,' which was set down by the Coast and Geodesic Survey many years ago."[25] Instead of subsidizing the surveyors, this clever home-owner lays a rough measurement from his home to a near-by bench mark, and then consults a map to provide the distance between one benchmark and the one outside 1600 Pennsylvania Avenue. The resulting uncertainty in this shrewd procedure is surprisingly small: the error in the rough local measurement is far outweighed by the precision of the much greater distance between the local benchmark and the White House. (The reader is presumed to live somewhere far from Washington D. C.; Berkeley, for example.)

Just like Calkin, the Alvarez group fastened on the control of labor costs. But as soon as the physicists began to take apart the act of photographic reading, it became apparent that the purely economic could not be isolated from the empyrean realms of pure physics. It was obvious, Alvarez pointed out, that the full cost of analysis would depend on the fraction of "interesting events" that "needed" to be scanned. Less evidently, as Alvarez saw it, the number of interesting events that merited scanning depended on cost:

> We have fallen into the habit of speaking of 'interesting events,' as though there was an absolute measure of 'interest.' If we look at the situation more realistically, we find that our definition of 'interest' is exactly tailored to the measuring capability we either have, or expect to have in the near future! This is not at all surprising, since in planning an experiment, a physicist must decide what events he should measure.[26]

Nothing mattered more than a crucial ratio: the rate of picture production from the bubble chamber (per year) divided by the rate of measurement of these pictures (per year). At the end of 1960, this ratio was about 60 for the 72″-chamber; when all of the Franckensteins went on-line, this ratio would fall to about 15. Alvarez wrote: "Since N is 60, we can only measure one event per 60 frames now, but we hope to measure 1 event in 15 frames, when 'the experiment is finished'; we therefore define an interesting event as one which occurs every 20 frames!"[27]

For example, two different event types give information on the interaction of pion-pion interactions in which a beam of negative pions (π^-) was shot into the hydrogen bubble chamber.

$$\pi^- + p \rightarrow \pi^- + p + \pi^0 \text{ and}$$
$$\pi^- + p \rightarrow \pi^- + n + \pi^+.$$

While the first set of reactions had been extensively studied, the second had not. Purely because of the lack of scanning tables, an entire set of events had been rendered "uninteresting."[28]

Less abstractly, Alvarez turned to the material aspects of the reading machine. Following the distance calculation from Berkeley to Washington, the machine proceeded in two steps. First, a small (1 cm) window in a spring-loaded shutter was placed over the desired section of the projected bubble chamber image. (The projected image falls on a glass plate; underneath the plate is a photomultiplier.) Reading the position of the window automatically, the computer stored the identity of the fiducial mark visible in the window. Next the "local measurement" (distance of a point on a track to the fiducial mark) took place by vibrating the glass plate on which the fiducial point is etched. When the benchmark crosses the image of the track, the light transmitted to the photocell diminishes, and a circuit signals the position of the moving benchmark to the computer.[29]

Practically, what happens is this: the operator moves the window along the track and, as often as desired, sets the machine to work at measuring the precise position of the track. Since measuring and scanning take place in one and the same motion of the window over the track, the human and machine are in constant joint action. The method easily lends itself to modification in which the machine asks for a remeasure (for example if, upon computer analysis, a track is physically "impossible" the computer can ask the operator to check again to see if perhaps what was thought to be one track was in fact a juxtaposition of two). But since the human is necessarily active at every stage, the method only awkwardly participates in a broader program of full automation. This did not disturb Alvarez's vision of the bubble chamber "fraternity:"

> Much of the thinking in the past has been flavored by thoughts of thousands of technicians. No doubt automatic pattern recognition will be one of the most important technological developments of the 20th century, but it is hard to see just how the bubble chamber fraternity could be justified in spending appreciable sums of money to bring it into being at a date much earlier than it will appear for other important reasons.[30]

If Alvarez's Physics Note 223 was the vision of interactionism in hardware, J. N. Snyder's of 25 August 1961 did the same for software. For hardware alone could not abstract the "interesting" bubbles from the prosaic ones in real-world film. Noise, in the form of dirt, had to be filtered out; spurious tracks from beam particles had to be eliminated; and the confusion resulting from track crossings had to be resolved. Just as Alvarez resisted pushing too far into the general problem of pattern recognition in the construction of hardware, so Snyder held back in software design. "Progress," he asserted, "is never made in such uncharted areas [as pattern recognition] by over-generalizing in a vacuum of ignorance; rather one first picks some specific problems and tries to solve them. The things learned can then be extended. An attempt to create an automated, computer-controlled, scanning, measuring, and analysis system for bubble chamber film is one example of such a specific problem."[31] Participants in SMP development held a "greater personal interest" in particle physics, "hence it has been chosen as a representative of the larger field." But no attempt would be made to leapfrog directly into the general problem of non-numerical information processing.[32]

Alvarez and his team wanted the "human operator" to be "the black box pattern recognizer." According to Snyder, "from this point of view one does not regard the system as a human scanner and an SMP with an on-line computer to do the computational part of the work on request, but rather views it as a comprehensive analysis program in the computer with an on-line human-and-SMP which can be interrogated when necessary for those visual and pattern recognizing tasks which are not yet automated and which a human does so well."[33] The competition (the HPD) by contrast detached the human from the machine processing, so that all human scanning functions took place before the machine took over; and once the machine had hold of the information, recourse to the human became a more difficult and time-consuming affair. While the CERN fantasy pursued total automation, the Berkeley imagination clasped a highly mechanized but still interactive process of reading pictures. One is a classical robot, the other a cyborg.[34] The output from a future Alvarez group analysis might go to an oscilloscope, asking the operator for more information. As Snyder reasoned:

For example, the following type of question might be asked: 'This track (arrow points to one of the tracks in the scope reconstruction) gave a sloppy fit, re-search it for a kink'; or, 'This V fitted badly, (arrow) search for a recoil from a possible neutral scatter in this (arrow or box) area.' etc.

While this scenario might seem far-fetched, Snyder maintained that it was not wildly beyond current air traffic control or teaching machines, and would eventually find its place under the broader rubric of pattern recognition.

The competition between human-centered and fully automated readers roused passion. Kowarski, summarizing the state of analysis in 1960, insisted that the choice for automation was unambiguous.

Clearly the problem is that of speed, and since human attention and action introduce a rock-bottom bottleneck, speed can be achieved either by pouring in parallel through many bottlenecks, or by eliminating them altogether. Either vast armies of slaves armed with templates and desk calculators – maybe even strings of beads – for a few people operating a lot of discriminating and thinking machinery. The evolution is towards the elimination of humans, function by function.[35]

Franckensteins meant slavery; the HPD was the harbinger of "discriminating and thinking machinery." If, for Kowarski, liberty meant freedom from human involvement in reading pictures, then it should hardly be surprising that he would consider Alvarez's invention of the SMP to have left slavery intact. Indeed, three years later, in 1963, Kowarski summarized advances in data analysis that had occurred over the previous year, and the contrast between the Alvarez approach and the HPD was as strong as ever. At stake, he believed, was the fundamental division between humans and machines, a division that had evolved in four stages. First, Kowarski contended, human observers, aided by nothing but an inch rule, had probed, explored and discovered physics amidst the slow output of the grand old cloud chambers. Workers proceeded with no speed constraint whatsoever. A second stage followed in 1956 in Berkeley under the watchful eye of Alvarez. This was the age of the Franckenstein (and CERN's equivalent device, the IEP) that drove production rates to the order of 100,000 pictures per year. Only recently had the third epoch begun, the age of a million events, ushered in on one side by Alvarez's SMP, and on the other by Hough with his innovative hardware. As Kowarski put it, though both still required humans, their philosophies were utterly at variance:

On one end, the Hough–Powell system segregates the machine from the human; it puts the human operation in its allotted corner and lets most of the machine operation work without the human hampering it. On the contrary, in the Alvarez system, the human operator still is in the center of things but the machine is arranged so that it speeds up the human operation, and corrects its inherent lack of accuracy.[36]

For Kowarski, the HPD's tentative steps towards total automation was just what was needed for the fourth stage, a coming generation of experiments – presumably involving tens of millions of events – that would *only* be accessible to an inhuman reader. Here too the road of development would bifurcate. On the one side, HPD-like devices would be readied with a "philosophy" of "tricky hardware": a computer would drive the spot in ways that reflected the physics interest of the experiment. On the other side, the tricky software solution was to write selectivity into computer programs that would extract information from the experiment without ever producing a photograph.[37]

Kowarski's "fourth stage" did not move the Alvarez group, even as an ideal. For as the Alvarez group penned software and crafted hardware, its members hailed the interactive feature of their "design philosophy" as a virtue, not a vice. If the "interrogation" of the human by the machine got too complex for the scanner, then so much the better. As four Berkeley software writers pointed out, then the physicists could be part of the reading (of) machines:

> if no hypothesis contained in the KICK [kinematic analysis] programs sufficed to fit the event, then additional hypotheses to try could either be generated automatically or requested of the operator. However, if the dialogue demands too involved decisions or too much physical understanding from the operator then it will be necessary to use more highly trained personnel on the SMP tables. This in turn opens a whole new vista of using the SMP and the dialogue concept as tool for understanding and processing very complicated or very recalcitrant events by manning such tables with highly trained experimental physicists.[38]

Nothing like these remarks can be found among the programmatic statements of the HPD enthusiasts. If "scanning girls" were necessary in the early stages of road preparation, so be it. But everything in the CERN program (and its BNL and LBL links) was designed to *obviate* the need for a physicist to sit behind a computer. Reading, in the HPD mode, was entirely without the hermeneutic pause desired by Alvarez's group. Alvarez, by contrast, saw the "scanning girl's" pattern recognition as a capability unmatched (and probably unmatchable) by electronics. If the physicist too was induced to interact with the computer in the exploration of "recalcitrant" images, so much the better. For at every level of analysis, these authors argued, reliability would be *enhanced* not reduced by the human/computer interaction. One group member commented: "Immediate feedback of results to the SMP operator enhances the operating efficiency and reliability of the analysis system."[39]

The commitment of the Alvarez group to the symbiosis of computer and human extended all the way from the identification of tracks to the legitimation of a histogram for publication. At every level, the team wrote computer routines to test both humans and machines. One program titled FAKE used a Monte Carlo routine to generate simulated bubble chamber tracks "faking" real events. Such collections of simulated images could then be used to test the analysis and track reconstruction programs against a set of events with specified characteristics and statistical distributions. Along with this event simulator, the Berkeley team developed a second program, GAME, that produced artificial histograms – reduced data in precisely the form that would normally appear in *Physical Review*. Such phony "final" data had a variety of applications, one of which was to test the reliability of the human eye as it cast its gaze not on individual events, but on the laboratory's end product. Here is what one of the lead programmers, Arthur Rosenfeld, had to say about FAKE in 1963:

> As an example of its application let us consider [a particular histogram] and ask the following questions. "Is the peak above [one of the bins] a resonance or a statistical fluctuation?" "What is the Poisson probability that these adjacent bins will be over populated in so striking a way?" It is difficult to formulate a definition of "striking" so as to answer this question with a chi-squared tale (thus χ^2 does not distinguish between under- and over-population).

To probe the robustness of the physicists' judgment, the programming team would draw a curve through the (real) data, and feed it to GAME. The program then produced 100 fake distributions of this smooth curve modified by Poisson fluctuations. These 100 printed histograms, the real data and 99 fakes, were then presented to the physicists, who are asked to rank them from the one they considered to most forcefully indicate a bump (new particle) to the one that least smacked of new physics. If the real McCoy is *not* the top contender, then the team agreed that an equally "striking" peak would present itself in 100 runs by Poisson statistics alone. "GAME," Rosenfeld concluded, "has protected us from publishing statistical fluctuations instead of resonances."[40]

An example from the published literature can be found in the pages of *Physical Review* from 1962, where the Alvarez group reported on pion-pion interactions, resulting from the collision of antiprotons with protons. In particular, when studying the reaction

$$\bar{p} + p \rightarrow 2\pi^+ + n\pi^0,$$

the bubble chamber experimenters were divided about whether to interpret their results as indicating the existence of one new particle or two. All hinged on the interpretation of Figure 5 – the plots in the left hand column run a single-peak hypothesis curve through the data, and the right hand column indicate the double-peak hypothesis. Again making use of the FAKE/GAME procedure, 20 experimental physicists were

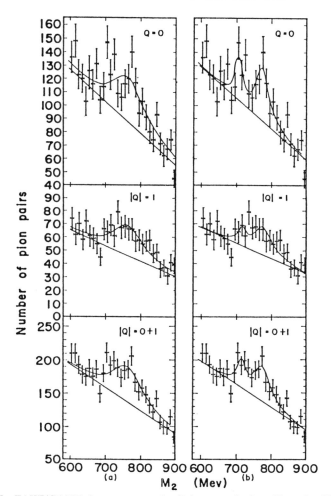

Fig. 5. FAKE/GAME. In contrast to the flying spot device effort, the Alvarez group remained staunchly committed to the idea of maintaining a central role for human judgment in the assessment of data. For example, when the data seemed ambiguous between 'two peaks' (column b) and 'one peak' (column a), the Alvarez group generated Monte Carlo data displaying statistical fluctuations from a distribution with only one 'true' peak. These pictures were then interspersed with real data, and physicists were asked to rank the images in two-peakedness, from the most two-peak-like to the least. When the real data failed to emerge as the clear victor, the physicists withheld the accolade of reality from the second peak. *Source*: Button *et al.*, 'Pion-Pion Interaction,' *Physical Review* **126** (1962): 1158–1863 on page 1861.

shown 72 Monte Carlo histograms of the invariant mass distribution for the pions. Each was asked to identify the one "most suggestive of a double peak." The real data was chosen the second most frequently; six picked it as "most suggestive" of a double peak. "Although such a test is an extremely subjective one," the authors surmised, "we can conclude from it that there is about a 1 to 3% chance of obtaining data . . . which are as suggestive of a double peak as are our data when only a single resonance exists."[41] Humans checked computers with their pattern-recognizing capability; computers checked humans' inclination to see patterns where there were none.

Film, kilometers of it, continued to pour from the bubble chamber cameras, and the debate over reading regimens raged. In 1963, Alvarez intervened in a discussion of the HPD to comment on the relative merits of the competing systems, when stacked against the SMP. Both machines, he observed, cost on the order of $35,000. In terms of speed, he confessed, the HPD held an advantage, as its proponents had repeatedly pointed out, taking 15 seconds per event, whereas on the SMP, an event occupied the machine and its operator for 4 minutes. In part this was due to the radically different procedures. Alvarez:

On the SMP a cursor goes along the whole track, whereas on the FSD [Flying Spot Device, another name for HPD] only three points are measured, but the three points are measured more accurately. . . . At the end of the measuring, the event is completely measured and analysed on the SMP. With the FSD it now goes onto tape, and a day or so later it goes to a $250,000 machine where it is then measured. I think with this method of comparison that the FSD is slower and more expensive. However, when a pattern recogniser is built into the FSD, everything I have said must be reexamined, and the FSD clearly has a potential advantage. One must then reconsider the question, taking into account economic factors, and one's basic philosophy about the interaction of physicists with their own data.[42]

Alvarez's caveats were two. First, as he had six years earlier, Alvarez reiterated his pessimism about even the medium-term prospects for pattern recognition: "I do not think that automatic scanning of bubble chamber film is going to come soon enough to contribute to our present campaign in particle physics. Perhaps it will help in some renaissance, twenty-five years from now, just as the laser has brought new vigor into the field of optical spectroscopy, after it had lain dormant for almost twenty-five years."[43] But his real objection lay in the second admonition, in the loaded phrase about "taking into account . . . one's basic philosophy about the interaction of physicists with their own data." As we have already begun to see, Alvarez saw far more in the choice of reading regimens than the selection of this or that projector design: at stake was the nature of discovery and the defining features of what it meant to be a physicist.

4. READING AND THE NATURE OF DISCOVERY

Computers were in no position to discover the fluke event, the tracks that fit no prior hypotheses and conformed to no previously-known entity. Already in 1960, the Alvarez group stressed the importance of the search for the individual occurrence:

As in cloud chamber work, it was considered very important to examine and keep a record of each event . . . . A large part of the interest was in studying rare, new, or unexpected events . . . . [T]his is still considered very important today, though some experiments involve so many thousand events that individual scrutiny is almost impossible.[44]

Dubbed zoo-ons as symbols of their oddness, Alvarez and his group found these beasts to be the heart and soul of their enterprise.

I should like to say a few things about zoo-ons because I think that they are exceedingly important. At present I believe we miss many of them because physicists no longer do the scanning. There is a hope that this will not be the case when the SMP's are in operation, where the computer will be able to help the scanner or the physicist in interpreting an odd event.[45]

By contrast, the HPD, at least in its full-fledged automatic form, would focus on reducing masses of data *without* any human intervention. Kowarski ceded the point: "These devices [such as the HPD] are therefore promising mostly in statistical experiments in which the chamber acts as a 'seeing counter'; for rarer and more sophisticated events, in which human search and evaluation have a greater role, their promise is less immediate."[46] Thus even within the enterprise of reading bubble chamber photographs, we find the clash between the visual and the logical traditions. Now, though, it is the *statistical* form of the HPD search that accords pictures the role more usually played by Geiger counters and coincidence circuits. The strategy of automating the reading process bypasses the epistemic role of the human search for the golden event, and in so doing has thrown one style of picture analysis out of the more usual image tradition, and into that of the logical counts of electronic gear. In Kowarski's words, HPD-style of reading has effectively reconstructed the automated reader of bubble chamber pictures into a "seeing counter."

Anticipating demurral from an audience well schooled in the great photographic triumphs of the past, Kowarski then moved to challenge the unwritten supposition that the golden event exceeded the statistical discovery in importance or value. One might take it as perfectly obvious that rare events – in physics or elsewhere – were more precious than those appearing in multitudes. This, Kowarski insisted, was unbridled prejudice.

> [I]f a 100,000-picture run contains information on several kinds of physical processes, one of which will require a measurement on only one picture out of ten, and another will call for measuring them all, it will be only human to find that the first process is more interesting and rewarding than the second. The physics of rare events thus gets, for purely material reasons, preference over the physics of high-number statistics. . . .[47]

Kowarski acknowledged the widespread preference for the rare over the common. Such choices had led to the conclusion that physics should be left to physicists; they were the ones who could fully extricate the golden event from those of fool's gold. This was specious reasoning, Kowarski maintained. But in Kowarski's book, three arguments militated against the supervaluation of the extraordinary event. First, physicists often found themselves in situations when "there are millions of events to be processed, and where zoology plays only a small role." Second, physics did not exist in isolation, and the techniques of automated analysis, far more than the interactive SMP, could help advance the cause, for example, of microbiology. Third, and finally, automated pattern recognition was so manifestly important to defense analysts (for example, in the analysis of aerial reconnaissance) that the military would surely be developing techniques relevant to the search for a fully-automated flying spot. And "after a suitable time lag to let the secrecy abate, it may well happen that fundamental science will yet profit from gadgets developed for defense purposes." One such "gadget" (not yet a full pattern recognition device) had, he reported, already come on the market.[48]

CERN's vaulting ambition, however, also hampered its progress. As historians John Krige and Dominique Pestre have pointed out,[49] many commentators have wrongly censored CERN for not engineering enough to keep up with the Americans. Quite the contrary, if CERN lagged behind the United States (especially Brookhaven and the LRL) it was principally because they *over-* rather than under-designed apparatus. Data analysis was no exception. As Kowarski quipped: "So Berkeley lacks universality, Brookhaven lacks automaticity, and CERN, which proudly refuses to sacrifice either universality or automaticity, so far lacks successful achievement in bubble chamber physics."[50]

Even if the glory of new discoveries eluded them for the moment, CERN's drive for ever-larger bubble chambers and more sophisticated and faster analysis machines pressed ahead. The next year (1964) at the Karlsruhe meeting on Automatic Data Acquisition, Kowarski repeatedly extolled the virtue of garnering massive quantities of data and then analyzing them with various automated techniques. His audience, clearly disquieted, responded with a mixture of shock and resistance to the notion of discovery that seemed implicit in the new experimental order. The following dense exchange is transcribed from the taped discussion that followed the talk:

K. Ekberg: I would like to ask a question of principle which touched on the point that one might wish to do the first scanning of these pictures automatically. Now it is surely so that many important discoveries have been made because the scientists have noticed something which they did not expect. Something new which they could not explain from their previous knowledge. Now we surely can't programme computers and automatic devices for this kind of thing. Do you think there is any danger here?

Kowarski: Well, first of all you must know what you want. You obviously cannot produce in a year several million or several tens of million pictures and at the same time explore each picture in an adventurous way as Blackett used to treat his cloud chamber pictures back in 1934. It just cannot be done. These millions will be taken mainly for kinds of experiments in which you know more or less in advance what kind of information you want to extract. In some intermediate cases there might perhaps be not millions but shall we say a hundred thousand pictures, which are worth to look at not in a statistical way but in a more enquiring mood. . . .

H. Schopper: This point of view frightens me a little because it would mean that in a few years if one wants to do a high energy experiment one would not go to start a new experiment but would just go into the archives, get a few magnetic tapes[,] and start to scan the tapes from a new point of view [–] that would be the experiment.

Kowarski: If you want to measure something in . . . space, for example, about ionized belts or something like that you do not immediately jump and go up there. You send a satellite, usually an unmanned one. You take the information collected by the satellite and then you process the information in your laboratory and I think that this kind of attitude will more and more prevail in high energy physics. . . .

W. W. Havens: I would like to point out that looking for the unexpected and hidden is not characteristic for these times . . . . It is important for the physicist to examine his data to look for the unexpected. If the results come only exactly as planned very little is learned.[51]

In the midst of a technical conference on data reduction, something has broken the tone and rhythm of discussion. Suddenly, modifications of software routines such as PANG and KICK are pushed aside; hardware innovations in the faster scan fall from view. What up to then has been implicit becomes explicit: reading photographic evidence one way rather than another, entrains one scientific life rather than another. This is why Ekberg brings to the floor "a question of principle," concerning the possible and dangerous squelching of novelty by automatic scanning. It is why Schopper is "frightened" at the loss of a laboratory for an archive (and the dreadful transformation from scientist to scholar that such a spatial switch implies). It is why Havens disparages the character of the times, times that compare poorly with a lost and better world that preceded them, earlier times when the search for the unexpected and hidden was the rule. And it is why Kowarski himself half laments and half celebrates the passing of the "adventurous" era of Blackett, the old days of viewing pictures in an "enquiring mood."

5. FORTRAN, PHYSICS, AND HUMAN NATURE

At one level, Alvarez's argument resided in engineering practice. The HPD approach overstepped the engineer's reluctance to find solutions far beyond what was "good enough" to solve the problem at hand. At the beginning of a major roundtable discussion at SLAC in 1966, Alvarez suggested that the participants would do well to be "more concerned with engineering and production than with instrumentation." Engineering and production meant an economically-informed activity. Citing an inaugural lecture at the United States Academy of Engineering the previous year, Alvarez paraphrased: "Engineering of necessity deals with economics, and any engineering-like activity that does not deal with economic realities is, in fact, not engineering at all." Instrumentation, which Alvarez referred to as the "delightful activity" without pecuniary considerations, did not qualify. Student shops, department shops, laboratory shops – even professional instrument-making shops – were not engineering. In the problem at hand, thinking like an engineer meant making certain choices that improved the number of "events per dollar" rather than perfecting a piece of hardware or software that would marginally assist in the analysis of a handful of events.[52]

For example, certain "fitting programs" straightened out a track degraded by thermal gradients or turbulent hydrogen flow within the chamber. These, Alvarez contended, cleaned up the data, but at the cost of slowing down the analysis far too much. To make his analysis more systematic, he then turned to the techniques of operational analysis. Operations analysis was a new field, inaugurated during the Second World War to recommend battle strategies for a variety of objectives, most notably the location and destruction of enemy submarines, and the most effective bombing runs. After the war, many of the key figures in operations analysis turned to economics as well as military matters. Key to all their work, and to Alvarez's use of it, was the breakdown of a goal into its constituent parts, each part describable in terms of partial derivatives. In the case at hand, the goal was the production of scanned and measured pictures, and the key constituent quantities were the specific scanning rate for an individual and the scanning rate per person for the whole group:

$\partial R/R \cong$ fractional increase in individual measuring rate
$\partial r/r \cong$ fractional increase in group specific rate arising from $\partial R/R$.

In 1966, $R \approx 100$ per hour, and $r \approx 5$ per hour (remember, r is the rate per total employee hours, and many of the employees are supervising, accounting etc., and not scanning.) The crucial quantity then becomes the ratio of these two dimensionless fractions; that is,

$$\frac{\partial rR}{\partial Rr} \equiv k.$$

Alvarez argued that the fact that k is near unity implies that the group rate depends almost entirely on the individual scanning rate. Other factors were negligible.[53]

Following this operational analysis led the group, for example, to focus attention on variables that were leading to the greatest change in $\partial RX/\partial XR$. These included disambiguating tracks that crossed at a small angle, locating the fiducial marks, or the measurement of the scanning stage to mark the end of short tracks. To Alvarez, the HPD-users' focus on the rate, f, at which events are digitized is irrelevant since it has nothing to do with the more fundamental quantities r (the group specific rate which depends only on the number of road makers) and R (the individual rate of production). In particular, $\partial r/\partial f = 0$. Alvarez conceded that this did not prove that the HPD rate r is not higher than his group's r –

but he did conclude that they were not using the same operations analytical to the field.⁵⁴

On first glance, the Alvarez group seems absolutely dominated by pragmatic concerns. But despite the massive attention to supervisors, accounting, and engineering; despite the reams of paper devoted to production per operator-hour, machine-hour, team hour, I suspect that the real animus behind their opposition to the HPD had little to do with partial derivatives of production. We have seen hints of this earlier – when Alvarez said in 1963 that even if the HPD speed increased their would be differences in philosophy of how physicists interacted with their data. Now, this deep concern emerged more strongly than every: "[M]ore important than [my] negative reaction to the versatile pattern recognition abilities of digital computers is my strong positive feeling that human beings have remarkable inherent scanning abilities. I believe these abilities should be used because they are better than anything that can be built into a computer."⁵⁵

In these remarks lies a belief about a peculiarly human capacity to unravel pictures. It is, I would argue, at the root of the many-sided development of the Alvarez's reading regime in all its many facets. It is the basis for making all of the apparatus interactive, rather than segregated. It is behind all the talk of zoo-ons, and a vision of discovery that is inextricably attached to the human eye.

To bolster his position, Alvarez invoked a hypothetical cautionary tale. An astronomer had found Pluto by using a blink comparator. Flicking into focus first one picture of a star field and then another taken sometime later, the discoverer had noticed a single, crucial spot of light that had changed. What, Alvarez rhetorically asked, would have been the outcome if astronomers had searched with full automation?

> One could certainly scan a star field with an FSD-like device, and store the coordinates and intensities of all stars in some memory system. Then he could repeat this measurement on a star plate taken at a different time, and again store the coordinates and intensities. And finally, he could perform the arithmetic comparisons between the two lists, and throw out all stars that had duplicate images on the other plate to within some "least count."⁵⁶

"Certainly," he concludes, it would have been a "mistake" to try "to beat the human with a computer." Everything, even evolutionary theory itself pointed to the "extraordinary" capacities of the eye-brain system, a system not easily duplicated by machines, however complex. Pluto stood as a warning to those hoping to push the human eye aside.

I hope that we never go so far in automatic scanning that we do in a similar tedious and expensive way something that can be done so easily by a human scanner. I'm sure I can't discourage anyone who really wants to replace scanners by machines. But I don't think that such an effort is usefully related to bubble chamber physics.[57]

To maintain the humanity of reading, Alvarez was willing to impose industrial production as a work regime. By contrast the HPD advocates were horrified at the style of work ("slavery") imposed by the interactive mode, and in exchange were willing to sacrifice the human as part of the regimen of discovery. Alvarez brushed aside anyone who was "horrified to hear any scientist setting such unscientific goals, let me remind you that I have my 'engineering hat' on at the moment so I have no apologies."[58] K. Gottstein, by contrast, held out the precise hope that the HPD reading machine would alleviate the oppression of routinized round-the-clock labor. Perhaps physics could be restored to physicists. At the 1967 Munich meeting on HPD's, Gottstein sketched a Dickensian picture of film reading in the age of million-photograph experiments:

We are now still used to a situation in which the accelerators and their bubble and spark chambers turn out data faster than the physicists and their assistants can analyze them. As a consequence, many laboratories which analyze ... bubble chamber film, work around the clock, in shifts, day and night and weekends, like a blast-furnace or the fire-brigade, and the lights never go out. Now the day may not be so far on which with help of computers, flying spots, etc. the data can be analysed faster than they are produced. This may mean that our laboratories could return to a normal day-time routine of operation which would certainly be more appropriate to the natural rhythm of human life. ....[59]

To Kowarski, the struggle against the human was one marked, as we saw, by an almost aesthetic pleasure in isolating the machine from the human. With a palpable frustration he ended a 1964 conference with the impression that "no matter how automatic any given system claims to be, the human element comes galloping back and the system begins to incorporate again some sort of human guidance." Brookhaven's "gadget" for spark chamber pictures might aspire to the "completely inhuman," but ended up employing "quite a bit of human guidance;" even the HPD spark chamber analyzer which momentarily "achieved complete inhumanity" fell back upon human intervention.[60]

Summarizing one of his HPD associates at their 1964 conference, Kowarski noted that the term "interface" had been used as but "another word for humanity's ugly face showing up again in some way, so that the system is not fully automatic after all."[61] Perhaps discouraged by the eternal return of this repressed humanity, Kowarski concluded the last of the dedicated HPD meetings with the concession that "The real purpose of the whole HPD development, including its recent cathode-ray tube sequels, may turn out to be definable not exclusively in relation to a certain class of problems arising from high-energy physics, but rather in terms of a simpler and wider ambition to provide a computer with an eye."[62]

In a sense, the reading machine debate of the 1960s pitted two equally, though differently situated, ambitions against one another. In the Alvarez group at Berkeley, the ambition for the discovery of novelty, classically defined in terms of a startling golden event, was the order of the day. And they did find such objects, indeed several of them. Muon-catalyzed fusion was one, in which a muon caused two hydrogen atoms to fuse at liquid-hydrogen temperatures. For a fleeting moment, long before the 1980s, the excitement of cold fusion swept into physics. Another was the discovery of the cascade zero, a single remarkable event which only painstaking analysis could transform into an argument. At CERN, the ambition was equally grand, to do new physics with the massive statistical sweep previously possessed only by the competing tradition of electronic counters. More than that, the CERN physicists and their American allies hoped their new device would find application far beyond particle physics, from chromosomal analysis to aerial photography – the dream was to build a device that would read photographic evidence like a phone book.

To a certain extent some of the hopes and most of the fears of both sides were realized in the coming decades of the 1970s and 1980s. Experiments grew vastly larger, leaving teams of Alvarez size (20, 30 or 40 PhD's) looking minuscule by comparison. Electronic devices took over, making the individual scrutiny of events by humans less and less feasible. Yet at the same time the pattern recognition device, like the automatic language translator and so much else in the burgeoning field of artificial intelligence, had to compromise its goals in order to realize them.

As he surveyed the field for the last time in 1967, Kowarski had to admit that "the goal of complete automation is slightly receding." And if neither elegant electromechanical devices nor fancy programming could blot out the human as a reader, Kowarski brought back the reader in the form she had originally taken. Alluding to his oft-repeated dichotomy between the two paths reading might take, he now added (or returned) to a third: "Should we add to tricky hardware and tricky software also tricky girlware?"[63] At the end of the day, both in Berkeley and in Geneva, there was a woman tracing particle trajectories.

Departments of Philosophy and Physics
Stanford University

NOTES

[1] On the history of bubble chambers themselves, see P. Galison, 'Bubble Chambers and the Experimental Workplace,' in P. Achinstein and O. Hannaway, eds., *Observation, Experiment and Hypothesis in Modern Physical Science* (Cambridge: MIT Press, 1985); also see Armin Hermann, John Krige, Ulrike Mersits, Dominique Pestre and a contribution by Laura Weiss, *History of CERN, Volume II: Building and Running the Laboratory, 1954–1965* (Amsterdam, Oxford, New York, Tokyo: North-Holland, 1990), especially chapters 6, 8, 9 – chapter 9 subsection 6 includes the best history of CERN's efforts in track-chamber picture handling facilities. I have profited greatly from this and from Krige's preliminary version in the preprint CERN History Series,CHS-20.
[2] As far as I know, the first systematic exploration of the concept of reading machines as text is in an unpublished paper of Michael Mahoney, 'Reading A Machine: The Products of Technology as Texts for Humanistic Study' (1983); more recent relevant work on this subject can be found in W. Bernard Carlson and Michael E. Gorman, 'Understanding Invention as a Cognitive Process: The Case of Thomas Edison and Early Motion Pictures, 1888–91,' *Social Studies of Science* **20**(3) (1990): 387–430, which focuses on the inventor's use of mental models in the construction of new devices. Also of interest in the "reading" of images are the essays in Michael Lynch and Steve Woolgar, eds., *Representation in Scientific Practice* (Cambridge and London: MIT Press, 1990).
[3] L. Alvarez, 'The Bubble Chamber Program at UCRL.' Stenciled typescript dated 18 April 1955, unpublished but widely circulated.
[4] Jack V. Franck, interview with author, 4 September 1991.
[5] On the Franckenstein as integrated into the full analysis system, see Hugh Brandner, 'Problems and Techniques in the Analysis of Bubble Chamber Photographs,' UCRL-9104 (1960). Also, Jack Franck, interview with author, 4 September 1991.
[6] J. Krige, 'The Development of Techniques for the Analysis of Track-Chamber Pictures at CERN,' CHS-20 (Geneva: CERN, 1987), introduction.
[7] On women's work in the analysis of star charts, see Pamela Mack, 'Straying from Their Orbits,' in G. Kass-Simon and P. Farnes, eds., *Women of Science: Righting the Record* (Bloomington: Indiana University Press, 1990), 72–116, and see Margaret Rossiter's discussion of women's work in her *Women Scientists in America* (Baltimore and London: The Johns Hopkins University Press, 1982), esp. chapter 3.
[8] In nuclear physics, women often worked as the observers of scintillation screens. See Roger Stuewer, 'Artificial Disintegration and the Cambridge-Vienna Controversy,' in P. Achinstein and O. Hannaway, eds., *Observation, Experiment, and Hypothesis in Modern Physical Science* (Cambridge: MIT Press, 1985), 239–307.

[9] See P. Galison, 'The Anxiety of the Experimenter,' in *Image and Logic* (in preparation), chapter 4.

[10] L. Kowarski, 'Concluding Remarks,' in 'Informal Meeting on Track Data Processing,' held at CERN, 19 July 1962 (CERN Yellow Report, CERN 62-37, Geneva: CERN 1962), 99.

[11] L. Alvarez, 'Round Table Discussion on Bubble Chambers' in *Proceedings of the 1966 International Conference on Instrumentation for High Energy Physics*, SLAC September 9–10, 1966 (Springfield, Virginia: National Bureau of Standards, 1966), 271–295, on 276 (hereafter, 'Round Table,' SLAC 1966).

[12] E.g., M. L. Stevenson, 'The Elementary Particles as Seen in the Hydrogen Bubble Chamber,' Lawrence Radiation Laboratory Physics Notes 327, 28 August 1961.

[13] Ed Hoedemaker, 'Alvarez Group Scanning Training Film,' October 1968, Lawrence Radiation Laboratory (LRL) Physics Note 595, 11. For an earlier manual see M. L. Stevenson, 'The Elementary Particles as Seen in the Hydrogen Bubble Chamber,' LRL Physics Note 327, 28 August 1961; and M. L. Stevenson, 'Reaction Dynamics for Scanners,' LRL Physics Note 300, 16 June 1961.

[14] 'Alvarez Group Scanning Training Film' (1968), 28.

[15] *Ibid.*, 128.

[16] *Ibid.*, 66.

[17] *Ibid.*, 71.

[18] *Ibid.*, 75–77.

[19] *Ibid.*, 78.

[20] *Ibid.*, 78.

[21] P. Hough and B. Powell, 'A Method for Faster Analysis of Bubble Chamber Photographs,' in *Proceedings of an International Conference on Instrumentation for High Energy Physics*, September 1960 (New York, London: Interscience, 1960), 245.

[22] P. Hough and B. Powell, 'A Method for Faster Analysis of Bubble Chamber Photographs,' *Nuovo Cimento* **18** (1960), 1184–91, on 1186.

[23] J. Calkin, 'A Mathematician Looks at Bubble and Spark Chamber Data Processing,' in J. Howie, S. McCarroll, B. Powell, and A. Wilson, eds., *Programming for HPD and Other Flying Spot Devices*, Collège de France, Paris, 21–23 August 1963, CERN Yellow Report (CERN 63-34), 193 (hereafter: 'Mathematician,' in Paris 1963), 191–200 or 193.

[24] J. Calkin, 'Mathematician,' in Paris 1963 (ref. 23), 192.

[25] L. Alvarez, 'A Proposal for the Rapid Measurement of Bubble Chamber Film,' LRL Physics Note 223, 8 November 1960, 4.

[26] *Ibid.*, 20.

[27] *Ibid.*, 21.

[28] *Ibid.*, 21.

[29] More specifically, the huge Berkeley bubble chamber was 72″ long, and was typically photographed on film with a useful area of 12 cm by 3 cm. An accuracy of 4 microns ($4\mu = 4 \times 10^{-5}$ cm) on the film was equivalent to a 60μ uncertainty in the chamber. In the long direction this amounted to a fractional error of 4/120,000 or 1/30,000. Measuring one part in 3 times 10^5 is no mean feat, and at root it was this difficulty that led to the high cost of the Franckenstein. Put into binary form 30,000 is $2*10^5$ or 15 bits; similarly

13 bits of information were required in the x-direction. Since the team was already using projectors that magnified ten times, it was practical to adapt these to the new reader; in terms of the projected image, an accuracy of 40μ would be needed. The analogue of the Coast and Geodetic Survey benchmarks would be frosted circles, 300μ in diameter, etched onto a glass projection screen in a square array of one centimeter by one centimeter: 7 bits of information would be needed to specify the "name" of the x benchmark in binary, since the projected length of the chamber would be 120 cm, which is nearly $2^7 = 128$ cm. Bubble position measurement in x-direction then reduces to two measurements. First, one provides the "name" of the x-benchmark that has come into view, e.g., 0100 110, and then a fraction of the distance between benchmarks that corresponds to 40μ ≈ 1/256 cm, or 8 bits: e.g., 0011 1101.

[30] L. Alvarez, 'A Proposal for the Rapid Measurement of Bubble Chamber Film,' LRL Physics Note 223 (1960) (ref. 25), 24.

[31] J. Snyder, 'Some Remarks on a Data Analysis System Based Upon the Scanning-Measuring Projector (SMP),' LRL Physics Note 326, 25 August 1961, 2–3.

[32] In Snyder's analysis, the goal of pattern recognition lay behind two great barriers. First, one needed a "front end" pattern recognizer which, given the location of points, would decide where the points constituted line segments. Second, given the line segments, there were the equally daunting sequential tasks of putting these segments into particle tracks, eliminating noise, discarding "uninteresting tracks," recording the topological features of the interesting ones, and executing the geometrical reconstruction of the tracks in space, and performing the kinematic analysis (determination of the particles' energy and momentum). For Snyder, even if a "front end" pattern recognizer existed as a black box, the project of full automation was far from complete. J. Snyder, 'Some Remarks on a Data Analysis System Based Upon the Scanning-Measuring Projector (SMP),' LRL Physics Note 326, 25 August 1961, 3–4.

[33] J. Snyder, 'Some Remarks on a Data Analysis System Based Upon the Scanning-Measurement Projector (SMP),' LRL Physics Notes 326, August 1961, 4.

[34] On Cyborgs, see Donna Haraway, e.g., *Primate Visions: Gender, Race and Nature in the World of Modern Science* (New York: Routledge, 1989).

[35] L. Kowarski, 'VIa Introduction,' in *Proceedings of an International Conference on Instrumentation for High-Energy Physics*, held at the Ernest O. Lawrence Radiation Laboratory, Berkeley, California, 12–14 September 1960 (New York and London: Interscience 1961), 223 (hereafter: 'Introduction,' in Berkeley 1960), 223–234 on 223.

[36] L. Kowarski, 'Introduction,' in Paris 1963 (ref. 23), 2–3.

[37] *Ibid.*, 4.

[38] J. Snyder, R. Hulsizer, J. Munson, and H. Schneider, 'Bubble Chamber Data Analysis Using a Scanning and Measuring Projector (SMP) On-Line to a Digital Computer,' in K. Beckurts, W. Gläser, and G. Krüger, eds., *Conference on the Automatic Acquisition and Reduction of Nuclear Data*, Proceedings of a Conference, Karlsruhe, 13–16 July 1964, (Karlsruhe: Gesellschaft für Kernforschung m. b. H. Karlsruhe, 1964), 239–248 on 243–44.

[39] R. Hulsizer, J. Munson and J. Snyder, 'A System for the Analysis of Bubble Chamber Film Based upon the Scanning and Measuring Projector (SMP),' *Methods in Computational Physics* **5** (1966), 157–211 on 159.

[40] A. Rosenfeld, 'Current Performance of the Alvarez-Group Data Processing System,' *Nucl. Inst. and Meth.* **20** (1963), 422–34, on 433.
[41] Button et al., *Phys. Rev* **126** (1962), 1858–63, on 1863.
[42] L. Alvarez, comment in *Nucl. Inst. and Methods* **20** (1963), 382.
[43] L. Alvarez, 'Round Table Discussion on Bubble Chambers,' in *Proceedings of the 1966 International Conference on Instrumentation for High Energy Physics* (Springfield, Virginia: National Bureau of Standards, U.S. Department of Commerce, n.d.), 294 (hereafter: 'Round Table,' in SLAC 1966).
[44] H. Bradner, 'Capabilities and Limitations of Present Data-Reduction Systems,' in Berkeley 1960, 225-228 on 225.
[45] L. Alvarez, comment, *Nucl. Inst. and Methods* **20** (1963), 383.
[46] L. Kowarski, 'Introduction,' in Berkeley 1960 (1961) (ref. 35), 224.
[47] L. Kowarski, 'General Survey: Automatic Data Handling in High Energy Physics,' in Karlsruhe 1964 (Karlsruhe: Gesellschaft für Kernforschung m. b. H., 1964) 26–38 on 26 (Hereafter Karlsruhe 1964.)
[48] L. Kowarski, 'Concluding Remarks,' in Paris 1963 (ref. 23), 241.
[49] D. Pestre and J. Krige, 'Some Thoughts on the History of CERN,' in P. Galison and B. Hevly, eds., *Big Science: The Growth of Large-Scale Research* (Stanford: Stanford University Press, 1992).
[50] L. Kowarski, 'Concluding Remarks,' in Paris 1963 (ref. 23), 238.
[51] L. Kowarski, 'General Survey: Automatic Data Handling in High Energy Physics,' in Karlsruhe 1964 (ref. 47), 26–40 on 39–40.
[52] L. Alvarez, 'Round Table,' in SLAC 1966 (ref. 43), 271–72.
[53] *Ibid.*, 290–91.
[54] *Ibid.*, 292–93.
[55] *Ibid.*, 294.
[56] *Ibid.*, 294.
[57] *Ibid.*, 295.
[58] *Ibid.*, 289.
[59] K. Gottstein, 'Introductory Remarks', in B. Powell and P. Seyboth, eds., *Programming for Flying Spot Devices*, Conference held at the Max-Planck-Institut für Physik und Astrophysik, Munich, 18–20 January 1967 (Munich, 1967), 1–4 on 3.
[60] L. Kowarski, 'Concluding Remarks', in W. Moorhead and B. Powell, eds., *Programming for Flying Spot Devices* (CERN Yellow Report 65–11, 26 March 1965). Proceedings of a Conference held at the Centro Nazionale Analisi Fotogrammi, I.N.F.N., Bologna, on 7–9 October 1964, 259–266 on 264–265 (hereafter: 'Concluding Remarks,' Bologna, 1964).
[61] L. Kowarski, 'Concluding Remarks', in Bologna 1964 (ref. 60), 259–266 on 264.
[62] L. Kowarski, 'Concluding Remarks', in Munich 1967 (ref. 59), 409–416 on 415.
[63] *Ibid.*, 409–416 on 414.

APPENDIX I

ERWIN N. HIEBERT'S DOCTORAL STUDENTS AND THEIR DISSERTATIONS

I. UNIVERSITY OF WISCONSIN

Frederick A. White (1959), 'Significant Contributions of American Industrial Research Laboratories in the Development of Analytical Instruments for the Physical Sciences 1900–1950.'

Ollin J. Drennan (1961), 'Electrolytic Solution Theory: Foundations of Some Thermodynamical Considerations.'

Bernard S. Finn (1963), 'Developments in Thermoelectricity 1860–1920.'

J. Brookes Spencer (1964), 'An Historical Investigation of the Zeeman Effect 1896–1913.'

Clifford L. Maier (1964), 'The Role of Spectroscopy in the Acceptance of the Internally Structured Atom 1860–1920.'

Michael J. Crowe (1965), 'The History of the Idea of a Vectorial System to 1910.'

Edward E. Daub (1966), 'Rudolf Clausius and the 19th Century Theory of Heat.'

Joan L. Bromberg (1967), 'Maxwell's Concept of Electric Displacement' (with Robert S. Cohen).

Carolyn Merchant Iltis (1967), 'The Controversy over Living Force: Leibniz to D'Alembert.'

Thomas W. Hawkins (1968), 'The Origin and Development of Lebesgue's Theory of Integration.'

Roger H. Stuewer (1968), 'The History of the Compton Effect.'

Robert J. McRae (1969), 'The Origin of the Conception of the Continuous Spectrum of Heat and Light.'

Mary Jo Nye (1970), 'Jean Perrin and Molecular Reality.'

Henry G. Small (1971), 'The Helium Atom in the Old Quantum Theory.'

Thaddeus J. Trenn (1971), 'The Rise and Early Development of the Disintegration Theory of Radiation.'

Donald F. Moyer (1973), 'The Use of Dynamics as the Basis of Physical Theory by British Theoretical Physicists in the Latter Half of the 19th Century.'

Gisela Kutzbach (1973), 'The Genesis of the Thermal Theory of Cyclones.'

David Hargreave (1973), 'Thomas Young's Theory of Vision: Its Roots, Development and Acceptance by the British Scientific Community.'

II. HARVARD UNIVERSITY

Joseph W. Dauben (1972), 'The Early Development of Cantorian Set Theory.'

Frederick G. Gregory (1973), 'Scientific Materialism in 19th Century Germany.'

Jed Z. Buchwald (1974), 'Matter, the Medium and the Electric Current: A History of Electricity and Magnetism from 1842 to 1895.'

Susan Presswood Wright (1975), 'Henri Poincaré: A Developmental Study of the Philosophical and Scientific Thought.'

Keith A. Nier (1975), 'The Emergence of Physics in 19th Century Britain as a Socially Organized Category of Knowledge: Preliminary Studies.'

Craig Zwerling (1976), 'The Emergence of the Ecole Normale Supérieure as a Center of Scientific Education in 19th Century France.'

Lorraine J. Daston (1979), 'The Reasonable Calculus: Classical Probability Theory 1650–1840.'

Barbara Reeves Buck (1980), 'Italian Physicists and Their Institutions 1861–1911.'

Joan L. Richards (1980), 'Non-Euclidean Geometry in 19th Century England: A Study in Changing Perceptions of Mathematical Truth.'

Peter L. Galison (1981), 'How Experiments End: Three Case Studies on the Interaction of Experiment and Theory in 20th Century Physics.'

Neil Wasserman (1981), 'The Bohr–Kramers–Slater Paper and the Development of the Quantum Theory of Radiation in the Work of Niels Bohr.'

Seyyed Mohammed Haghi (1983), 'Matter, Life, Consciousness: The Interrelation of Science, Religion and Reality in the Philosophy of Henri Bergson' (with Wilfred C. Smith).

Richard L. Kremer (1984), 'The Thermodynamics of Life and Experimental Physiology 1770–1880.'

Maila K. Walter (1985), 'Science and Cultural Crisis: An Intellectual Biography of Percy Williams Bridgman.'

Gilbert F. Whittemore, Sr. (1986), 'The National Committee on Radiation Protection 1928–1960: From Professional Guidelines to Government Regulation.'

Sara S. Genuth (1988), 'From Monstrous Signs to Natural Causes: The Assimilation of Comet Lore into Natural Philosophy' (with I. Bernard Cohen and Owen J. Gingerich).

Diana K. Barkan (1990), 'Walther Nernst and the Transition to Modern Physical Chemistry.'

Dong-Won Kim (1991), 'The Emergence of the Cavendish School: An Early History of the Cavendish Laboratory 1871–1900' (with Silvan S. Schweber).

Skuli Sigurdsson (1991), 'Hermann Weyl, Mathematics and Physics 1900–1927' (with Silvan S. Schweber).

APPENDIX II

ERWIN N. HIEBERT
SELECTED LIST OF PUBLICATIONS

I. BOOKS AND ARTICLES

'The Experimental Basis of Kekulé's Valence Theory,' *Journal of Chemical Education* **36** (1959), 320–327.
The Impact of Atomic Energy. A History of Responses by Governments, Scientists, and Religious Groups (Newton, Kansas, 1961).
Historical Roots of the Principle of Conservation of Energy (Madison, 1962; reissued New York: Arno Press, 1981).
'Historical Remarks on the Discovery of Argon: The First Noble Gas,' in Herbert H. Hyman, ed., *Noble-Gas Compounds* (University of Chicago Press, 1963), 1–20.
'The Concept of Chemical Affinity in Thermodynamics,' *Proceedings of the Tenth International Congress of History of Science* (Ithaca, 1962) **2** (1964), 871–873.
'The Problem of Organic Analysis,' in I. Bernard Cohen and René Taton, eds., *Mélanges Alexandre Koyré*, Vol. I, *L'Aventure de la Science* (Paris: 1964), 303–325.
'The Uses and Abuses of Thermodynamics in Religion,' *Daedalus* **95** (1966), 1046–1080.
'Thermodynamics and Religion,' in F. J. Crosson, ed., *Science and Contemporary Society* (Notre Dame: University of Notre Dame Press, 1967), 57–104.
The Conception of Thermodynamics in the Scientific Thought of Mach and Planck (Freiburg i. Br.: Wissenschaftlicher Bericht, Ernst-Mach-Institut der Fraunhofer-Gesellschaft zur Förderung der angewandten Physik, 1968); also in Frank Kerkhoff, ed., *Symposium aus Anlass des 50. Todestages von Ernst Mach* (Freiburg i. Br.: 1967).
'The Role of Mechanics in Chemistry,' *Proceedings of the Eleventh International Congress of History of Science* (Warsaw) **3** (1968), 402–405.
'The Genesis of Mach's Early Views on Atomism,' in *Ernst Mach Physicist and Philosopher*, Boston Studies in the Philosophy of Science, Vol. 6 (Dordrecht: Reidel, 1970), 79–106.
'Niels Janniksen Bjerrum,' *Dictionary of Scientific Biography* **2** (1970), 169–171.
'Mach's Philosophical Use of the History of Science,' in Roger H. Stuewer, ed., *Historical and Philosophical Perspectives of Science*, Minnesota Studies in the Philosophy of Science, Vol. 5 (Minneapolis: University of Minnesota Press, 1970), 184–203.
'The Energetics Controversy and the New Thermodynamics,' in Duane H. D. Roller, ed., *Perspectives in the History of Science and Technology* (Norman: University of Oklahoma Press, 1971), 67–86.
'Ernst Mach,' *Dictionary of Scientific Biography* **8** (1974), 595–607.
'Mach's Conception of Thought Experiments in the Natural Sciences,' in Yehuda Elkana, ed., *The Interaction between Science and Philosophy* (Atlantic Highlands, N.J.: Humanities Press, 1975), 329–348.
'An Appraisal of the Work of Ernst Mach: Scientist-Historian-Philosopher,' in P. K.

Machamer and R. G. Turnbull, eds., *Motion and Time, Space and Matter* (Columbus: Ohio State University Press, 1976), 360–388.

'Introduction' to Ernst Mach's *Knowledge and Error: Sketches on the Psychology of Enquiry* (Dordecht: Reidel, 1976), xi–xxx.

'Chemical Thermodynamics and the Third Law: 1884–1914,' in *Proceedings of the Fifteenth International Congress of the History of Science* (Edinburgh, 1978), 305–313.

'Hermann Walther Nernst,' *Dictionary of Scientific Biography* 15 (1978), 432–453.

'Wilhelm Friedrich Ostwald,' *Dictionary of Scientific Biography* 15 (1978), 455–469 (with Hans-Günther Körber).

'Nernst and Electrochemistry,' in George Dubpernell *et al.*, eds., *Proceedings of the Symposium on Selected Topics in the History of Electrochemistry* (Princeton: The Electrochemical Society, 1978), 180–200.

'The State of Physics at the Turn of the Century,' in Mario Bunge and William R. Shea, eds., *Rutherford and Physics at the Turn of the Century* (New York: Science History Publications, 1979), 3–22; Chinese translation 'Shangshijimo he benshijichude wulixue zhuanghkuang,' *Kexue Shi Yicong* 1 (1983), 70–82.

'The Integration of Revealed Religion and Scientific Materialism in the Thought of Joseph Priestley,' in Lester Kieft and Bennett R. Willeford, eds., *Joseph Priestley: Scientist, Theologian and Metaphysician* (London: Associated University Presses, 1980), 27–61.

'Boltzmann's Conception of Theory Construction: The Promotion of Pluralism, Provisionalism and Pragmatic Realism,' in J. Hintikka *et al.*, eds., *Probabilistic Thinking, Thermodynamics, and the Interaction of the History and Philosophy of Science*, Pisa Conference Proceedings, Vol. 2 (Dordrecht: Reidel, 1981), 175–198.

'Einstein as a Philosopher of Science,' in C. M. Kinnon *et al.*, eds., *The Impact of Modern Scientific Ideas on Society*, In Commemoration of Einstein (Dordrecht: Reidel, 1981), 83–99.

'The Changing Relation of the Theoretical and Practical Aspects of Science,' *Acta historiae rerum naturalium necnon technicarum*, Czechoslovak Studies in the History of Science, Special Issue 13 (1982), 507–531.

'Report on Current Activities in the History of Science in the United States,' *Schriftenreihe für Geschichte der Naturwissenschaften, Technik und Medizin*, Leipzig 19 (1982), 12–15; Chinese translation: 'Guanyu Meigno kexueshi yanjiu gongzuo muqian dongxiang baodao,' *Kexue Shi Yicong* 1 (1983), 111–113.

'O sovremennom sostaianii istorii nauki v SShA Kratkoe soobshchenie,' *Voprosy Istorii Estestvoznaniia i Tekhniki*, Istoriko-nauchnye Issledovaniia za Rubezhom, Akademiya Nauk SSSR, Moskva 1 (1982), 105–107.

'Developments in Physical Chemistry at the Turn of the Century,' in C. G. Bernhard, E. Crawford, and P. Sorbom, eds., *Science, Technology and Society in the Time of Alfred Nobel* (Oxford: 1982), 97–118.

'Walther Nernst and the Application of Physics to Chemistry,' in R. Aris, H. T. Davis, and R. H. Stuewer, eds., *Springs of Scientific Creativity* (Minneapolis: University of Minnesota Press, 1983), 203–231.

'Priestley Found Harmony in "Work" and "Works" of God,' *Unitarian Universalist World* 14 (1983), 4.

'Einstein's Image of Himself as a Philosopher of Science,' in Everett Mendelsohn, ed.,

Transformation and Tradition in the Sciences, Essays in Honor of I. Bernard Cohen (Cambridge: 1984), 175–190.

'The Influence of Mach's Thought on Science,' *Philosophia Naturalis* **21** (1984), 598–615.

'Modern Physics and Christian Faith,' in Ronald Numbers and David Lindberg, eds., *God and Nature* (Berkeley: University of California Press, 1986), 423–447.

'The Scientist as Philosopher of Science,' *Schriftenreihe für Geschichte der Naturwissenschaften, Technik und Medizin*, Leipzig **24** (1987), 7–17.

'The Role of Experiment and Theory in the Development of Nuclear Physics in the Early 1930s,' in Diderik Batens and Jean Paul van Bendegem, eds., *Theory and Experiment* (Dordrecht: Reidel, 1988), 55–76.

'The Role of Experiment and Theory in the Genesis of Nuclear Physics,' in W. Muschik and E. Scheibe, eds., *Philosophie, Physik, Wissenschaftsgeschichte*, TUB-Documentation Kongresse und Tagungen, Heft 37 (Berlin, 1988), 141–147.

'The Transformation of Physics,' in Mikulás Teich and Roy Porter, eds., *Fin de Siècle and Its Legacy* (Cambridge: 1990), 236–253.

'Reflections on the Origin and Verification of the Third Law of Thermodynamics,' in K. Martinás, L. Ropolyi, and P. Szegedi, eds., *Thermodynamics: History and Philosophy: Facts, Trends, Debates* (Singapore: 1991), 90–138.

'The Discovery of Nuclear Physics in China,' in Yehuda Elkana and Erwin N. Hiebert, eds., *Ideas in Context: History of Nuclear Physics* (Cambridge University Press, forthcoming) (with Shuping Yao).

II. BOOK REVIEWS

Dirk Struik, *Het Land van Stevin en Huygens* (Amsterdam: 1958), in *Scripta Mathematica* **24** (1959), 325–326.

K. Przibram, ed., *Letters on Wave Mechanics: Schrödinger, Planck, Einstein, Lorentz* (New York: Philosophical Library, 1967), in *Science* **160** (1968), 664–666.

Russell McCormmach, ed., *Historical Studies in the Physical Sciences*, Vol. 1 (Philadelphia: University of Pennsylvania Press, 1969), in *Science* **168** (1970), 735–736.

J. Bradley, *Mach's Philosophy of Science* (London: Athlone, 1971), in *Isis* **65** (1974), 253–255.

Stephen G. Brush, *The Kind of Motion We Call Heat*, 2 Vols. (Amsterdam: North-Holland, 1976), in *American Journal of Physics* **45** (1977), 1130–1131.

Sadi Carnot, *Réflexions sur la puissance motrice du feu*, Critical Edition by Robert Fox (Paris: 1978), in *Isis* **71** (1980), 259.

F. I. Ordway and M. R. Sharpe, *The Rocket Team* (New York: Cromwell, 1979), in *Sci Quest* **53** (1980), 30–31.

Rudolf Schmitz, *Die Naturwissenschaften an der Philipps-Universität Marburg, 1527–1977* (Marburg: 1978), in *Isis* **72** (1981), 293–294.

Frederick V. Hunt, *Origins in Acoustics: The Science of Sound from Antiquity to the Age of Newton* (New Haven: 1978), in *Annals of Science* **38** (1981), 367–370.

Sadi Carnot et L'Essor de la Thermodynamique, Table Ronde du CNRS (Paris: 1976), in *Historia Mathematica* **8** (1981), 375–377.

I. Prigogine, *From Being to Becoming: Time and Complexity in the Physical Sciences* (San Francisco: Freeman, 1980), in *Physics Today* (January 1982), 69–70.

Stephen G. Brush, *Statistical Physics and the Atomic Theory of Matter* (Princeton: Princeton University Press, 1988), in *Archives Internationale d'Histoire des Sciences* **38** (1988), 342–343.

Werner Heisenberg, *Collected Works*, ed. W. Blum et al., Series A, Part 1, Wissenschaftliche Originalarbeiten (Berlin: Springer, 1985), in *Archives International d'Histoire des Sciences* **38** (1988), 346–347.

Hilde Levi, *George de Hevesy, Life and Work* (Bristol and Boston: Hilger, 1985), in *Archives International d'Histoire des Sciences* **38** (1988), 374–375.

Ernst Mach, *The Principles of the Theory of Heat*, ed. Brian McGuinness (Dordrecht: Reidel, 1987), in *Isis* **80** (1989), 160–161.

Arthur Fine, *The Shaky Game: Einstein, Realism, and the Quantum Theory* (Chicago: University of Chicago Press, 1986), in *American Historical Review* (December 1989), 1340.

Emil Fischer, *Aus meinem Leben*, ed. M. Bergmann (Berlin: Springer, 1987), in *Isis* **81** (1990), 584–585.

G. E. Bacon, ed., *Fifty Years of Neutron Diffraction: The Advent of Nuclear Scattering* (Bristol: Hilger, 1987), in *Archives International d'Histoire des Sciences* **40** (1990), 129–130.

C. Habfast, *Grossforschung mit kleinen Teilchen: Das Deutsche Elektronen-Synchrotron DESY 1956–1970* (Berlin: 1989), in *Archives International d'Histoire des Sciences* **40** (1990), 382.

M. J. van Lieburg and H. A. M. Snelders, eds., *De bevordering en volmaking der proefondervindelijke wijsbegeerte* (Amsterdam: 1989), forthcoming in *Isis*.

Tore Frängsmyr, ed., *Science in Sweden: The Royal Swedish Academy of Sciences 1739–1989* (Canton, Mass.: Science History Publications, 1989), forthcoming in *Isis*.

Ludwig Boltzmann, *Principien der Naturfilosofi*, ed. Ilse Fasol-Boltzmann (Berlin: 1990), forthcoming in *Isis*.

Thomas Hapke, *Die Zeitschrift für physikalische Chemie* (Herzberg: 1990), forthcoming in *Centaurus*.

NOTES ON CONTRIBUTORS

DIANA KORMOS BARKAN is Assistant Professor of History at the California Institute of Technology and a member of the School of Social Sciences at the Institute for Advanced Study during 1992–1993. She is completing a book, *Physicists as Chemists: Helmholtz, Planck, and Nernst and the Appeal of Multiple Methods*.

JED Z. BUCHWALD has been Professor of the History of Science at the University of Toronto. He is now Dibner Professor of History of Science and first Director of the Dibner Institute for the History and Philosophy of Science and Technology at the Massachusetts Institute of Technology. He is the author of several articles and two books: *From Maxwell to Microphysics* (1985) and *The Rise of the Wave Theory of Light* (1989). The article in the present collection draws on his next book: *The Creation of Scientific Effect: Heinrich Hertz and the Discovery of Electric Waves*, forthcoming in 1993.

LORRAINE J. DASTON has been Professor of the History of Science at the University of Göttingen, Germany. She is now Professor of History at the University of Chicago and at the Morris Fishbein Center for the History of Science and Medicine. She is the author of *Classical Probability in the Enlightenment* (1988), and is currently at work on a study of the history of the ideals and practices of scientific objectivity, 1600–1925.

PETER GALISON has been Professor in the departments of Philosophy and Physics at Stanford University and co-chair of the Program in the History of Science. He is now Professor of the History of Science at Harvard. His primary interest is in the history and philosophy of experimentation, the subject of his *How Experiments End* (1987) and *Big Science: The Growth of Large-Scale Research*, edited with Bruce Hevly (1992). His current project is entitled *Image and Logic: The Material Culture of Modern Physics*.

FREDERICK GREGORY is Chairperson of the Department of History at the University of Florida. His research has focussed on science in Germany during the eighteenth and nineteenth centuries. Recently he has published *Nature Lost? Natural Science and the German Theological Traditions of the Nineteenth Century* (1992). He continues work on an intellectual biography of the German romantic philosopher/physicist Jakob Fries.

RICHARD L. KREMER is Assistant Professor of History at Dartmouth College. He has published articles on Helmholtz, color theory and experimental physiology, has edited *Letters of Hermann von Helmholtz to His Wife, 1847–1859* (1990), and is completing a monograph on the "culture of experiment" in physiology at German universities between 1810 and 1865.

MARY JO NYE is George Lynn Cross Research Professor of the History of Science at the University of Oklahoma in Norman, Oklahoma. She is a past president of the History of Science Society and the author of *Molecular Reality: A Perspective on the Scientific Work of Jean Perrin* (1972) and *Science in the Provinces: Scientific Communities and Provincial Leadership in France, 1860–1930* (1986). She is completing a book on the history of chemical philosophy and "theoretical" chemistry since the eighteenth century.

JOAN L. RICHARDS is an Associate Professor in the History Department of Brown University. She has published *Mathematical Visions: The Pursuit of Geometry in Victorian England* (1988). At present, she is working on a cross-cultural study of mathematics in England and France at the beginning of the nineteenth century.

SARA SCHECHNER GENUTH teaches in the Department of Science, Technology, and Society at Sarah Lawrence College. Her current research treats the relationship of high and low culture, the history and philosophy of the physical sciences, and early scientific instruments. As the 1991–1992 recipient of the Herbert C. Pollock Award for Research in the History of Astronomy and Astrophysics, she is writing a book provisionally entitled, *Comets, Popular Culture, and the Birth of Modern Cosmology*.

SKULI SIGURDSSON holds a B.A. in mathematics and physics from Brandeis University (1981) and a Ph.D. in the history of science from Harvard University (1991). He was a Walther Rathenau Fellow in Berlin during the academic year 1990–1991, and during 1992 he is an Alexander von Humboldt Fellow at the University of Göttingen.

ROGER H. STUEWER is Professor of the History of Science and Technology at the University of Minnesota. He is the author of *The Compton Effect: Turning Point in Physics* (1975); editor of *Historical and Philosophical Perspectives of Science* (1970; reprinted 1989), and of *Nuclear Physics in Retrospect: Proceedings of a Symposium on the 1930s* (1979); co-editor of *Springs of Scientific Creativity* (1983), and *The Michelson Era in American Science 1870–1930* (1988).

NAME INDEX

Agassiz, Louis, 88–89
Albisetti, J. C., 123
d'Alembert, Jean, 28, 61
Alexander, xii
Allmann, 103
Altholtz, Joseph, 54
Alvarez, Luis, xx, 226–31, 235–37, 240–56
Ampère, André-Marie, 184
Apelt, Ernst, 86
Arbuthnot, John, 44
Aristotle, 30, 33, 36, 41, 207, 210
Arnold, Matthew, 73
Arrhenius, Svante, 175, 182, 185, 191, 193, 194, 212, 213
Aubert, Hermann, 165, 166
Austen, Jane, 55

Babbage, Charles, 54, 55, 63, 68, 69, 70, 72, 74
Bacon, Francis, 31, 32, 33, 35, 45
Baeyer, Adolf von, 212
Bancroft, Wilder D., 188
Barkan, Diana, xxii, xxviii, xxx
Barrow, Isaac, 103
Barschall, H. H., xi, xv
Barth, Karl, 93
Bayes, Thomas, 39
Beetz, Wilhelm, 130
Bentley, Richard, 6
Berlin, Isaiah, 85
Bernoulli, Jakob, xxvii, 27–30, 35, 37, 38, 39, 40, 41–4, 46, 47
Bernoulli, Nicholas, 44
Berthelot, Marcellin, 191, 211, 212
Berthollet, Claude Louis, 185
Berzelius, Jöns Jacob, 209
Bezold, Wilhelm von, 130
Blackett, 251, 252

Boettger, Rudolph, 128
Bohr, Niels, xx
Boltzmann, Ludwig, 109, 177, 183, 190
Borchardt [Prof.], 136
Boring, Edwin, 148
Born, Max, 205, 214
Boyle, Robert, 34, 61, 209
Braunmühl, 103
Brettschneider, 103
Brewster, David, 98, 163
Bridgewater, Francis Henry, Earl of, 57
Buchwald, Jed, xxix
Buddenbrook, Johann, 126
Buffon, Georges-Louis Leclerc, Comte de, 53
Bultmann, Rudolf, 93
Bunsen, Robert, 191
Brücke, Ernst, 155
Byles, Mather, 9

Calkin, J. W., 239, 240, 241
Calvin, Jean, 30
Campbell, xii
Cantor, Moritz Benedikt, xxviii, 97, 100–1, 103, 104, 109, 112
Cardano, Girolamo, 28, 38–44, 45
Carnap, Rudolf, 43
Cassirer, Ernst, xii
Cauchy, Augustin, 64, 66, 67, 184
Chalmers, Thomas, 57, 58, 59
Chasles, Michel, 103
Cheyne, George, 7–8, 9, 20, 21
Clairault, Alexis, 61
Clausius, Rudolf, 185
Clifford, William, 72, 74
Cohen, Bob, xvii
Colebrooke, 103
Comte, Auguste, 178
Condorcet, 28, 29

Copernicus, Nicholas, 61
Coulomb, Charles, 184
Coulson, Charles Alfred, xxxi–xxxii, 215–16
Curtze, 103
Cuvier, Georges, 211

Dalton, John, 209, 210
Daniels, Farrington, xiii
Darwin, Charles, xxviii, 55, 56, 72, 88, 89, 90
Daston, Lorraine, xxvii, xxviii
Daub, Edward, xii, xviii
Dauben, Joseph, xviii
Daudel, Raymond, 217
Davy, Humphry, 180
De Candolle, A. P., 211
De Moivre, Abraham, xxvii, 28, 29, 41, 44, 46, 47
De Morgan, Augustus, 28, 55, 64, 65, 66, 67, 71
Derham, William, 6, 7, 9, 20, 21, 46
Descartes, René, 10, 31, 32–33, 34, 35, 45, 51, 162
Deville, Henri Sainte-Claire, 211
Diderot, Denis, 53
Dirac, Paul, 205, 214, 215
Donovan, Arthur, xii
Doppler, Christian, 162
Draper, John, 93
Du Bois-Reymond, Emil, 152–53
Duhem, Pierre, xii, xx, xxix, xxx, 183–87, 189, 191, 192, 212
Dukas, Helen, xvi
Dumas, Jean Baptiste, 210
Durkheim, xx

Edwards, Henry Milne, 211
Einstein, xii, xvi, xx
Ekberg, K., 251, 252
Eneström, 103
Enriques, xii
Euler, Leonhard, 61

Faraday, Michael, 180, 209
Fechner, Gustav, 149, 152, 153, 154, 155, 157, 158, 160, 161, 163, 164, 165, 167, 168
Ferguson, James, 10–12, 20, 21
Feuerbach, Ludwig, 89, 90
Feynman, Richard, 111
Fick, Adolf, 159, 160, 164
Feigl, Herbert, xv
Flürscheim, Bernard, 214
Fontenelle, Bernard le Bovier de, 3–4, 6, 10, 21, 35
Fourcroy, 185
Franck, Jack V., 228
Frank, xii
Fries, Jakob, 86

Galileo, 31, 34, 35, 36, 38, 43, 61
Galison, Peter, xxx, xxxi
Genuth, Sara, xxvii
Gibbs, J. W., 183, 185, 192, 212
Goethe, Johann von, 147
Gordan, Paul, xxvi
Gottstein, K., 255
Grant, Edward, xiii
Gregory, Frederick, xxviii
Guyton de Morveau, Louis B., 184, 210

Hacking, Ian, 30, 112
Hale, George Ellery, 206
Hall, A. Rupert, 98, 99
Halley, Edmond, 7, 16
Hamermesh, Morton, xv
Hankel, 103
Hardy, Godfrey Harold, 99, 109
Havens, W. W., 251, 252
Hawkins, Thomas, xii
Hegel, Georg Wilhelm Friedrich, 86, 87
Heiberg, 103
Heitler, W., 215
Helmholtz, Hermann von, xx, xxix, 120, 125, 130–40, 147–55, 157, 160, 161, 162, 163, 164, 182, 185, 191, 211
Herbart, Johann Friedrich, 155–58, 162, 167, 168
Herburg, 149
Hering, Ewald, xxix, 147–55, 158, 160, 161, 164, 165, 166, 167, 168

NAME INDEX

Herrmann, Wilhelm, 92, 93
Herschel, John, xxviii, 54–5, 58–60, 61, 63, 66, 67, 70, 71, 72, 74
Hertz, Anna Elisabeth, 120–23, 128
Hertz, Gustav, 120–24, 126, 128
Hertz, Heinrich, xii, xxix, 119–40
Hewit, Beau, 45
Hiebert, Elfrieda, xi–xxiv
Hiebert, Erwin N., xi–xxiv, xxxi, xxxii, 109, 214
Hilbert, David, xxvi, 108
Hobbes, Thomas, 30, 43
Hobsbawm, Eric, 175
Høffding, 103
Hoffman, Roald, 217, 218
Holbach, P. H., Baron d', 30
Holmes, Frederic L., 209
Hough, Paul, 226, 237, 239, 244
Hubbard, Frank, xx
Hultsch, 103
Humboldt, Alexander von, 184
Humboldt, Wilhelm von, 123
Hume, David, 31, 56, 68
Hund, Friedrich, 215
Huxley, Thomas, 73
Huygens, Christiaan, 6, 46, 103

Iltis, Carolyn Merchant, xii
Ingold, Christopher K., 195, 214

Jacob, James R., 98
Jacob, Margaret C., 98
James, William, xx, 87, 99
Jansenius, 30
Jellinek, Karl, 193–94
Jolly, Philip von, 129
Jones, H. C., 188

Kant, Immanuel, 86, 128, 211
Kayser, Heinrich, 138
Kekulé, Auguste, 188, 189, 217
Kepler, Johannes, 61
Keynes, John Maynard, 43
Kingsley, Charles, 53
Kirchhoff, Gustav, 131, 132, 138, 139, 140

Kitts, David B., 219
Klein, Felix, 135
Knight, Fanny, 55
Kohlrausch, Friedrich Wilhelm, 131, 132
Kohut, Adolph, 89
Köstlin [Dr.], 124
Kowarski, Lew, xxxi, 226, 228, 230, 239, 243–45, 249–52, 255–57
Koyré, Alexandre, xiii, 98
Kremer, Richard, xxii, xxix
Krige, John, 250
Kronecker, Leopold, 139, 140
Kuhn, Thomas, 82, 109, 111–12

La Mettrie, Julien Offray de, 30, 53
Lacroix, Sylvester-François, 63, 64
Ladenburg, Alfred, 188
Lagrange, Joseph-Louis, 61, 63, 64, 129, 130
Lakatos, Imre, xvii
Lambert, Johann, 12, 14, 15, 18, 20, 21
Lange, Richard, 121, 122
Laplace, Pierre-Simon, xxvii, 18–19, 27–29, 40, 47, 61, 62, 101, 184, 185
Lapworth, Arthur, 214
Laudan, Rachel, 219
Lavoisier, Antoine Laurent, 18, 182, 185, 191, 194, 208–09, 210
Leibniz, Gottfried Wilhelm, 31, 34–35, 37–38, 42, 43, 44, 98, 99, 100, 101, 102, 104, 105, 107, 110
Lenin, xii
Lespieau, Robert, 190, 218
Lewis, G. N., 188, 212
Lindberg, David, xv
Locke, John, 31, 45, 51, 53, 58, 62, 68
Lodge, Oliver, 119
Loeb, Jacques, 188
London, F., 215
Longuet-Higgins, Christopher, 217
Lorentz, 190
Loria, 103
Lotze, Hermann, 86
Lovejoy, Arthur, 4
Lowry, Thomas, 213
Luther, Martin, 88

NAME INDEX

Mach, Ernst, xii, xx, xxix, 93, 101, 102, 103, 147, 148, 151, 155–68, 177, 179
Macquer, P. J., 184, 185, 207
Mahoney, Michael, 225
Manuel, Frank, 98
Marconi, Guglielmo, 119
Marx, Karl, 178
Mather, Cotton, 8, 9, 20, 21
Maupertuis, Pierre Louis Moreau de, 12
Maxwell, James Clerk, 47, 211
McClelland, C. E., 134
McRae, Robert, xii, xiv
Meitner, Lisa, xxiv
Melhado, Evan, 209
Mersenne, Marin, 35, 37
Merton, Robert K., 98
Meyerson, xii
Mill, John Stuart, 67
Moleschott, Jakob, 85
Monge, Gaspard, 185
Montesquieu, Charles de, 32
Montucla, Jean Étienne, 100, 129, 130
Müller, Johannes, 166
Mulliken, Robert, 215, 217
Murdoch, John, xiii, xvii

Nernst, Walther, xxx, 178, 182–83, 188, 191, 194, 213, 214–15
Newton, Isaac, 4, 6, 7, 15, 20, 31, 34, 61, 98, 99, 100–105, 110–11, 147, 162, 164, 166, 185, 207–08, 209–10
Nietzsche, Friedrich, 89
Noether, Max, 108
Novick, Peter, 81
Noyes, A., 188
Nye, Mary Jo, xxx, xxxi, xxxii

Ohm, G. S., 180
Oldenburg, Henry, 98
Oliver, Andrew, 13–14, 16–18, 20, 21
Oresme, Nicole, 32
Ostwald, Wilhelm, xxx, 175, 178–82, 183, 184, 185, 187, 188, 189, 190, 191, 193, 212, 213
Oven, Emil von, 127

Paley, William, 56, 57, 60
Paracelsus, 208
Pascal, xxvii, 35, 38, 43, 46
Pauling, Linus, 215, 216, 217
Peacock, George, 54–55, 63–64, 65, 67, 71
Peirce, Charles Sanders, xii, 87
Perrin, Jean, 186, 213–14
Pestre, Dominique, 250
Petzval, Joseph, 162
Pfefferkorn, Johann, 120
Pilate, Pontius, 81
Planck, Max, xii, 132–33, 183, 189, 192, 194
Plato, 85, 99
Poincaré, Henri, xii, 179
Poisson, Siméon-Denis, 44, 66, 130, 184
Pope, Alexander, 59
Powell, Brian, 226, 237, 239
Powell, Hough, 230
Priestley, Joseph, 17–18
Primas, Hans, 218
Prior, A. N., 86
Proust, Louis Joseph, 185
Pullman, Alberte, 216, 217
Pullman, Bernard, 217
Pyenson, Lewis, 135
Pynchon, Thomas, 209

Quetelet, Adolphe, 157

Ramus, Peter, 34
Ranke, Leopold, 81
Ray, John, 6
Richards, Joan L., xxvii, xxviii
Richards, Joseph W., 195
Richards, T. W., 188
Rickert, Heinrich, 177
Riecke, Eduard, 131
Riemann, Bernhard, 72
Ritschl, Albrecht, 87, 91, 93
Ritter, 103
Robinson, Robert, 214
Röntgen, Wilhelm, 119
Rosenfeld, Arthur, 228, 246
Rouelle, G. F., 185, 207

NAME INDEX

Rutherford, Ernest, 206

Schiller, 129
Schleiermacher, Friedrich, 87
Schlottke, [Herr] F., 124
Schmid, Rudolf, 90, 91
Schneider, Ivo, 110
Schopenhauer, Arthur, 166
Schopper, H., 251, 252
Schultz, 124
Schuster, Arthur, 131, 132
Schweber, Silvan S., 217
Shapin, Steven, 98
Sigurdsson, Skuli, xxviii
Slater, John C., 205, 206, 215
Smollett, Tobias, 45
Snyder, J. N., 242–43
Sommerfeld, Arnold, xvi
Spencer, Herbert, 178
Spinoza, Benedict, 31
Stahl, Georg, 207
Stallo, xii
Stauffer, Robert C., xi
Stent, Gunther, 196
Strauss, David Friedrich, 89–90
Struik, Dirk J., 100
Süssmilch, Johann, 44, 46
Swindon, Tobias, 8

Tannery, Paul, 103, 109
Thackray, Arnold, 210
Thiele, Johannes, 216
Thomsen, Julius, 191, 212
Thomson, William, 110
Trevor, J.-E., 188
Trevor-Roper, Hugh, 175
Tyndall, John, 128

Urey, Harold C., 195–96

Vaihinger, Hans, 93
van 't Hoff, Jacobus Hendricus, xxx, 175, 178, 182, 185, 187–93, 194, 212

Van Vleck, John, 215
van der Waals, 190
Venel, G. F., 207–08
Vinci, Leonardo da, 165
Volta, Alessandro, 180, 181

Walker, Adam, 207
Wall, William, 8–9, 21
Weber, Eduard, 165
Weber, Wilhelm, 131–32, 133, 158, 159, 161
Wegener, 219
Weinberg, Steven, xxxii
Weininger, Stephen, 218
Werner, Alfred, 214
Wertheim, 103
Wesley, John, 55
Whewell, William, xxviii, 53, 54–55, 60–62, 64, 67, 69, 70, 71, 72, 74
Whiston, William, 7–8, 9, 14, 20, 21
White, Andrew Dickson, 93
Whiteside, D. T., 102
Williamson, Alexander, 213, 217
Williamson, Hugh, 15–16, 17, 18, 20, 21
Willstätter, R., 211
Wilson, E. Bright, 217
Windelband, Wilhelm, 177
Winthrop, John, 12, 17
Woepcke, 103
Woolley, R. G., 218
Wüllner, Adolf, 128, 129
Wundt, Wilhelm, 157, 158
Wurtz, Adolphe, 189
Wussing, Hans, xxiv

Young, Thomas, 152, 162, 163, 164, 165, 166

Zeuthen, Hieronymous Georg, xxviii, 97, 99, 102–10, 112
Zöckler, Otto, 88–89, 90

Boston Studies in the Philosophy of Science

96. G. Márkus: *Language and Production.* A Critique of the Paradigms. Translated from French. 1986 ISBN 90-277-2169-6
97. F. Amrine, F.J. Zucker and H. Wheeler (eds.): *Goethe and the Sciences: A Reappraisal.* 1987 ISBN 90-277-2265-X; Pb 90-277-2400-8
98. J.C. Pitt and M. Pera (eds.): *Rational Changes in Science.* Essays on Scientific Reasoning. Translated from Italian. 1987 ISBN 90-277-2417-2
99. O. Costa de Beauregard: *Time, the Physical Magnitude.* 1987 ISBN 90-277-2444-X
100. A. Shimony and D. Nails (eds.): *Naturalistic Epistemology.* A Symposium of Two Decades. 1987 ISBN 90-277-2337-0
101. N. Rotenstreich: *Time and Meaning in History.* 1987 ISBN 90-277-2467-9
102. D.B. Zilberman: *The Birth of Meaning in Hindu Thought.* Edited by R.S. Cohen. 1988 ISBN 90-277-2497-0
103. T.F. Glick (ed.): *The Comparative Reception of Relativity.* 1987 ISBN 90-277-2498-9
104. Z. Harris, M. Gottfried, T. Ryckman, P. Mattick Jr, A. Daladier, T.N. Harris and S. Harris: *The Form of Information in Science.* Analysis of an Immunology Sublanguage. With a Preface by Hilary Putnam. 1989 ISBN 90-277-2516-0
105. F. Burwick (ed.): *Approaches to Organic Form.* Permutations in Science and Culture. 1987 ISBN 90-277-2541-1
106. M. Almási: *The Philosophy of Appearances.* Translated from Hungarian. 1989 ISBN 90-277-2150-5
107. S. Hook, W.L. O'Neill and R. O'Toole (eds.): *Philosophy, History and Social Action.* Essays in Honor of Lewis Feuer. With an Autobiographical Essay by L. Feuer. 1988 ISBN 90-277-2644-2
108. I. Hronszky, M. Fehér and B. Dajka: *Scientific Knowledge Socialized.* Selected Proceedings of the 5th Joint International Conference on the History and Philosophy of Science organized by the IUHPS (Veszprém, Hungary, 1984). 1988 ISBN 90-277-2284-6
109. P. Tillers and E.D. Green (eds.): *Probability and Inference in the Law of Evidence.* The Uses and Limits of Bayesianism. 1988 ISBN 90-277-2689-2
110. E. Ullmann-Margalit (ed.): *Science in Reflection.* The Israel Colloquium: Studies in History, Philosophy, and Sociology of Science, Vol. III. 1988 ISBN 90-277-2712-0; Pb 90-277-2713-9

See also Volumes 94 and 95.

111. K. Gavroglu, Y. Goudaroulis and P. Nicolacopoulos (eds.): *Imre Lakatos and Theories of Scientific Change.* 1989 ISBN 90-277-2766-X
112. B. Glassner and J.D. Moreno (eds.): *The Qualitative- Quantitative Distinction in the Social Sciences.* 1989 ISBN 90-277-2829-1
113. K. Arens: *Structures of Knowing.* Psychologies of the 19th Century. 1989 ISBN 0-7923-0009-2

Boston Studies in the Philosophy of Science

114. A. Janik: *Style, Politics and the Future of Philosophy.* 1989
 ISBN 0-7923-0056-4
115. F. Amrine (ed.): *Literature and Science as Modes of Expression.* With an Introduction by S. Weininger. 1989 ISBN 0-7923-0133-1
116. J.R. Brown and J. Mittelstrass (eds.): *An Intimate Relation.* Studies in the History and Philosophy of Science. Presented to Robert E. Butts on His 60th Birthday. 1989 ISBN 0-7923-0169-2
117. F. D'Agostino and I.C. Jarvie (eds.): *Freedom and Rationality.* Essays in Honor of John Watkins. 1989 ISBN 0-7923-0264-8
118. D. Zolo: *Reflexive Epistemology.* The Philosophical Legacy of Otto Neurath. 1989 ISBN 0-7923-0320-2
119. M. Kearn, B.S. Philips and R.S. Cohen (eds.): *Georg Simmel and Contemporary Sociology.* 1989 ISBN 0-7923-0407-1
120. T.H. Levere and W.R. Shea (eds.): *Nature, Experiment and the Science.* Essays on Galileo and the Nature of Science. In Honour of Stillman Drake. 1989
 ISBN 0-7923-0420-9
121. P. Nicolacopoulos (ed.): *Greek Studies in the Philosophy and History of Science.* 1990 ISBN 0-7923-0717-8
122. R. Cooke and D. Costantini (eds.): *Statistics in Science.* The Foundations of Statistical Methods in Biology, Physics and Economics. 1990
 ISBN 0-7923-0797-6
123. P. Duhem: *The Origins of Statics.* Translated from French by G.F. Leneaux, V.N. Vagliente and G.H. Wagner. With an Introduction by S.L. Jaki. 1991
 ISBN 0-7923-0898-0
124. H. Kamerlingh Onnes: *Through Measurement to Knowledge.* The Selected Papers, 1853-1926. Edited and with an Introduction by K. Gavroglu and Y. Goudaroulis. 1991 ISBN 0-7923-0825-5
125. M. Čapek: *The New Aspects of Time: Its Continuity and Novelties.* Selected Papers in the Philosophy of Science. 1991 ISBN 0-7923-0911-1
126. S. Unguru (ed.): *Physics, Cosmology and Astronomy, 1300- 1700.* Tension and Accomodation. 1991 ISBN 0-7923-1022-5
127. Z. Bechler: *Newton's Physics on the Conceptual Structure of the Scientific Revolution.* 1991 ISBN 0-7923-1054-3
128. É. Meyerson: *Explanation in the Sciences.* Translated from French by M-A. Siple and D.A. Siple. 1991 ISBN 0-7923-1129-9
129. A.I. Tauber (ed.): *Organism and the Origins of Self.* 1991
 ISBN 0-7923-1185-X
130. F.J. Varela and J-P. Dupuy (eds.): *Understanding Origins.* Contemporary Views on the Origin of Life, Mind and Society. 1992 ISBN 0-7923-1251-1
131. G.L. Pandit: *Methodological Variance.* Essays in Epistemological Ontology and the Methodology of Science. 1991 ISBN 0-7923-1263-5

Boston Studies in the Philosophy of Science

132. G. Munévar (ed.): *Beyond Reason.* Essays on the Philosophy of Paul Feyerabend. 1991 ISBN 0-7923-1272-4
133. T.E. Uebel (ed.): *Rediscovering the Forgotten Vienne Circle.* Austrian Studies on Otto Neurath and the Vienna Circle. Partly translated from German. 1991
ISBN 0-7923-1276-7
134. W.R. Woodward and R.S. Cohen (eds.): *World Views and Scientific Discipline Formation.* Science Studies in the [former] German Democratic Republic. Partly translated from German by W.R. Woodward. 1991 ISBN 0-7923-1286-4
135. P. Zambelli: *The Speculum Astronomiae and its Enigma.* Astrology, Theology and Science in Albertus Magnus and his Contemporaries. 1992
ISBN 0-7923-1380-1
136. P. Petitjean, C. Jami and A.M. Moulin (eds.): *Science and Empires.* Historical Studies about Scientific Development and European Expansion.
ISBN 0-7923-1518-9
137. W.A. Wallace: *Galileo's Logic of Discovery and Proof.* The Background, Content, and Use of His Appropriated Treatises on Aristotle's *Posterior Analytics.* 1992 ISBN 0-7923-1577-4
138. W.A. Wallace: *Galileo's Logical Treatises.* A Translation, with Notes and Commentary, of His Appropriated Latin Questions on Aristotle's *Posterior Analytics.* 1992 ISBN 0-7923-1578-2
Set (137 + 138) ISBN 0-7923-1579-0
139. M.J. Nye, J.L. Richards and R.H. Stuewer (eds.), *The Invention of Physical Science.* Intersections of Mathematics, Theology and Natural Philosophy since the Seventeenth Century. Essays in Honor of Erwin N. Hiebert. 1992
ISBN 0-7923-1753-X
140. G. Corsi, M.L. dalla Chiara and G.C. Ghirardi (eds.), *Bridging the Gap: Philosophy, Mathematics and Physics.* Lectures on the Foundations of Science. 1992 ISBN 0-7923-1761-0
141. C.-H. Lin and D. Fu (eds.), *Philosophy and Conceptual History of Science in Taiwan.* 1992 ISBN 0-7923-1766-1
142. S. Sarkar (ed.), *The Founders of Evolutionary Genetics.* A Centenary Reappraisal. 1992 ISBN 0-7923-1777-7
143. J. Blackmore (ed.), *Ernst Mach – A Deeper Look.* Documents and New Perspectives. 1992 ISBN 0-7923-1853-6
144. P. Kroes and M. Bakker (eds), *Technological Development and Science in the Industrial Age.* New Perspectives on the Science – Technology Relationship. 1992 ISBN 0-7923-1898-6

Previous volumes are still available.

KLUWER ACADEMIC PUBLISHERS – DORDRECHT / BOSTON / LONDON